Lecture Notes in Mathematics

Volume 2315

This series reports on new developments in all areas of mathematics and their applications - quickly, informally and at a high level. Mathematical texts analysing new developments in modelling and numerical simulation are welcome. The type of material considered for publication includes:

1. Research monographs
2. Lectures on a new field or presentations of a new angle in a classical field
3. Summer schools and intensive courses on topics of current research.

Texts which are out of print but still in demand may also be considered if they fall within these categories. The timeliness of a manuscript is sometimes more important than its form, which may be preliminary or tentative.

Titles from this series are indexed by Scopus, Web of Science, Mathematical Reviews, and zbMATH.

Rúben Sousa • Manuel Guerra •
Semyon Yakubovich

Convolution-like Structures, Differential Operators and Diffusion Processes

Springer

Rúben Sousa
Department of Mathematics, Faculty of Sciences
University of Porto
Porto, Portugal

Manuel Guerra
Department of Mathematics, ISEG
Universidade de Lisboa
Lisbon, Portugal

Semyon Yakubovich
Department of Mathematics, Faculty of Sciences
University of Porto
Porto, Portugal

This work was supported by Fundação para a Ciência e a Tecnologia (Portugal) (UIDB/05069/2020, PD/BI/128072/2016, PD/BD/135281/2017).

ISSN 0075-8434 ISSN 1617-9692 (electronic)
Lecture Notes in Mathematics
ISBN 978-3-031-05295-8 ISBN 978-3-031-05296-5 (eBook)
https://doi.org/10.1007/978-3-031-05296-5

Mathematics Subject Classification: 60G53, 60G51, 47F10, 33C10, 33C15, 44A15

This Springer imprint is published by the registered company Springer Nature Switzerland AG
The registered company address is: Gewerbestrasse 11, 6330 Cham, Switzerland

Preface

It is well-known that convolutions, differential operators and diffusion processes are interconnected subjects: the ordinary convolution commutes with the Laplacian, and the law of Brownian motion has a convolution semigroup property with respect to the ordinary convolution. If we seek to generalize this useful connection so as to cover other differential operators and diffusion processes, we are naturally led to the notion of a convolution-like operator—i.e. a bilinear operator with respect to which a given diffusion (other than the Brownian motion) has the convolution semigroup property, and which commutes with the generator of the given diffusion. The study of this and other related concepts in generalized harmonic analysis has, since the 1930s, attracted the attention of many researchers, most notably Delsarte, Levitan, Berezansky, Urbanik, Chebli, Heyer and Trimèche, Zeuner, among others

A few books have been published on the theory of generalized harmonic analysis, but a general, systematic presentation of the problem of constructing convolution-like operators and of its applications to stochastic processes and differential equations is missing. The goal of this book is to try to fill in this gap, while at the same time providing an accessible introduction to recent developments on this topic. The book also intends to draw attention to a wide range of questions which still remain open in this area of research. We are confident that this book will be a valuable resource for graduate students and researchers interested on the intersections between harmonic analysis, probability theory and differential equations.

We are deeply thankful to all our friends and fellow researchers who have contributed to this project with comments, conversations or encouragement. We are also thankful to the organizers of the various conferences where our work has been featured and where many fruitful interactions took place, namely the *12th Vilnius Conference on Probability Theory and Statistics* (Vilnius, Lithuania, 2018), the *WSMC12 Workshop on Statistics, Mathematics and Computation* (Covilhã, Portugal, 2018), the *Probability and Analysis Conference* (Bedlewo, Poland, 2019), the *IWOTA 2019* (Lisbon, Portugal, 2019), the *12th ISAAC Congress* (Aveiro, Portugal, 2019) and the *ÖMG Conference* (Dornbirn, Austria, 2019).

The work of the first and third authors was partially supported by CMUP, which is financed by national funds through FCT—Fundação para a Ciência e a Tecnologia, I.P., under the project with reference UIDB/00144/2020. The first author was also supported by the grants PD/BI/128072/2016 and PD/BD/135281/2017, under the FCT PhD Programme UC|UP MATH PhD Program. The work of the second author was partially supported by the project CEMAPRE/REM—UIDB/05069/2020—financed by FCT through national funds.

Porto, Portugal — Rúben Sousa
Lisbon, Portugal — Manuel Guerra
Porto, Portugal — Semyon Yakubovich
September 2020

Contents

List of Symbols

$\mathbb{1}(\cdot)$	function identically equal to one
$\mathbb{1}_B(\cdot)$	indicator function of the subset B
$(a)_n$	Pochhammer symbol, 55
$\mathcal{A}_\alpha$	generator of the Shiryaev process, 51
$\mathrm{AC}_{\mathrm{loc}}(a, b)$	$\{f : (a, b) \longrightarrow \mathbb{C} \mid f \text{ is locally absolutely continuous}\}$
$\mathrm{B}_{\mathrm{b}}(E)$	$\{f : E \longrightarrow \mathbb{C} \text{ measurable} \mid \sup_{x \in E} \lvert f(x)\rvert < \infty\}$
$\mathbb{B}(x, \varepsilon)$	ball centred at x with radius ε
$\mathrm{C}(E)$	$\{f : E \longrightarrow \mathbb{C} \mid f \text{ is continuous}\}$
$\mathrm{C}_0(E)$	$\{f \in \mathrm{C}(E) \mid f \text{ vanishes at infinity}\}$
$\mathrm{C}_{\mathrm{b}}(E)$	$\{f \in \mathrm{C}(E) \mid f \text{ is bounded}\}$
$\mathrm{C}_{\mathrm{c}}(E)$	$\{f \in \mathrm{C}(E) \mid f \text{ has compact support}\}$
$\mathrm{C}^k(E)$	$\{f \in \mathrm{C}(E) \mid f \text{ is } k \text{ times continuously differentiable}\}$
$\mathrm{C}^k_{\mathrm{c}}(E)$	$\mathrm{C}_{\mathrm{c}}(E) \cap \mathrm{C}^k(E)$
$\mathrm{C}^\infty_{\mathrm{c,even}}$	$\{f : [0, \infty) \longrightarrow \mathbb{C} \mid f$ is the restriction of an even $\mathrm{C}^\infty_{\mathrm{c}}(\mathbb{R})$ $-$function$\}$
$\mathbf{d}(\cdot, \cdot)$	distance function, 11, 187
$D_\mu(z)$	parabolic cylinder function, 56
$\mathcal{D}(\mathcal{L})$	domain of the operator $\mathcal{L}$
δ_x	Dirac measure at the point x
$\mathbf{e}(\mu), \mathbf{e}_\alpha(\mu), \mathbf{e}_k(\mu)$	Poisson(-like) measure associated with μ, 85, 157, 230
$\mathcal{E}, \mathcal{E}_{\mathcal{L}}, \mathcal{E}_N$	sesquilinear forms, 29, 185
$\mathbb{E}_{x_0}$	expectation for a time-homogeneous Markov process started at x_0
$\mathcal{F}$	Sturm-Liouville integral transform, 23
${}_2F_1(a, b; c; z)$	Gauss hypergeometric function, 31
$\mathcal{G}, \mathcal{G}^{(0)}, \mathcal{G}^{(b)}, \mathcal{G}^{(2)}$	infinitesimal generators, 10, 13, 185
$\Gamma(z)$	Gamma function, 44
$H^k(E)$	Sobolev space, 186
$\mathcal{H}$	Hankel transform, 44
$I_\eta(z)$	modified Bessel function of the first kind, 44

$J_\eta(z)$	Bessel function of the first kind, 44
$\boldsymbol{J}_\eta(z)$	normalized Bessel function of the first kind, 44
$k_\alpha(x, y, \xi)$	kernel of the Whittaker product formula, 63
$K_\eta(z)$	modified Bessel function of the second kind, 55
ℓ, ℓ	Sturm-Liouville operators
$L^p(E, \mu)$	$\{f : E \longrightarrow \mathbb{C}$ measurable $\mid \int_E \lvert f\rvert^p d\mu < \infty\}$
$\mathcal{L}, \mathcal{L}^{(2)}$	realization of a Sturm-Liouville operator
$M_{\kappa,\nu}(z)$	Whittaker function of the first kind, 32
$\mathcal{M}_{\mathbb{C}}(E)$	space of finite complex Borel measures on E
$\mathcal{M}_+(E)$	space of finite positive Borel measures on E
$\mu^{*n}, \mu^{\diamond n}$	n-fold convolution of μ with itself
$\mu_n \xrightarrow{v} \mu$	vague convergence of measures, 10
$\mu_n \xrightarrow{w} \mu$	weak convergence of measures, 10
Ω_g	Riemannian volume on a manifold with cone-like metric, 216
$p_{t,x}(dy),\ p(t, x, y)$	transition probabilities, transition density of a diffusion process
P_{x_0}	distribution of a time-homogeneous Markov process started at x_0
$\mathcal{P}(E)$	space of probability measures on E
$\mathcal{P}_{\mathrm{id}}, \mathcal{P}_{\alpha,\mathrm{id}}$	subset of infinitely divisible distributions, 82, 156
$\Psi(a, b; z)$	confluent hypergeometric function of the second kind, 57
$q_t(x, y, \xi)$	kernel of the regularized product formula, 145, 192, 208
$R_n^{(\alpha,\beta)}(x)$	Jacobi polynomials, 31
$\mathbb{R}^+$	open half-line $(0, \infty)$
$\mathbb{R}_0^+$	closed half-line $[0, \infty)$
$\boldsymbol{\rho}_{\mathcal{L}}$	spectral measure for the operator $\mathcal{L}$, 23
$T_t, T_t^{(p)}$	strongly continuous one-parameter semigroups
$\mathcal{T}^x, \mathcal{T}^\mu$	generalized translation operator
$V_{k,\lambda}(\cdot), V_{\lambda,\eta}^{\wp}(\cdot)$	generalized eigenfunctions on spaces with cone-like metrics, 218, 240
$w_\lambda(\cdot)$	solution of the Sturm-Liouville boundary value problem, 18
$W_{\alpha,\nu}(z)$	Whittaker function of the second kind, 32
$\boldsymbol{W}_{\alpha,\nu}(z)$	normalized Whittaker function of the second kind, 54
$\mathcal{W}_\alpha$	index Whittaker transform, 71
$\mathfrak{X}$	domain for the Whittaker translation, 98
$Z_X^{A,f}$	Dynkin martingale, 95

Chapter 1
Introduction

The goal of this book is to provide an introduction to recent developments in the theory of generalized harmonic analysis and its applications to the study of differential operators and diffusion processes.

Throughout the book, we shall provide answers to the following research question, which triggered our interest in this topic: *given a diffusion process* $\{X_t\}_{t\geq 0}$ *on a metric space* E*, can we construct a convolution-like operator* $*$ *on the space of probability measures on* E *with respect to which the law of* X_t *has the* $*$*-convolution semigroup property, i.e. can be written as* $P[X_t \in \cdot] = \mu_t * \delta_{X_0}$*, where the measures* μ_t *are such that* $\mu_{t+s} = \mu_t * \mu_s$ *for all* $t, s \geq 0$*?*

The significance of this problem stems both from its interpretation as a generalization of classical harmonic analysis and from its probabilistic applications. These motivations are discussed in the next section, where we highlight the connections between the construction of convolution-like structures and disciplines such as stochastic processes, ordinary and partial differential equations, spectral theory, special functions, and integral transforms. In Sect. 1.2 we describe the logical sequence of the chapters and provide a summary of the main results presented in this volume.

1.1 Motivation and Scope

Harmonic analysis for elliptic differential operators

A first motivation for the problem formulated above comes from the fact that the existence of a generalized convolution structure for the diffusion process generated by a given elliptic differential operator $\mathcal{L}$ puts at our disposal a valuable tool for the study of elliptic and parabolic partial differential equations determined by $\mathcal{L}$.

R. Sousa et al., *Convolution-like Structures, Differential Operators and Diffusion Processes*, Lecture Notes in Mathematics 2315,
https://doi.org/10.1007/978-3-031-05296-5_1

Indeed, the most straightforward way to investigate the properties of various heat-type equations (and their nonlocal counterparts) is, in many cases, by making use of techniques from (standard) harmonic analysis [6, 129, 151, 160]. If the properties of the convolution-like operator are similar to those of the ordinary convolution, then the resulting algebraic structure allows one to develop the basic notions of harmonic analysis in parallel with the standard theory [15, 127]; therefore, it is natural to expect that a positive answer to our problem will lead to a better understanding of the properties of the corresponding differential operators and the associated potential-theoretic objects. We note that the problem of constructing a generalized convolution can be formulated for a large class of operators which includes, in particular, the (Dirichlet, Neumann, Robin) Laplacian on Euclidean domains and Riemannian manifolds.

This interplay between convolutions and elliptic operators originates in the observation that the existence of a convolution-like operator for the diffusion $\{X_t\}$ is closely related to the properties of the *generalized eigenfunctions* of its (infinitesimal) generator $\mathcal{L}$, i.e. of the solutions of the elliptic equation $\mathcal{L}u = \lambda u$ ($\lambda \in \mathbb{C}$). Indeed, suppose that $*$ is a bilinear operator on the set $\mathcal{P}(E)$ of probability measures on E satisfying the conditions

C1. (*Convolution semigroup property*) $P_x[X_t \in \cdot] = \mu_t * \delta_x$ $(t > 0, x \in E)$, where $\{\mu_t\}_{t\geq 0} \subset \mathcal{P}(E)$ is such that $\mu_{t+s} = \mu_t * \mu_s$ for all $t, s \geq 0$, and P_x stands for the distribution of $\{X_t\}$ started at x;

C2. There exists a family Θ of bounded continuous functions such that

$$\int_E \vartheta \, d(\mu * \nu) = \left(\int_E \vartheta \, d\mu\right) \cdot \left(\int_E \vartheta \, d\nu\right) \qquad \text{for all } \vartheta \in \Theta \text{ and } \mu, \nu \in \mathcal{P}(E). \tag{1.1}$$

Notice that (1.1) holds if $*$ is the ordinary convolution and $\vartheta(x) = e^{\lambda x}$ with $\lambda \in \mathbb{C}$; condition C2 can thus be interpreted as a general formulation of a trivialization property similar to that of the Fourier transform with respect to the ordinary convolution. From C1 and C2 it is not difficult to deduce (cf. Chap. 5) that each $\vartheta \in \Theta$ is a generalized eigenfunction of the transition semigroup of $\{X_t\}$ and, consequently, a generalized eigenfunction of the elliptic operator $\mathcal{L}$. Replacing μ and ν by Dirac measures in (1.1), we find that there exists a family of measures $\boldsymbol{\nu}_{x,y} \in \mathcal{P}(E)$ such that the *probabilistic product formula*

$$\int_E \vartheta_\lambda \, d\boldsymbol{\nu}_{x,y} = \vartheta_\lambda(x)\, \vartheta_\lambda(y), \qquad x, y \in E \tag{1.2}$$

holds for bounded solutions ϑ_λ of $\mathcal{L}\,\vartheta_\lambda = \lambda \vartheta_\lambda$. (We use the word 'probabilistic' in order to emphasize that $\{\boldsymbol{\nu}_{x,y}\}$ is a family of probability measures.) Conversely, if a probabilistic product formula of the form (1.2) holds for a sufficiently large family of generalized eigenfunctions of $\mathcal{L}$, then the generalized convolution operator defined as $(\mu * \nu)(d\xi) = \iint \boldsymbol{\nu}_{x,y}(d\xi)\, \mu(dx)\, \nu(dy)$ is such that the $*$-convolution semigroup property C1 holds for the distribution of $\{X_t\}$.

The historical development of the topic of generalized harmonic analysis began with the seminal works of Delsarte [45] and Levitan [117], where it was first noticed that product formulas are the key ingredient for the construction of such convolution-like structures. The nontrivial motivating example came from the Bessel differential operator, for which the existence of the product formula (1.2) follows from a classical result on the Bessel function. (An overview on this motivating example will be presented in Sect. 2.4.) This led, on the one hand, to the proposal of axiomatic structures—often referred to as generalized convolutions, generalized translations, hypercomplex systems or hypergroups—which aimed to identify the essential features which allow one to derive analogues of the basic facts of classical harmonic analysis. In Sect. 2.3 we provide some historical background on the development of such axiomatic theories, whose range extends far beyond the particular case of structures associated with diffusion processes or elliptic operators.

On the other hand, there has been a continuous interest in finding additional examples of nontrivial product formulas associated with Sturm–Liouville and elliptic operators. Besides the Bessel example, other product formulas have been obtained by exploiting the properties of special functions of hypergeometric type [70, 73, 105]. An alternative strategy relies on the fact that certain differential operators are related with topological groups [105]. Yet another approach, which (unlike the former techniques) is applicable to one-dimensional operators with general coefficients, is to rely on the *associated hyperbolic PDE*: if $\mathcal{L}$ is the Sturm–Liouville operator $\frac{1}{r}\big[-(pu')' + qu\big]$ and ϑ_λ are the generalized eigenfunctions satisfying the boundary condition $\vartheta_\lambda(a) = 1$, then the product $f(x,y) = \vartheta_\lambda(x)\vartheta_\lambda(y)$ is a solution of the hyperbolic PDE

$$\begin{aligned}\frac{1}{r(x)}\Big\{-\partial_x\big[p(x)\,\partial_x f(x,y)\big] + q(x) f(x,y)\Big\}\\ = \frac{1}{r(y)}\Big\{-\partial_y\big[p(y)\,\partial_y f(x,y)\big] + q(y) f(x,y)\Big\}.\end{aligned}$$

satisfying the boundary condition $f(x,a) = \vartheta_\lambda(x)$; studying the properties of this PDE is therefore a natural strategy for proving the existence of a product formula of the form (1.2) and extracting information about the measure $\boldsymbol{\nu}_{x,y}$ [30, 38, 120, 210].

The existing theory on convolution-like operators associated with elliptic operators is mostly limited to the one-dimensional (Sturm–Liouville) case. One of the reasons for this is the fact that there is a well-developed spectral theory for Sturm–Liouville operators which, in particular, ensures that (under suitable boundary conditions) the corresponding generalized eigenfunctions are the kernel of a Sturm–Liouville type integral transform $(\mathcal{F}h)(\lambda) := \int_I h(x)\, w_\lambda(x)\, \mathrm{m}(dx)$ which, similar to the Fourier transform, defines an isometric isomorphism between L^2 spaces. This class of transformations includes many common integral transforms (Hankel, Kontorovich–Lebedev, Mehler–Fock, Jacobi, Laguerre, etc.). The construction of

Sturm–Liouville convolutions satisfying the trivialization identity $\mathcal{F}(h * g) = (\mathcal{F}h)\cdot(\mathcal{F}g)$ triggers a better understanding of the mapping properties of such Sturm–Liouville integral transforms.

Due to our probabilistic motivations, the present discussion focuses on convolution-like structures where the convolution is a bilinear operator acting on finite complex measures. We observe, however, that in the theory of integral transforms and special functions it is more common to define a convolution (say, associated with a given integral transform) as an operator acting on suitable spaces of integrable functions [73, 207]. In this context, it usually becomes less of a concern whether or not the convolution preserves properties such as positivity or boundedness.

Construction of Lévy-like processes

Lévy processes are a very important class of Markov processes. By definition, they are stochastically continuous and have stationary and independent increments. Lévy processes are a versatile class of processes with jumps whose continuous representatives are the drifted Brownian motions (in the sense that any Lévy process with continuous paths is a drifted Brownian motion); therefore, they can be seen as a natural generalization of Brownian motion. Replacing Brownian motions by Lévy processes with jumps is a common strategy for obtaining models with greater flexibility in mathematical finance and other applications [6, 143].

The Brownian motion is the most famous diffusion process, but many other diffusion processes also find diverse applications in a wide range of fields. One such field is mathematical finance, where one-dimensional diffusions such as the Bessel, Ornstein–Uhlenbeck and Shiryaev processes are often used in the modelling of the underlying financial variables, while two-dimensional diffusion processes have been extensively applied in the context of stochastic volatility models [122, 143]. It is relevant to ask whether these other diffusion processes can also be generalized into a class of processes characterized by some analogue of the notions of stationarity and independence.

By a well-known characterization, Lévy processes can be equivalently defined as Feller processes whose law satisfies the convolution semigroup property (as stated in condition C1) with respect to the usual convolution. It is thus natural to generalize the notion of Lévy process by replacing the requirement of stationarity and independence by the convolution semigroup property with respect to any convolution-like operator with suitable properties. This provides us with a recipe for defining a class of Lévy-like processes associated with a given diffusion process: as prescribed in the problem above, one should construct a convolution-like operator such that condition C1 holds for the law of the given diffusion. This generalized notion of Lévy process has been proposed in various papers [23, 83, 163]; however, the class of diffusions which have been proved to admit such an associated family of Lévy-like processes is still very limited.

The notion of a convolution semigroup is closely related with that of an infinitely divisible distribution. In the case of the usual convolution, a central role is played by the Lévy–Khintchine theorem which provides a complete description of the set of infinitely divisible distributions; in addition, laws of large numbers and other limit theorems have been established for random walks (the discrete analogues of Lévy processes). It is, of course, desirable to determine what are the properties which ensure that analogues of those fundamental results hold for convolution-like operators constructed via the above procedure.

Scope of this book

Following the motivations above, this book gives an account of the construction of generalized convolution operators and measure algebras that are natural to a given diffusion or equivalently, to a given elliptic operator. This includes presentation of the corresponding analogues of classic harmonic analysis and Lévy processes. However, the presentation of these topics is kept at a fairly basic level. Harmonic analysis is mostly limited to the construction of analogues of Fourier transforms, while the discussion of Lévy processes is centered on Lévy–Khintchine-like representations, the characterization of certain processes playing a role analogous to Brownian motion in classic stochastic calculus, and basic sample paths properties. This is due to two main reasons: First, harmonic analysis and Lévy processes are very rich fields, with many ramifications into other fields of mathematics. Thus, a deeper study of these topics would require a much larger volume. Stark and possibly arbitrary choices in selection of subtopics would be required. Second, and most important, a large part of such study is not yet done. We trust that our presentation of the core building blocks will convince the reader that rather complete analogues to classical theories are not only feasible but also desirable.

1.2 Organization of the Book

Unlike previous monographs on generalized harmonic analysis [15, 19, 184], the presentation of the material in this book proceeds from the particular to the general. In our view, this reflects the natural flow of progression in mathematical reasoning, and stimulates the reader to pursue further generalizations of the results stated throughout the text.

To make this book accessible to a wider audience, in Chap. 2 we start by presenting a comprehensive summary of background material on stochastic processes, harmonic analysis and Sturm–Liouville theory.

Chapter 3 is devoted to the construction of the *Whittaker convolution*, a convolution-like structure associated with the Shiryaev process, whose infinitesimal generator is the Sturm–Liouville operator $x^2 \frac{d^2}{dx^2} + (1 + 2(1 - \alpha)x)\frac{d}{dx}$, where α is a real parameter. We open the chapter by explaining why this diffusion process is a natural starting point for our exposition. As we will see, this has to do not only with the importance and the numerous applications of this diffusion, but also with the fact that its connection with the Kontorovich-Lebedev transform provides us with the required product formula for the particular case $\alpha = 0$.

In the general case, the generalized eigenfunctions of the generator are written in terms of the Whittaker W function with first parameter α. In Chap. 3 we state and prove a product formula for the Whittaker W function whose kernel does not depend on the second parameter and is given in closed form in terms of the parabolic cylinder function. Our method is based on special function theory and standard integral transform techniques. Furthermore, we show that if $\alpha < \frac{1}{2}$ then the convolution-like operator induced by the product formula has the property that the convolution of probability measures is a probability measure, and therefore defines a measure algebra in which the Shiryaev process becomes a Lévy-like process. We also provide a Lévy–Khintchine type theorem which describes the general form of an index Whittaker transform of a Lévy-like process, and we show that the Shiryaev process admits a martingale characterization analogous to Lévy's characterization of Brownian motion.

Chapter 3 also contains three other results of independent interest: an integral representation for the Whittaker function as a Fourier transform of a parabolic cylinder function, an analogue of the Wiener-Lévy theorem for the index Whittaker transform, and an existence and uniqueness theorem for a family of convolution-type integral equations. An example is provided where the existence and uniqueness theorem yields an explicit expression for the solution of an integral equation with the Whittaker function in the kernel.

A property of the Whittaker convolution is that the support of the measures of the underlying product formula is the whole half-line. This fact is remarkable because, in contrast, the so-called Sturm–Liouville hypergroups—a class of convolutions associated with one-dimensional diffusions which have been extensively studied in previous literature [19, 163, 210], and whose general construction is based on the associated hyperbolic PDE introduced above—have the property that the measures of the underlying product formula have compact support. This distinction raises a natural question, namely whether it is possible to construct other one-dimensional convolutions where the measures of underlying product formula do not have compact support.

A positive answer is given in Chap. 4, where we develop a unified approach for constructing Sturm–Liouville convolutions whose supports can be either compact or noncompact. Our technique is based on the hyperbolic PDE approach, which is shown to be extensible to a larger class of Sturm–Liouville operators whose associated hyperbolic equations are possibly degenerate at the initial line. The extension relies on an existence and uniqueness theorem for hyperbolic Cauchy problems which is useful in itself, as it covers many parabolically degenerate cases which are outside the scope of the classical theory and for which it is not even clear whether the Cauchy problem is well-posed or not. We also introduce a regularization method which makes use of the properties of the diffusion semigroup to construct a sequence of regularized product formulas, from which the desired product formula is obtained via a weak convergence argument. Many probabilistic properties of the Whittaker convolution, such as the interpretation of the associated diffusion as a

Lévy-like process or the Lévy–Khintchine type theorem, extend to this general family of Sturm–Liouville convolutions.

The convolutions constructed in Chap. 4 satisfy the compactness axiom if and only if the hyperbolic equation determined by the Sturm–Liouville operator is uniformly hyperbolic on its domain. If this is the case, then one can check that the convolution satisfies all the axioms of hypergroups; this leads to an improvement of previous existence theorems for Sturm–Liouville hypergroups. In turn, the case where the hyperbolic PDE is parabolically degenerate yields a general family of degenerate Sturm–Liouville hypergroups which includes the Whittaker convolution as a particular case.

The results described thus far are restricted to convolution structures for one-dimensional diffusions, but our opening discussion makes it clear that the problem of constructing convolution-like operators associated with diffusion processes is meaningful in a much more general framework. In Chap. 5 we study the construction of convolutions for diffusions on a general locally compact separable metric space. We start by identifying the requirements that such a convolution should satisfy in order to allow for the development of the basic notions of probabilistic harmonic analysis, and we then determine necessary and sufficient conditions which relate the existence of the convolution structure with certain properties of the eigenfunctions of the generator. One of the necessary conditions is that the eigenfunctions should have a common maximizer, which is quite restrictive in dimension greater than 1; this explains in part why a significant part of the literature on generalized harmonic analysis is devoted to problems on one dimensional spaces.

Using standard results on spectral theory of differential operators, one can prove that the common maximizer property does not hold for reflected Brownian motions on smooth domains of $\mathbb{R}^d$ ($d \geq 2$) or on compact Riemannian manifolds; this leads to a nonexistence theorem for convolutions on such domains. Going back to the one-dimensional problem, (the failure of) the common maximizer property will also be shown to yield nonexistence theorems for some one-dimensional diffusions which are not covered in the preceding chapter.

Another difficulty which arises in the multidimensional setting is that the associated hyperbolic equation becomes ultrahyperbolic, and therefore the PDE approach for the construction of convolutions is only applicable if the elliptic operator admits separation of variables. This is a significant limitation, but it does not hinder the construction of nontrivial multidimensional convolutions, as there are many elliptic operators which admit separation of variables but do not decompose trivially into a product of one-dimensional operators. In the final section of Chap. 5 we discuss the interesting example of the Laplace–Beltrami operator on a general class of two-dimensional manifolds endowed with cone-like metrics. The product formula for the generalized eigenfunctions is shown to depend on one of the two spectral parameters; accordingly, it induces a family of convolution operators (rather than a single convolution). This structure gives rise to a Lévy-like representation for the reflected Brownian motion on the manifold, together with other analogues of the one-dimensional results.

Appendix A collects some open problems which naturally arise from the present work.

Part of the results presented in this book have appeared in our papers [172–176], but are here presented in a streamlined and more self-contained way, often with improvements to the original formulation of the theorems.

Chapter 2
Preliminaries

Our opening discussion in the introductory chapter sketched some connections between the problem of constructing generalized convolutions and various fields of mathematics such as harmonic analysis, stochastic processes, differential equations, spectral theory and special functions. In the first three sections of this chapter, we review some prerequisite notions and facts from these fields. Section 2.4 closes the chapter with an example of a generalized convolution (the Kingman convolution), providing motivation and a benchmark for the results in further chapters.

For brevity, our treatment is limited to the essential. Proofs are omited, being replaced by appropriate bibliographic references, whenever such references exist and are easily accessible. Since the topics are dispersed among various fields, there is no single work, or a small number of works, covering all the topics in this chapter. To help those readers wishing to study parts of this background in some depth, we provide some guidance: For Markov processes, associated operators and semigroups, the classical book by S. Ethier and T. Kurtz [57] is an excellent source. For one-dimensional diffusions, we suggest M. Fukushima's paper [65], and for infinitely divisible distributions and Lévy processes K. Sato's classical book [166]. G. Teschl's book [185] contains the essentials of Sturm-Liouville theory in a compact and accessible presentation, but readers may need to complement it with papers [53] and [122]. Concerning harmonic analysis on generalized convolution structures, the most important background material for this book can be found in W.R. Bloom and H. Heyer's monography [19].

2.1 Continuous-Time Markov Processes

In what follows we write P_{x_0} for the distribution of a given time-homogeneous Markov process started at the point x_0 and $\mathbb{E}_{x_0}$ for the associated expectation operator. A sequence of finite complex measures μ_n on a locally compact separable

R. Sousa et al., *Convolution-like Structures, Differential Operators and Diffusion Processes*, Lecture Notes in Mathematics 2315,
https://doi.org/10.1007/978-3-031-05296-5_2

metric space E is said to *converge weakly* (respectively, *vaguely*) to the finite complex measure μ if $\lim_n \int_E g(\xi)\mu_n(d\xi) = \int_E g(\xi)\mu(d\xi)$ for all $g \in \mathrm{C_b}(E)$ (respectively, for all $g \in \mathrm{C_c}(E)$). It is clear that weak convergence implies vague convergence. The converse is true if the sequence of total variations $\|\mu_n\|$ converges to the total variation of μ (for example, if μ_n and μ are probability measures), but may fail if μ_n is a generic sequence.

Feller Semigroups and Processes

A family $\{T_t\}_{t\geq 0}$ of linear operators from a Banach space $\mathcal{V}$ to itself is said to be a *semigroup* if $T_0 = \mathrm{Id}$ (the identity operator on $\mathcal{V}$) and $T_{t+s} = T_t T_s$ for all $t, s \geq 0$. The semigroup is said to be a *contraction semigroup* if $\|T_t\| \leq 1$ for all t and *strongly continuous* if for all $f \in \mathcal{V}$ we have $\|T_t f - f\| \longrightarrow 0$ as $t \downarrow 0$. A *Feller semigroup* on a locally compact separable metric space E is a strongly continuous contraction semigroup of positive operators on the $\mathrm{C_0}(E)$ provided with the supremum norm. A time-homogeneous Markov process $\{X_t\}_{t\geq 0}$ with state space E is called a *Feller process* if its transition semigroup $(T_t f)(x) := \mathbb{E}_x[f(X_t)]$ $(f \in \mathrm{C_0}(E))$ is a Feller semigroup.

Given a Feller semigroup $\{T_t\}_{t\geq 0}$ on E, one can use the Riesz representation theorem [10, §29] to write it as $(T_t f)(x) = \int_E f(y)\, p_{t,x}(dy)$, where $\{p_{t,x}(\cdot)\}$ is a uniquely defined family of sub-probability measures on E which is vaguely continuous in x (i.e. $p_{t,x_n}(\cdot) \xrightarrow{v} p_{t,x}(\cdot)$ whenever $x_n \to x$). We can then use the standard Kolmogorov consistency theorem to construct a Markov process $\{X_t\}$ on E such that $(T_t f)(x) := \mathbb{E}_x[f(X_t)]$ (cf. [9, §36]). Therefore, Feller processes are in one-to-one correspondence with Feller semigroups. Moreover, this representation allows us to define the natural extension of a Feller semigroup to a semigroup of operators on $\mathrm{B_b}(E)$ as

$$(T_t f)(x) = \int_E f(y)\, p_{t,x}(dy), \qquad f \in \mathrm{B_b}(E).$$

If this extension is such that $T_t\big(\mathrm{B_b}(E)\big) \subset \mathrm{C_b}(E)$ for all $t > 0$, then we say that $\{T_t\}$ is a *strong Feller semigroup* and $\{X_t\}$ is a *strong Feller process*.

A Feller semigroup $\{T_t\}_{t\geq 0}$ on E (and the associated Feller process) is said to be *conservative* if $T_t \mathbb{1} = \mathbb{1}$ for all t (where $\mathbb{1}$ denotes the function identically equal to one) or, equivalently, if $\{p_{t,x}(\cdot)\}$ is a family of probability measures on E. In this case the family $\{p_{t,x}(\cdot)\}$ is weakly continuous in x. If $\{T_t\}_{t\geq 0}$ is a conservative Feller semigroup, then the strong continuity on $\mathrm{C_0}(E)$ extends to local uniform continuity on $\mathrm{C_b}(E)$, i.e. we have $\lim_{t\downarrow 0} T_t f = f$ uniformly on compact sets for all $f \in \mathrm{C_b}(E)$.

The *(infinitesimal) generator* $\big(\mathcal{G}, \mathcal{D}(\mathcal{G})\big)$ of a strongly continuous semigroup $\{T_t\}_{t\geq 0}$ on a Banach space $\mathcal{V}$ is the (generally unbounded) operator

$$\mathcal{D}(\mathcal{G}) := \Big\{ f \in \mathcal{V} \;\Big|\; \lim_{t\downarrow 0} \tfrac{1}{t}(T_t f - f) \text{ exists in the topology of } \mathcal{V} \Big\},$$

$$\mathcal{G} : \mathcal{D}(\mathcal{G}) \subset \mathcal{V} \longrightarrow \mathcal{V}, \qquad \mathcal{G}f := \lim_{t\downarrow 0} \tfrac{1}{t}(T_t f - f).$$

In particular, the domain of the infinitesimal generator of a Feller semigroup $\{T_t\}_{t\geq 0}$ is the set $\mathcal{D}(\mathcal{G}^{(0)})$ of functions $f \in C_0(E)$ such that $\lim_{t\downarrow 0} \frac{1}{t}(T_t f(x) - f(x))$ exists as a uniform limit. Infinitesimal generators have the following property:

Proposition 2.1 ([57, Chapter 1, Corollary 1.6]) *The infinitesimal generator of any strongly continuous semigroup on a Banach space is a densely defined closed linear operator.*

Further, using Theorem 1.24 in Davies' book [42], it is possible to show that in the case of Feller semigroups, pointwise convergence of $\frac{1}{t}(T_t f(x) - f(x))$ can be substituted for uniform convergence:

Proposition 2.2 *Let $\{T_t\}$ be a Feller semigroup. The domain $\mathcal{D}(\mathcal{G}^{(0)})$ of its infinitesimal generator is the subspace*

$$\mathcal{D}(\mathcal{G}^{(0)}) = \left\{ f \in C_0(E) \,\middle|\, \begin{array}{l} \exists g \in C_0(E) \text{ such that} \\ g(x) = \lim_{t\downarrow 0} \frac{1}{t}\big((T_t f)(x) - f(x)\big) \text{ for all } x \in E \end{array} \right\}.$$

Concerning the sample path properties of Feller processes, it is clear that strong continuity of the Feller semigroup in $C_0(E)$ implies stochastic continuity of the corresponding Feller process $\{X_t\}_{t\geq 0}$, i.e., for every $t \geq 0$, we have

$$\lim_{s\to t} P\left[\mathbf{d}(X_s, X_t) > \varepsilon\right] = 0 \qquad \forall \varepsilon > 0,$$

where $\mathbf{d}(\cdot,\cdot)$ is the distance on E. Further, the following important regularity result holds:

Proposition 2.3 ([57, Chapter 4, Theorem 2.7]) *Let $\{T_t\}_{t\geq 0}$ be a Feller semigroup on E. For each $\mu \in \mathcal{P}(E)$, there exists a càdlàg Feller process $\{X_t\}_{t\geq 0}$ corresponding to $\{T_t\}$ with initial distribution μ. If $\{\boldsymbol{F}_t^X\}_{t\geq 0}$ is the filtration generated by the process $\{X_t\}$, then $\{X_t\}$ is strong Markov with respect to the filtration $\{\boldsymbol{F}_{t+}^X = \bigcap_{s>t} \boldsymbol{F}_s^X\}_{t\geq 0}$.*

In particular, every Feller process $\{X_t\}$ has a càdlàg modification, i.e. there exists a Feller process $\{\widetilde{X}_t\}$ such that $P[\widetilde{X}_t = X_t] = 1$ for all $t \geq 0$, and the sample path $t \mapsto \widetilde{X}_t(\omega)$ is for almost every (a.e.) ω right continuous with finite left-hand limits.

The following is a useful criterion for continuity of sample paths.

Proposition 2.4 ([57, Chapter 4, Proposition 2.9 and Remark 2.10]) *Let $\{X_t\}_{t\geq 0}$ be a càdlàg Feller process on E. If for all $\varepsilon > 0$, $x \in E$ we have*

$$P_x[X_t \in E \setminus \mathbb{B}(x,\varepsilon)] = o(t) \qquad \text{as } t \downarrow 0 \tag{2.1}$$

then the paths $t \mapsto X_t(\omega)$ are continuous for a.e. ω.

In particular, if the domain $\mathcal{D}(\mathcal{G}^{(0)})$ of the infinitesimal generator of $\{X_t\}$ is such that for all $\varepsilon > 0$, $x \in E$ there exists $f \in \mathcal{D}(\mathcal{G}^{(0)})$ such that $f(x) = \|f\|$, $\sup_{y\in E\setminus\mathbb{B}(x,\varepsilon)} f(y) < \|f\|$ and $\mathcal{G}^{(0)} f(x) = 0$, then (2.1) holds and consequently the Feller process $\{X_t\}$ has almost surely (a.s.) continuous paths.

Conversely, it is possible to prove some properties of infinitesimal generators starting from sample path properties of the corresponding Feller process. We will use the following:

Proposition 2.5 ([25, Theorem 1.40]) *Let* $\{X_t\}_{t\geq 0}$ *be a Feller process on* $\mathbb{R}^d$ *whose paths are a.s. continuous. Then the infinitesimal generator* $(\mathcal{G}^{(0)}, \mathcal{D}(\mathcal{G}^{(0)}))$ *is a local operator, i.e. we have* $(\mathcal{G}^{(0)} f_1)(x) = (\mathcal{G}^{(0)} f_2)(x)$ *whenever* $f_1, f_2 \in \mathcal{D}(\mathcal{G}^{(0)})$ *and* $f_1|_{\mathbb{B}(x,\varepsilon)} = f_2|_{\mathbb{B}(x,\varepsilon)}$ *for some* $\varepsilon > 0$.

The following criterion provides a sufficient condition for a Feller semigroup to have the strong Feller property.

Proposition 2.6 ([25, Theorem 1.14]) *Let* $\{T_t\}_{t\geq 0}$ *be a conservative Feller semigroup, and assume that the kernels* $p_{t,x}$ *can be written as*

$$p_{t,x}(dy) = p_t(x, y)\mu(dy),$$

where μ *is a positive Borel measure on* E *and the function* $p_t(\cdot, \cdot)$ *is locally bounded on* $E \times E$ *for each* $t > 0$. *Then* $\{T_t\}_{t\geq 0}$ *is a strong Feller semigroup.*

Martingales and Local Martingales

An adapted integrable stochastic process $\{X_t\}_{t\geq 0}$ on a filtered probability space $(\Omega, \mathbf{F}, \{\mathbf{F}_t\}, P)$ is called a *martingale* if

$$\mathbb{E}[|X_t|] < \infty, \qquad \mathbb{E}[X_t|\mathbf{F}_s] = X_s, \qquad 0 \leq s \leq t \leq \infty.$$

The systematic study of connections between martingales and Feller processes was initiated by D.Stroock and S.Varadhan in the 1960s [181]. Here, we give the following theorem:

Theorem 2.1 ([57, Chapter 4, Proposition 1.7 and Theorem 4.1]) *Let* $\{X_t\}_{t\geq 0}$ *be a càdlàg Feller process on a locally compact separable metric space E with initial distribution* $\mu = P[X_0 \in \cdot]$ *and let* $(\mathcal{G}^{(0)}, \mathcal{D}(\mathcal{G}^{(0)}))$ *be its generator. Let* $\mathcal{D}$ *be a* core *of* $\mathcal{D}(\mathcal{G}^{(0)})$, *i.e. a subset* $\mathcal{D} \subset \mathcal{D}(\mathcal{G}^{(0)})$ *such that* $(\mathcal{G}^{(0)}, \mathcal{D}(\mathcal{G}^{(0)}))$ *is the closure of the operator* $(\mathcal{G}^{(0)}, \mathcal{D})$. *For each* $f \in \mathcal{D}$, *the process*

$$f(X_t) - f(X_0) - \int_0^t (\mathcal{G}^{(0)} f)(X_s)\, ds, \qquad t \geq 0 \tag{2.2}$$

is a martingale with respect to the same filtration for which $\{X_t\}$ *is a Markov process. Moreover,* $\{X_t\}$ *is the unique (in distribution) E-valued stochastic process with initial distribution* μ *such that the process defined by* (2.2) *is a martingale for any* $f \in \mathcal{D}$.

A *stopping time* on the filtered probability space $(\Omega, \mathbf{F}, \{\mathbf{F}_t\}, P)$ is a random variable $\tau : \Omega \mapsto [0, \infty]$ such that $\{\tau \leq t\} \in \mathbf{F}_t$ for all $t \geq 0$. A stochastic process $\{X_t\}_{t\geq 0}$ on the space $(\Omega, \mathbf{F}, \{\mathbf{F}_t\}, P)$ is said to be a *local martingale* if there exists an increasing sequence $\{\tau_n\}_{n\in\mathbb{N}}$ of stopping times with $\lim_n \tau_n = +\infty$

a.s. and such that for every $n \in \mathbb{N}$ the process $\{X_{t\wedge\tau_n}\}_{t\geq 0}$ is a martingale, where $X_{t\wedge\tau_n} = X_t \mathbb{1}_{\{\tau_n \geq t\}} + X_{\tau_n}\mathbb{1}_{\{\tau_n < t\}}$.

The fundamental martingale characterization of Brownian motion is stated below. Here and later, we denote by $\{[X]_t\}_{t\geq 0}$ the *quadratic variation* of a stochastic process $\{X_t\}_{t\geq 0}$, defined as $[X]_t = \lim \sum_{j=0}^{m(\pi^n)-1}(X_{t_{j+1}^n} - X_{t_j^n})^2$, where the limit is in probability, taken over all sequences of partitions $\pi^n = \{0 = t_0^n < t_1^n < \ldots < t_{m(\pi^n)}^n = t\}$ such that $\max_j(t_{j+1}^n - t_j^n) \to 0$ as $n \to \infty$.

Theorem 2.2 (Lévy's Characterization of Brownian Motion [102, Theorem 25.28]) *Let $\{X_t\}_{t\geq 0}$ be a local martingale on $\mathbb{R}$ with a.s. continuous paths and $X_0 = 0$. The following are equivalent:*

- *(i) $\{X_t\}$ is a standard Brownian motion (i.e. for each $s < t$, the random variable $X_t - X_s$ is normally distributed with mean zero and variance $t - s$ and is independent of $\{X_u : u \leq s\}$);*
- *(ii) $\{X_t^2 - t\}_{t\geq 0}$ is a local martingale;*
- *(iii) $[X]_t = t$ for all $t \geq 0$.*

One-Dimensional Diffusion Processes

One-dimensional diffusion processes and associated Sturm-liouville operators play a central role in the contents of this book.

A *diffusion process* on an open interval $I \subset \mathbb{R}$ is a strong Markov process $\{X_t\}_{t\geq 0}$ with state space I and such that $t \mapsto X_t(\omega)$ is continuous for a.e. ω. We say that $\{X_t\}_{t\geq 0}$ is an *irreducible diffusion* if, in addition, we have $P_x(\tau_y < \infty) > 0$ for all $x, y \in I$, where $\tau_y = \inf\{t \geq 0 | X_t = y\}$.

An irreducible diffusion process $\{X_t\}$ on an open interval $I \subset \mathbb{R}$ is not in general a Feller process. However, it is $\mathrm{C_b}$-*Feller* in the sense that its transition semigroup $(T_t f)(x) = \mathbb{E}[f(X_t)]$ is such that $T_t\bigl(\mathrm{C_b}(I)\bigr) \subset \mathrm{C_b}(I)$. We can therefore define the *η-resolvent operator* of $\{X_t\}$ as

$$\mathcal{R}_\eta : \mathrm{C_b}(I) \longrightarrow \mathrm{C_b}(I), \qquad \mathcal{R}_\eta f := \int_0^\infty e^{-\eta t}\, T_t f\, dt \qquad (\eta > 0) \tag{2.3}$$

and define the $\mathrm{C_b}$-*generator* $\bigl(\mathcal{G}^{(b)}, \mathcal{D}(\mathcal{G}^{(b)})\bigr)$ of $\{X_t\}$ as the operator with domain $\mathcal{D}(\mathcal{G}^{(b)}) = \mathcal{R}_\eta\bigl(\mathrm{C_b}(I)\bigr)$, given by

$$(\mathcal{G}^{(b)}u)(x) = \eta u(x) - g(x) \qquad \text{for } u = \mathcal{R}_\eta g,\ \ g \in \mathrm{C_b}(I),\ \ x \in I. \tag{2.4}$$

It is possible to show that $\mathcal{G}^{(b)}$ is independent of η, (cf. [65, p. 295]). Further, if $\{T_t\}$ is a Feller semigroup with generator $\bigl(\mathcal{G}^{(0)}, \mathcal{D}(\mathcal{G}^{(0)})\bigr)$, then we have $\mathcal{D}(\mathcal{G}^{(0)}) = \mathcal{R}_\eta\bigl(\mathrm{C_0}(I)\bigr)$ and $\mathcal{G}^{(0)} f = \mathcal{G}^{(b)} f$ for all $f \in \mathcal{D}(\mathcal{G}^{(0)})$.

A *canonical scale* s is a strictly increasing continuous function $s : I \longrightarrow \mathbb{R}$. A *speed measure* m is a positive Radon measure on I with support $\mathrm{supp}(m) = I$. We say that (s, m, k) is a *canonical triplet* if s is a canonical scale, m is a speed measure and k is a positive Radon measure on I (called the *killing measure*).

The following theorem provides a correspondence between (generators of) one-dimensional diffusions and canonical triplets:

Theorem 2.3 ([65, Section 2.2]) *If $\{X_t\}$ is an irreducible diffusion process on the open interval I, then there exists a canonical triplet (s, m, k) on I such that*

$$(\mathcal{G}^{(b)} f)(x) = \frac{dD_s f - f dk}{dm}(x), \qquad f \in \mathcal{D}(\mathcal{G}^{(b)}), \; x \in I \tag{2.5}$$

(where $D_s f(x) := \lim_{\varepsilon \downarrow 0} \frac{f(x+\varepsilon)-f(x)}{s(x+\varepsilon)-s(x)}$) in the sense that the measure $dD_s f - f dk$ is absolutely continuous with respect to dm and the corresponding Radon-Nikodym derivative has a representative which belongs to $\mathrm{C_b}(I)$ and is equal to $\mathcal{G}^{(b)} f$.

Conversely, if (s, m, k) is an arbitrary canonical triplet on I, then there exists an irreducible diffusion process $\{X_t\}$ on I whose $\mathrm{C_b}$-generator is given by (2.5) *for all $f \in \mathcal{D}(\mathcal{G}^{(b)})$.*

Let $\{X_t\}$ be a diffusion on $I = (a, b)$, where $-\infty \le a < b \le +\infty$, and let (s, m, k) be its canonical triple. Write $j = m + k$, and for $c \in I$ consider the integrals

$$I_a = \int_a^c \int_a^y ds(x)\, j(dy), \qquad J_a = \int_a^c \int_y^c ds(x)\, j(dy),$$

$$I_b = \int_c^b \int_y^b ds(x)\, j(dy), \qquad J_b = \int_c^b \int_c^y ds(x)\, j(dy).$$

The endpoint $e \in \{a, b\}$ is said to be:

$$\begin{aligned} &\textit{regular} \text{ if } I_e < \infty,\ J_e < \infty; && \textit{entrance} \text{ if } I_e = \infty,\ J_e < \infty; \\ &\textit{exit} \quad\ \text{ if } I_e < \infty,\ J_e = \infty; && \textit{natural} \text{ if } I_e = \infty,\ J_e = \infty, \end{aligned} \tag{2.6}$$

this classification being independent of the choice of c. This is the so-called *Feller boundary classification* of the diffusion process $\{X_t\}$, and it determines the behaviour of $\{X_t\}$ near the endpoints of (a, b) (see [89] for details). In particular, if the endpoint e is entrance or natural then the process $\{X_t\}$ cannot reach e in finite time. Moreover, if neither a nor b is a regular endpoint then there exists a unique diffusion process $\{X_t\}$ on (a, b) with canonical triple (s, m, k). (Uniqueness here means that the domain $\mathcal{D}(\mathcal{G}^{(b)})$ of the $\mathrm{C_b}$-generator is uniquely determined by (s, m, k), see [65].)

Theorem 2.3 ensures, in particular, that each second-order differential operator of the form

$$\mathfrak{a}(x) \frac{d^2}{dx^2} + \mathfrak{b}(x) \frac{d}{dx} - \mathfrak{c}(x) \qquad (x \in I),$$

where $\mathfrak{a}, \mathfrak{b}, \mathfrak{c} \in C(I)$ with $\mathfrak{a} > 0$ and $\mathfrak{c} \geq 0$ on I, is the generator of an irreducible diffusion process $\{X_t\}_{t\geq 0}$. The associated canonical triplet is

$$s(x) = \int_{x_0}^{x} e^{-B(y)} dy, \qquad m(dx) = \frac{e^{B(x)}}{\mathfrak{a}(x)} dx, \qquad k(dx) = \mathfrak{c}(x) \frac{e^{B(x)}}{\mathfrak{a}(x)} dx,$$

where $B(x) := \int_{x_0}^{x} \frac{\mathfrak{b}(\xi)}{\mathfrak{a}(\xi)} d\xi$ and $x_0 \in I$ is arbitrary.

Consider the stochastic differential equation (SDE)

$$dX_t = \mathfrak{b}(X_t)\, dt + \sigma(X_t)\, dW_t, \tag{2.7}$$

where $\sigma(x) = \sqrt{2\mathfrak{a}(x)}$ and $\{W_t\}_{t\geq 0}$ is a standard Brownian motion. An I-valued stochastic process $\{X_t\}_{t\geq 0}$ is said to be a *solution* of the SDE (2.7) if it satisfies the integral equation $X_t = X_0 + \int_0^t \mathfrak{b}(X_s) ds + \int_0^t \sigma(X_s) dW_s$, where the latter term is a stochastic integral with respect to the standard Brownian motion. By [99, Chapter 5, Theorem 5.15], the SDE (2.7) has a unique weak solution $\{X_t\}$ up to a possibly finite lifetime $\zeta := \inf\{t \geq 0 \mid X_t \notin I\}$. If both endpoints a and b are entrance or natural, then it follows from [99, Chapter 5, Theorem 5.29] that $\{X_t\}$ is a diffusion process on I whose lifetime is infinite a.s. and whose generator is the differential operator $\mathfrak{a}(x)\frac{d^2}{dx^2} + \mathfrak{b}(x)\frac{d}{dx}$.

Even though diffusions are not always Feller processes on the open interval I, they become Feller processes after a suitable extension to the boundaries of the interval:

Proposition 2.7 ([65, Sections 4 and 6]) *Let $\{X_t\}$ be an irreducible diffusion on $I = (a, b)$, and let $\overline{I}$ be the interval obtained by attaching the regular or entrance endpoints of $\{X_t\}$ to the interval I. Then there exists a Feller process $\{\overline{X}_t\}$ with state space $\overline{I}$ satisfying the following conditions:*

- *The process $\{\overline{X}_t\}$ is an extension of $\{X_t\}$, in the sense that $X_t(\omega) = \overline{X}_t(\omega)$ for $0 \leq t \leq \tau_I(\omega) := \inf\{t \geq 0 \mid X_t(\omega) \notin I\}$;*
- *If $a \in \overline{I}$ (respectively $b \in \overline{I}$), then $\{\overline{X}_t\}$ is instantaneously reflecting at the endpoint a (resp. b), in the sense that we have $P_x[\overline{X}_t \neq a$ for a.e. $t \geq 0] = 1$ (resp. $P_x[\overline{X}_t \neq b$ for a.e. $t \geq 0] = 1$);*
- *The transition semigroup of $\{\overline{X}_t\}$ is a Feller semigroup whose infinitesimal generator is given by*

$$\mathcal{D}(\mathcal{G}^{(0)}) = \left\{ u \in C_0(\overline{I}) \;\middle|\; \begin{array}{l} \frac{dD_s f - f dk}{dm} \in C_0(\overline{I}) \\ D_s u(a) = 0 \text{ if the endpoint } a \text{ is regular or entrance} \\ D_s u(b) = 0 \text{ if the endpoint } b \text{ is regular or entrance} \end{array} \right\},$$

$$(\mathcal{G}^{(0)} f)(x) = \frac{dD_s f - f dk}{dm}(x) \qquad \big(f \in \mathcal{D}(\mathcal{G}^{(0)}),\ x \in I\big).$$

If $\{X_t\}$ has no regular endpoints, then $\{\overline{X}_t\}$ is the unique extension of $\{X_t\}$ to a strong Markov process with continuous paths on the interval $\overline{I}$.

Lévy Processes, Infinitely Divisible Distributions and Convolution Semigroups
A stochastic process $\{X_t\}_{t\geq 0}$ on $\mathbb{R}^d$ with $X_0 = 0$ is said to be a *Lévy process* if it is stochastically continuous, has independent increments (i.e. $X_t - X_s$ is independent of $\{X_u : u \leq s\}$ for all $s < t$) and has stationary increments (i.e. $X_{t+s} - X_s$ has the same distribution as $X_t - X_0$ for all $t, s \geq 0$).

It is clear from this definition that any drifted Brownian motion on $\mathbb{R}^d$ started at zero (i.e. any process of the form $B_t = \alpha t + A W_t$ with $\alpha \in \mathbb{R}^d$, A a symmetric nonnegative definite $d \times d$-matrix and $\{W_t\}$ a d-dimensional standard Brownian motion) is a Lévy process.

In the definition of Lévy process, some authors also require that $\{X_t\}$ is càdlàg. This is unimportant because of the following proposition:

Proposition 2.8 ([166, Theorem 11.5]) *If $\{X_t\}_{t\geq 0}$ is a Lévy process on $\mathbb{R}^d$, then:*

(a) *$\{X_t\}$ has a càdlàg modification;*
(b) *The transition semigroup $(T_t f)(x) := \mathbb{E}^x[f(X_t)] \equiv \mathbb{E}[f(X_t + x)]$ is a Feller semigroup on $\mathbb{R}^d$ (and, therefore, $\{X_t\}$ is a Feller process).*

There is a one-to-one correspondence between Lévy processes, convolution semigroups and infinitely divisible distributions. Before stating this result, we recall some notions. The *(ordinary) convolution* of two measures $\mu, \nu \in \mathcal{M}_\mathbb{C}(\mathbb{R}^d)$ is defined by $(\mu * \nu)(B) := \int_{\mathbb{R}^d} \int_{\mathbb{R}^d} \delta_{x+y}(B)\mu(dx)\nu(dy)$ for each Borel subset $B \subset \mathbb{R}^d$. A probability measure $\mu \in \mathcal{P}(\mathbb{R}^d)$ is said to be *infinitely divisible* if for each integer $n \in \mathbb{N}$, there exists a measure $\nu_n \in \mathcal{P}(\mathbb{R}^d)$ such that $\mu = \nu_n^{*n}$, where ν_n^{*n} denotes the n-fold convolution of ν_n with itself. A family $\{\mu_t\}_{t\geq 0} \subset \mathcal{P}(\mathbb{R}^d)$ is called a *convolution semigroup* if we have $\mu_s * \mu_t = \mu_{s+t}$ $(s, t \geq 0)$, $\mu_0 = \delta_0$ and $\mu_t \overset{w}{\longrightarrow} \delta_0$ as $t \downarrow 0$.

Proposition 2.9 ([166, Theorem 7.10]) *Let $\{X_t\}_{t\geq 0}$ be a Feller process on $\mathbb{R}^d$. The following assertions are equivalent:*

(i) *$\{X_t\}$ is a Lévy process;*
(ii) *There exists a convolution semigroup $\{\mu_t\}_{t\geq 0} \subset \mathcal{P}(\mathbb{R}^d)$ such that $\mathbb{E}[f(X_t)] = \int_{\mathbb{R}^d} f(y)\mu_t(dy)$ for each $f \in \mathrm{B_b}(\mathbb{R}^d)$.*

If these conditions hold then μ_t is, for all $t \geq 0$, an infinitely divisible measure. Conversely, if $\mu \in \mathcal{P}(\mathbb{R}^d)$ is an infinitely divisible measure then there exists a Lévy process $\{X_t\}$ such that $\mathbb{E}[f(X_1)] = \int_{\mathbb{R}^d} f(y)\mu(dy)$ for $f \in \mathrm{B_b}(\mathbb{R}^d)$.

Given a $\mathbb{R}^d$-valued random variable X with law $\mu = \mathbb{P}[X \in \cdot\,]$, the *Fourier transform* of μ (also called the *characteristic function* of X) is defined as $(\mathfrak{F}\mu)(z) := \mathbb{E}[e^{iz\cdot X}] \equiv \int_{\mathbb{R}^d} e^{iz\cdot x}\mu(dx)$ $(z \in \mathbb{R}^d)$. The celebrated Lévy-Khintchine formula provides an explicit characterization of the characteristic function (or Fourier transform of the law) of a Lévy process:

Theorem 2.4 (Lévy-Khintchine Representation [166, Theorem 8.1]) *Let $\{X_t\}_{t\geq 0}$ be a Lévy process and $\{\mu_t\} \subset \mathcal{P}(\mathbb{R}^d)$ the associated convolution semigroup. We have*

$$\mathbb{E}[e^{iz\cdot X_t}] \equiv (\mathfrak{F}\mu_t)(z) = e^{-t\phi(z)} \qquad (t \geq 0,\ z \in \mathbb{R}^d) \tag{2.8}$$

for some function $\phi(\cdot)$ of the form

$$\phi(z) = z \cdot Qz + i\alpha \cdot z + \int_{\mathbb{R}^d\setminus\{0\}} \biggl(1 - e^{iz\cdot y} + \frac{iz\cdot y}{1+|y|^2}\biggr)\nu(dy), \tag{2.9}$$

where Q is a symmetric nonnegative definite $d \times d$-matrix, $\alpha \in \mathbb{R}^d$ and ν is a Lévy measure *on $\mathbb{R}^d$, i.e. a positive measure on $\mathbb{R}^d \setminus \{0\}$ such that $\int_{\mathbb{R}^d\setminus\{0\}} \frac{|y|^2}{1+|y|^2}\nu(dy) < \infty$. Conversely, for any function $\phi(\cdot)$ of the form* (2.9) *there exists a convolution semigroup $\{\mu_t\} \subset \mathcal{P}(\mathbb{R}^d)$ with $(\mathfrak{F}\mu_t)(z) = e^{-t\phi(z)}$.*

The function $\phi(\cdot)$ in (2.9) is called the *Lévy symbol* of the process $\{X_t\}$. One can show that the integral term in the expression for the Lévy symbol is, for every Lévy measure ν, the characteristic function of a discontinuous Lévy process, and therefore the following result holds:

Proposition 2.10 *Let $\{X_t\}$ be a càdlàg Lévy process on $\mathbb{R}^d$ with Lévy-Khintchine representation* (2.8)–(2.9)*. The following are equivalent:*

(i) *$\{X_t\}$ has a.s. continuous paths;*
(ii) *$\nu = 0$;*
(iii) *$X_t = \alpha t + \sqrt{Q}\, W_t$, where $\{W_t\}$ is a standard Brownian motion on $\mathbb{R}^d$.*

2.2 Sturm-Liouville Theory

Sturm-Liouville operators with positive coefficients are infinitesimal generators of one-dimensional diffusions. In Chaps. 3 and 4 we will address the problem of constructing convolution-like operators associated with diffusions whose generators are (reducible to) Sturm-Liouville operators. This section collects the necessary background material from Sturm-Liouville theory.

We will consider the Sturm-Liouville expression

$$\ell(u)(x) := \frac{1}{r(x)}\Bigl(-(pu')'(x) + q(x)u(x)\Bigr), \qquad x \in (a,b) \subset \mathbb{R}, \tag{2.10}$$

where we assume that the coefficients are such that $p, r > 0$ and $q \geq 0$ on (a, b), p, r and q are locally integrable on (a, b) and

$$\int_a^c \int_y^c \frac{dx}{p(x)}\bigl(r(y) + q(y)\bigr)dy < \infty, \tag{2.11}$$

where $c \in (a, b)$ is arbitrary.

The Sturm-Liouville expression (2.10) is of the form $-\frac{dD_s f - f dk}{dm}$ with $s(x) = \int_{x_0}^x \frac{dy}{p(y)}$, $m(dx) = r(x)dx$ and $k(dx) = q(x)dx$. As in Sect. 2.1, an endpoint $e \in \{a, b\}$ is called regular, entrance, exit or natural according to the classification (2.6), where $I_a = \int_a^c \int_a^y \frac{dx}{p(x)}\bigl(r(y) + q(y)\bigr)dy$, $J_a = \int_a^c \int_y^c \frac{dx}{p(x)}\bigl(r(y) + q(y)\bigr)dy$, $I_b = \int_c^b \int_y^b \frac{dx}{p(x)}\bigl(r(y) + q(y)\bigr)dy$ and $J_b = \int_c^b \int_c^y \frac{dx}{p(x)}\bigl(r(y) + q(y)\bigr)dy$. The standing assumption that the coefficients $\{p, q, r\}$ satisfy (2.11) means that *the endpoint a is regular or entrance.*

2.2.1 Solutions of the Sturm-Liouville Equation

It is a basic fact from the theory of ordinary differential equations that the vector space of solutions of the Sturm-Liouville equation $\ell(u) = \lambda u$ is two-dimensional, and that a basis is formed by the (unique) solutions $u_{1,\lambda}(x)$, $u_{2,\lambda}(x)$ which satisfy the initial conditions

$$u_{1,\lambda}(c) = \sin\alpha, \quad u_{1,\lambda}'(c) = \cos\alpha, \qquad u_{2,\lambda}(c) = -\cos\alpha, \quad u_{2,\lambda}'(c) = \sin\alpha,$$

where $\alpha \in [0, \pi)$ and c is any (interior) point of the interval (a, b). When the initial conditions are instead given at an endpoint of the interval, the possibility of solving the Sturm-Liouville problem depends on the boundary classification for the set of coefficients $\{p, q, r\}$. Our starting lemma asserts that under the assumption (2.11) we have existence and uniqueness of solution for the Sturm-Liouville problem with Neumann-type condition at the left endpoint. Let us recall that an entire function $h : \mathbb{C} \longrightarrow \mathbb{C}$ is said to be of *exponential type* if there exist constants $c, M > 0$ such that $|h(z)| \leq Me^{c|z|}$ for all $z \in \mathbb{C}$.

Lemma 2.1 *The initial value problem*

$$\ell(w) = \lambda w \quad (a < x < b,\ \lambda \in \mathbb{C}), \qquad w(a) = 1, \qquad (pw')(a) = 0 \tag{2.12}$$

has a unique solution $w_\lambda(\cdot)$. Moreover, $\lambda \mapsto w_\lambda(x)$ is, for fixed x, an entire function of exponential type.

We emphasize that the boundary assumption (2.11) for this lemma includes Sturm-Liouville equations where the left endpoint can be either limit point or limit circle, while the usual existence and uniqueness theorems for Sturm-Liouville problems with initial condition at an endpoint rely on the assumption that the endpoint is limit circle (cf. e.g. [11, Section 5]). Here we recall the well-known Weyl limit point/limit circle endpoint classification: the endpoint a (respectively b) is called *limit point* if $\int_a^c |u_\lambda(x)|^2 r(x)dx = \infty$ (respectively $\int_c^b |u_\lambda(x)|^2 r(x)dx = \infty$) for some solution of $\ell(u) = \lambda u$ and *limit circle* if $\int_a^c |u_\lambda(x)|^2 r(x)dx < \infty$ (respectively $\int_c^b |u_\lambda(x)|^2 r(x)dx < \infty$) for all solutions of $\ell(u) = \lambda u$. A regular end point is always a limit circle endpoint, but an entrance endpoint can be either a limit point or a limit circle endpoint [54, Theorem 2.1]. Lemma 2.1 is not new—a special case is established in [96, Lemma 3]—but seems to be little known. We give a self-contained proof based on [96].

Proof of Lemma 2.1 Pick an arbitrary $\beta \in (a, b)$. Define $\mathfrak{s}(x) := \int_c^x \frac{d\xi}{p(\xi)}$ and $S(x) = \int_a^x (\mathfrak{s}(\beta) - \mathfrak{s}(\xi))(q(\xi) + r(\xi))d\xi$. From the boundary assumption (2.11) it follows that $0 \le S(x) \le S(\beta) < \infty$ for $x \in (a, \beta]$. Let

$$\eta_0(x;\lambda) = 1,$$

$$\eta_j(x;\lambda) = \int_a^x (\mathfrak{s}(x) - \mathfrak{s}(\xi))\eta_{j-1}(\xi;\lambda)(q(\xi) - \lambda r(\xi))d\xi \quad (j = 1, 2, \ldots). \tag{2.13}$$

One can check (using induction) that $|\eta_j(x;\lambda)| \le \frac{1}{j!}((1 + |\lambda|)S(x))^j$ for all j. Therefore, the function

$$w_\lambda(x) = \sum_{j=0}^{\infty} \eta_j(x;\lambda) \qquad (a < x \le \beta,\ \lambda \in \mathbb{C}) \tag{2.14}$$

is well-defined as an absolutely convergent series. The entireness of $\lambda \mapsto w_\lambda(x)$ follows at once from the Weierstrass theorem for compactly convergent series of holomorphic functions, and the estimate

$$|w_\lambda(x)| \le \sum_{j=0}^{\infty} \frac{1}{j!}((1 + |\lambda|)S(x))^j = e^{(1+|\lambda|)S(x)} \le e^{(1+|\lambda|)S(\beta)} \qquad (a < x \le \beta) \tag{2.15}$$

shows that $\lambda \mapsto w_\lambda(x)$ is of exponential type. For $a < x \le \beta$ we have

$$1 + \int_a^x \frac{1}{p(y)} \int_a^y w_\lambda(\xi)\,(q(\xi) - \lambda r(\xi))d\xi\,dy$$

$$= 1 + \int_a^x (\mathfrak{s}(x) - \mathfrak{s}(\xi))\,w_\lambda(\xi)(q(\xi) - \lambda r(\xi))d\xi$$

$$= 1 + \int_a^x (\mathfrak{s}(x) - \mathfrak{s}(\xi)) \Big(\sum_{j=0}^{\infty} \eta_j(\xi; \lambda) \Big) \big(q(\xi) - \lambda r(\xi)\big) d\xi$$

$$= 1 + \sum_{j=0}^{\infty} \int_a^x (\mathfrak{s}(x) - \mathfrak{s}(\xi)) \eta_j(\xi; \lambda) \big(q(\xi) - \lambda r(\xi)\big) d\xi$$

$$= 1 + \sum_{j=0}^{\infty} \eta_{j+1}(x; \lambda) \; = \; w_\lambda(x),$$

i.e., $w_\lambda(x)$ satisfies

$$w_\lambda(x) = 1 + \int_a^x \frac{1}{p(y)} \int_a^y w_\lambda(\xi) \big(q(\xi) - \lambda r(\xi)\big) d\xi \, dy. \tag{2.16}$$

This integral equation is equivalent to (2.12), so the proof is complete.

Corollary 2.1 *If* $\lambda < 0$*, then the solution of* (2.12) *is strictly increasing. If* $\lambda < 0$ *and the endpoint b is exit or natural, then the solution of* (2.12) *is unbounded.*

Proof Rewriting the functions $\eta_j(x; \lambda)$ from the proof of Lemma 2.1 as

$$\eta_j(x; \lambda) = \int_a^x \frac{1}{p(y)} \int_a^y \eta_{j-1}(\xi; \lambda) \big(q(\xi) - \lambda r(\xi)\big) d\xi \, dy,$$

we see at once (using induction on j) that each $\eta_j(\cdot; \lambda)$ is positive and strictly increasing, and therefore $w_\lambda(\cdot) = \sum_{j=0}^{\infty} \eta_j(\cdot; \lambda)$ is strictly increasing. Moreover, $\lim_{x \uparrow b} \eta_1(x; \lambda) = \int_a^b \frac{1}{p(y)} \int_a^y \big(q(\xi) - \lambda r(\xi)\big) d\xi \, dy \geq \min\{1, |\lambda|\} J_b = \infty$, hence w_λ is unbounded.

It is also worth noting that the following converse of Lemma 2.1 holds: *if* $\int_a^c \int_y^c \frac{dx}{p(x)} r(y) dy = \infty$ *(so that* (2.11) *fails to hold) then for* $\lambda < 0$ *there exists no solution of* $\ell(w) = \lambda w$ *satisfying the boundary conditions* $w(a) = 1$ and $(pw')(a) = 0$. Indeed, if the integral $\int_a^c \int_y^c \frac{dx}{p(x)} r(y) dy$ diverges, then it follows from [88, Sections 5.13–5.14] that any solution w of $\ell(w) = \lambda w$ $(\lambda < 0)$ either satisfies $w(a) = 0$ or $(pw')(a) = +\infty$, so in particular (2.12) cannot hold.

In the sequel, $\{a_m\}_{m \in \mathbb{N}}$ will denote a sequence $b > a_1 > a_2 > \ldots$ with $\lim a_m = a$. Next we verify that the solution $w_\lambda(\cdot)$ for the Sturm-Liouville equation on the interval (a, b) is approximated by the corresponding solutions on the intervals (a_m, b). We will use this fact on Chap. 4.

Lemma 2.2 *For* $m \in \mathbb{N}$ *and* $\lambda \in \mathbb{C}$*, let* $w_{\lambda,m}(x)$ *be the unique solution of the boundary value problem*

$$\ell(w) = \lambda w \quad (a_m < x < b), \qquad w(a_m) = 1, \qquad (pw')(a_m) = 0. \tag{2.17}$$

Then

$$\lim_{m\to\infty} w_{\lambda,m}(x) = w_\lambda(x) \qquad \textit{and} \qquad \lim_{m\to\infty} (pw'_{\lambda,m})(x) = (pw'_\lambda)(x) \tag{2.18}$$

pointwise for each $a < x < b$ *and* $\lambda \in \mathbb{C}$.

Proof In the same way as in the proof of Lemma 2.1 we can check that the solution of (2.17) is given by

$$w_{\lambda,m}(x) = \sum_{j=0}^{\infty} \eta_{j,m}(x;\lambda) \qquad (a_m < x < b,\ \lambda \in \mathbb{C}),$$

where $\eta_{0,m}(x;\lambda) = 1$ and $\eta_{j,m}(x;\lambda) = \int_{a_m}^{x} \big(\mathfrak{s}(x) - \mathfrak{s}(\xi)\big)\eta_{j-1,m}(\xi;\lambda)\big(q(\xi) - \lambda r(\xi)\big)d\xi$. As before we have $|\eta_{j,m}(x;\lambda)| \le \frac{1}{j!}\big((1+|\lambda|)S(x)\big)^j$ for $a_m < x \le \beta$ (where S is the function from the proof of Lemma 2.1). Using this estimate and induction on j, it is easy to see that $\eta_{j,m}(x;\lambda) \to \eta_j(x;\lambda)$ as $m \to \infty$ ($a < x \le \beta$, $\lambda \in \mathbb{C}$, $j = 0, 1, \ldots$). Noting that the estimate on $|\eta_{j,m}(x;\lambda)|$ allows us to take the limit under the summation sign, we conclude that $w_{\lambda,m}(x) \to w_\lambda(x)$ as $m \to \infty$ ($a < x \le \beta$). Finally, by (2.16) we have for $a < x \le \beta$

$$\begin{aligned}
\lim_{m\to\infty} (pw'_{\lambda,m})(x) &= -\lambda \lim_{m\to\infty} \int_{a_m}^{x} w_{\lambda,m}(\xi)\, r(\xi) d\xi \\
&= -\lambda \int_{a}^{x} w_\lambda(\xi)\, r(\xi) d\xi \\
&= (pw'_\lambda)(x),
\end{aligned}$$

using dominated convergence and the estimates $|w_{\lambda,m}(x)| \le e^{(1+|\lambda|)S(\beta)}$, $|w_\lambda(x)| \le e^{(1+|\lambda|)S(\beta)}$.

The following lemma provides a sufficient condition for the solution $w_\lambda(\cdot)$ to be uniformly bounded in the variables $x \in (a,b)$ and $\lambda \ge 0$:

Lemma 2.3 *If* $q \equiv 0$, $\lambda \ge 0$ *and* $x \mapsto p(x)r(x)$ *is an increasing function, then the solution of* (2.12) *is bounded:*

$$|w_\lambda(x)| \le 1 \qquad \textit{for all } a < x < b,\ \lambda \ge 0. \tag{2.19}$$

Proof (Adapted from [210, Proposition 4.3]) Let us start by assuming that $p(a)r(a) > 0$. For $\lambda = 0$ the result is trivial because $w_0(x) \equiv 1$. Fix $\lambda > 0$. Multiplying both sides of the differential equation $\ell(w_\lambda) = \lambda w_\lambda$ by $2pw'_\lambda$, we obtain $-\frac{1}{pr}[(pw_\lambda')^2]' = \lambda(w_\lambda^2)'$. Integrating the differential equation and then using integration by parts, we get

$$\begin{aligned}\lambda\bigl(1-w_\lambda(x)^2\bigr) &= \int_a^x \frac{1}{p(\xi)r(\xi)}\bigl((pw_\lambda')(\xi)^2\bigr)'d\xi \\ &= \frac{(pw_\lambda')(x)^2}{p(x)r(x)} + \int_a^x \bigl(p(\xi)r(\xi)\bigr)'\biggl(\frac{(pw_\lambda')(\xi)}{p(\xi)r(\xi)}\biggr)^2 d\xi, \qquad a<x<b,\end{aligned}$$

where we also used the fact that $(pw_\lambda')(a) = 0$ and the assumption that $p(a)r(a) > 0$. The right hand side is nonnegative, because $x \mapsto p(x)r(x)$ is increasing and therefore $(p(\xi)r(\xi))' \geq 0$. Given that $\lambda > 0$, it follows that $1 - w_\lambda(x)^2 \geq 0$, so that $|w_\lambda(x)| \leq 1$.

If $p(a)r(a) = 0$, the above proof can be used to show that the solution of (2.17) is such that $|w_{\lambda,m}(x)| \leq 1$ for all $a < x < b$, $\lambda \geq 0$ and $m \in \mathbb{N}$; then Lemma 2.2 yields the desired result.

2.2.2 *Eigenfunction Expansions*

Eigenfunction expansion theorems for ordinary and partial differential operators are a key tool for the construction of generalized convolutions. Under the running assumption that the left endpoint is regular or entrance, the Sturm-Liouville operator (2.10) has a self-adjoint realization with Neumann-type boundary conditions, and the corresponding spectral expansion gives rise to an invertible integral transform whose kernel is the solution $w_\lambda(\cdot)$: (For brevity we write $L^p(r) := L^p\bigl((a,b); r(x)dx\bigr)$.)

Theorem 2.5 *The operator*

$$\mathcal{L}^{(2)} : \mathcal{D}(\mathcal{L}^{(2)}) \subset L^2(r) \longrightarrow L^2(r), \qquad \mathcal{L}^{(2)}u = \ell(u),$$

where

$$\mathcal{D}(\mathcal{L}^{(2)}) := \begin{cases} \left\{ u \in L^2(r) \;\middle|\; \begin{array}{l} u, u' \in \mathrm{AC}_{\mathrm{loc}}(a,b),\ \ell(u) \in L^2(r), \\ (pu')(a) = 0 \end{array} \right\}, & \textit{if } b \textit{ is limit point,} \\ \left\{ u \in L^2(r) \;\middle|\; \begin{array}{l} u, u' \in \mathrm{AC}_{\mathrm{loc}}(a,b),\ \ell(u) \in L^2(r), \\ (pu')(a) = (pu')(b) = 0 \end{array} \right\}, & \textit{if } b \textit{ is limit circle} \end{cases} \tag{2.20}$$

is positive and self-adjoint. There exists a unique locally finite positive Borel measure $\boldsymbol{\rho}_{\mathcal{L}}$ on $\mathbb{R}$ such that the map $h \mapsto \mathcal{F}h$, where

$$(\mathcal{F}h)(\lambda) := \int_a^b h(x)\, w_\lambda(x)\, r(x) dx \qquad \big(h \in \mathrm{C_c}[a,b),\ \lambda \geq 0\big), \tag{2.21}$$

induces an isometric isomorphism $\mathcal{F} : L^2(r) \longrightarrow L^2(\mathbb{R}; \boldsymbol{\rho}_{\mathcal{L}})$ whose inverse is given by

$$(\mathcal{F}^{-1}\varphi)(x) = \int_{\mathbb{R}} \varphi(\lambda)\, w_\lambda(x)\, \boldsymbol{\rho}_{\mathcal{L}}(d\lambda), \tag{2.22}$$

the convergence of the latter integral being understood with respect to the norm of $L^2(r)$. The spectral measure $\boldsymbol{\rho}_{\mathcal{L}}$ is supported on $\mathbb{R}_0^+$. Moreover, the operator $\mathcal{F}$ is a spectral representation of $\mathcal{L}^{(2)}$, i.e. we have

$$\mathcal{D}(\mathcal{L}^{(2)}) = \Big\{u \in L^2(r) \;\Big|\; \lambda \cdot (\mathcal{F}u)(\lambda) \in L^2\big(\mathbb{R}_0^+, \boldsymbol{\rho}_{\mathcal{L}}\big)\Big\}, \tag{2.23}$$

$$\big(\mathcal{F}(\mathcal{L}^{(2)}h)\big)(\lambda) = \lambda \cdot (\mathcal{F}h)(\lambda), \qquad h \in \mathcal{D}(\mathcal{L}^{(2)}). \tag{2.24}$$

Proof The fact that $(\mathcal{L}^{(2)}, \mathcal{D}(\mathcal{L}^{(2)}))$ is a positive self-adjoint operator is a known result [185, Chapter 9]. The existence of a spectral transformation $\mathcal{F}$ associated with the operator $\mathcal{L}$ is a consequence of the standard Weyl-Titchmarsh-Kodaira theory of eigenfunction expansions of Sturm-Liouville operators (cf. [201, Chapter 8]).

In the general case the eigenfunction expansion is written in terms of two linearly independent solutions of $\ell(u) = \lambda u$ and a 2×2 matrix measure. However, by Lemma 2.1, the function $w_\lambda(x)$ is square-integrable near $x = 0$ with respect to the measure $r(x)dx$ and, for fixed x, it is an entire function of λ. Therefore, the possibility of writing the expansion in terms only of the eigenfunction $w_\lambda(x)$ follows from the results of [53, Sections 9 and 10].

The isometric integral transform $\mathcal{F}$ will be called the *$\mathcal{L}$-transform.*

Remark 2.1 Assume that the coefficients of the Sturm-Liouville expression (2.10) are such that p', r' are locally absolutely continuous on (a, b). Let u be a solution of $\ell(u) = \lambda u$, and consider a transformation of independent and dependent variables of the form

$$u = Z(x)v, \qquad y = \int_c^x H(\xi)\, d\xi,$$

where the functions Z, H are positive and sufficiently smooth. A straightforward computation yields that the function $v(y)$ is a solution of the Sturm-Liouville equation $\widetilde{\ell}(v)(y) := \frac{1}{\widetilde{r}(y)}\big(-(\widetilde{p}v')'(y) + \widetilde{q}(y)v(y)\big) = \lambda v(y)$, where

$$\widetilde{r} = \frac{rZ^2}{H}, \qquad \widetilde{p} = pHZ^2, \qquad \widetilde{q} = \frac{qZ^2}{H} + Z\frac{d}{dy}\Big(pH\frac{dZ}{dy}\Big).$$

We can write $\widetilde{\ell}(v) = \boldsymbol{U}^{-1}\ell(\boldsymbol{U}v)$, where $\boldsymbol{U} : L^2\big((\gamma^{-1}(a), \gamma^{-1}(b)), \widetilde{r}\big) \longrightarrow L^2(r)$ is the isometry defined by

$$(\boldsymbol{U}v)(x) = Z(x)\, v(\gamma(x)), \qquad (\boldsymbol{U}^{-1}u)(y) = \frac{u(\gamma^{-1}(y))}{Z(\gamma^{-1}(y))},$$

with $\gamma(x) := \int_c^x H(\xi)\,d\xi$ and γ^{-1} its inverse function. Therefore, the operator $\widetilde{\mathcal{F}}$ defined by

$$(\widetilde{\mathcal{F}}h)(\lambda) := \big(\mathcal{F}(\boldsymbol{U}h)\big)(\lambda) = \int_{\gamma^{-1}(a)}^{\gamma^{-1}(b)} h(y)\,\frac{w_\lambda(\gamma^{-1}(y))}{Z(\gamma^{-1}(y))}\,\widetilde{r}(y)dy,$$

$$(\widetilde{\mathcal{F}}^{-1}\varphi)(y) := \big(\boldsymbol{U}^{-1}(\mathcal{F}^{-1}\varphi)\big)(y)$$

is a spectral representation of the self-adjoint realization $\widetilde{\mathcal{L}} := \boldsymbol{U}^{-1}\mathcal{L}\boldsymbol{U}$ of the operator $\widetilde{\ell}$. Under suitable additional assumptions (for instance, if $Z(a) = 1$, $(pZ')(a) = 0$ and ℓ is limit point at b), one can check that this spectral representation coincides with that obtained by applying Theorem 2.5 to the transformed operator $\widetilde{\ell}$; in particular, the spectral measure given by Theorem 2.5 is invariant under such transformations of variable.

A special case is the so-called *Liouville transformation* $u = [p(x)r(x)]^{1/4}v$, $y = \int\sqrt{r(x)/p(x)}\,dx$. This choice yields a simplified operator $\widetilde{\ell}$ without first-order term, namely $\widetilde{\ell}(v)(y) = v''(y) + \widetilde{q}(y)v(y)$, where $\widetilde{q} = \frac{q}{r} + (pr)^{-\frac{1}{4}}\frac{d^2}{dy^2}[(pr)^{\frac{1}{4}}]$. This is called the *Liouville normal form* of the operator ℓ.

Theorem 2.5 establishes the existence of a spectral measure $\boldsymbol{\rho}_{\mathcal{L}}$ such that the $\mathcal{L}$-transform maps the space $L^2(r)$ isometrically onto $L^2(\mathbb{R}; \boldsymbol{\rho}_{\mathcal{L}})$, but it provides no information on how to compute the measure $\boldsymbol{\rho}_{\mathcal{L}}$. When the Sturm-Liouville operator has no natural endpoints, the spectral measure is discrete and can be obtained by determining the eigenvalues and the norms of the eigenfunctions:

Proposition 2.11 *Suppose that the endpoint b is regular, entrance or exit. Let* $\boldsymbol{u}_\lambda(\cdot)$ *be a nontrivial solution of* $\ell(\boldsymbol{u}) = \lambda\boldsymbol{u}$ $(\lambda \in \mathbb{C})$ *such that*

$$\boldsymbol{u}_\lambda(b) = 1, \quad (p\boldsymbol{u}_\lambda')(b) = 0, \qquad \textit{if } b \textit{ is regular or entrance,} \tag{2.25}$$

$$\boldsymbol{u}_\lambda \in L^2((c,b), r(x)dx) \textit{ for some } c \in (a,b), \qquad \textit{if } b \textit{ is exit} \tag{2.26}$$

and let $\mathrm{Wr}(w_\lambda, \boldsymbol{u}_\lambda) := p(w_\lambda \boldsymbol{u}'_\lambda - \boldsymbol{u}_\lambda w'_\lambda)$ *be the modified Wronskian of the solutions* w_λ *and* $\boldsymbol{u}_\lambda$. *Then* $\mathrm{Wr}(w_\lambda, \boldsymbol{u}_\lambda)$ *is independent of* x *and its positive zeros* $0 \leq \lambda_1 < \lambda_2 < \lambda_3 < \ldots \uparrow \infty$ *are eigenvalues of the self-adjoint operator* $\mathcal{L}$. *The spectrum of* $\mathcal{L}$ *is* $\{\lambda_k\}_{k\in\mathbb{N}}$, *and its (purely discrete) spectral measure is given by*

$$\boldsymbol{\rho}_{\mathcal{L}} = \sum_{k=1}^{\infty} \|w_{\lambda_k}\|_{L^2(r)}^{-2}\, \delta_{\lambda_k}.$$

Proof See [122, Section 5.1].

If the endpoint b is natural, the spectrum of $\mathcal{L}$ has, in general, a more complicated structure. We refer to the work of Linetsky [122] for a complete characterization of the structure of the spectrum of a large class of Sturm-Liouville operators with natural endpoints. The following results describe two approaches for computing the spectral measure of operators whose endpoint b is natural—the so-called real variable approach, where $\boldsymbol{\rho}_{\mathcal{L}}$ is obtained as a limit of discrete measures which correspond to eigenvalue problems on approximating intervals, and an alternative approach which relies on complex analysis and the so-called Weyl-Titchmarsh m-function:

Proposition 2.12 *Suppose that the endpoint* b *is natural. For* $\beta \in (a, b)$, *let* $0 \leq \lambda_{1,\beta} < \lambda_{2,\beta} < \ldots \uparrow \infty$ *be the zeros of the function* $\lambda \mapsto w_\lambda(\beta)$ *and let* $\boldsymbol{\rho}^{\beta}_{\mathcal{L}}$ *be the measure*

$$\boldsymbol{\rho}^{\beta}_{\mathcal{L}} = \sum_{k=1}^{\infty} \|w_{\lambda_{k,\beta}}\|_{2,\beta}^{-2}\, \delta_{\lambda_{k,\beta}}, \qquad \textit{where } \|f\|_{2,\beta} = \int_a^\beta |f(x)|^2 r(x)\,dx.$$

There exists a right continuous, monotone increasing function $\Psi(\cdot)$ *on* $\mathbb{R}$ *such that* $\lim_{\beta\uparrow b} \boldsymbol{\rho}^{\beta}_{\mathcal{L}}(-\infty, \lambda] = \Psi(\lambda)$ *at all points of continuity of* Ψ. *Moreover, the spectral measure of* $\mathcal{L}$ *is the Lebesgue-Stieltjes measure with distribution function* $\Psi(\lambda)$, *i.e. we have* $\boldsymbol{\rho}_{\mathcal{L}}(\lambda_1, \lambda_2] = \Psi(\lambda_2) - \Psi(\lambda_1)$ *for all* $\lambda_1, \lambda_2 \in \mathbb{R}$ *with* $\lambda_1 < \lambda_2$.

Proof See [122, Section 5.2].

Notice that Lemma 2.2 guarantees existence of a solution θ of $\ell(\theta) = \lambda\theta$ which is real entire in λ (i.e. for fixed $x \in (a, b)$ the function $\lambda \mapsto \theta_\lambda(x)$ is entire, and $\theta_\lambda(x) \in \mathbb{R}$ for $\lambda \in \mathbb{R}$), and such that $\mathrm{Wr}(w_\lambda, \theta_\lambda) = 1$. After this remark, we can formulate the following characterization of the spectral measure.

Proposition 2.13 *Suppose that the endpoint* b *is natural, and let* $\theta_\lambda(\cdot)$ *be a solution of* $\ell(\theta) = \lambda\theta$ *which is real entire in* λ. *There exists a function* $m : \mathbb{C} \setminus \mathbb{R} \longrightarrow \mathbb{C}$, *called the* Weyl-Titchmarsh m-function, *which is uniquely defined by the* *requirement that* $\psi_\lambda(x) := \theta_\lambda(x) + m(\lambda)w_\lambda(x)$ *belongs to* $L^2((c,b), r(x)dx)$ *for some* $c \in (a,b)$*. The spectral measure of the operator* $\mathcal{L}$ *is given by*

$$\boldsymbol{\rho}_{\mathcal{L}}(\lambda_1, \lambda_2] = \lim_{\delta \downarrow 0} \lim_{\varepsilon \downarrow 0} \frac{1}{\pi} \int_{\lambda_1+\delta}^{\lambda_2+\delta} \mathrm{Im}(m(\lambda + i\varepsilon))\, d\lambda \qquad (\lambda_1, \lambda_2 \in \mathbb{R},\ \lambda_1 < \lambda_2). \tag{2.27}$$

Proof See [53, Sections 9 and 10].

It is often important to know whether the inversion integral for the $\mathcal{L}$-transform is absolutely convergent. A sufficient condition, which is valid for any Sturm-Liouville operator satisfying the left boundary assumption (2.11), is given in the next lemma:

Lemma 2.4 *Set* $J = [a,b)$ *if* $\int_a^c \int_a^y \frac{dx}{p(x)} \bigl(q(y) + r(y)\bigr) dy < \infty$*, and* $J = (a,b)$ *otherwise. Then:*

(a) *For each* $\mu \in \mathbb{C} \setminus \mathbb{R}$*, the integrals*

$$\int_{\mathbb{R}_0^+} \frac{w_\lambda(x)\, w_\lambda(y)}{|\lambda - \mu|^2} \boldsymbol{\rho}_{\mathcal{L}}(d\lambda) \qquad \text{and} \qquad \int_{\mathbb{R}_0^+} \frac{(pw_\lambda')(x)\, (pw_\lambda')(y)}{|\lambda - \mu|^2} \boldsymbol{\rho}_{\mathcal{L}}(d\lambda) \tag{2.28}$$

converge uniformly on compact squares in $J \times J$*.*

(b) *If* $h \in \mathcal{D}(\mathcal{L}^{(2)})$*, then*

$$h(x) = \int_{\mathbb{R}_0^+} (\mathcal{F}h)(\lambda)\, w_\lambda(x)\, \boldsymbol{\rho}_{\mathcal{L}}(d\lambda), \tag{2.29}$$

$$(ph')(x) = \int_{\mathbb{R}_0^+} (\mathcal{F}h)(\lambda)\, (pw_\lambda')(x)\, \boldsymbol{\rho}_{\mathcal{L}}(d\lambda), \tag{2.30}$$

where the right-hand side integrals converge absolutely and uniformly on compact subsets in J*.*

Proof

(a) It is known that the resolvent of the Sturm-Liouville operator $(\mathcal{L}^{(2)}, \mathcal{D}(\mathcal{L}^{(2)}))$ is given by

$$(\mathcal{L}^{(2)} - \mu)^{-1} g(x) = \int_a^b g(y)\, G(x,y,\mu)\, r(y) dy, \qquad g \in L^2(r),\ \mu \in \mathbb{C} \setminus \mathbb{R},$$

where

$$G(x,y,\mu) = \begin{cases} \frac{1}{\mathrm{Wr}(w_\mu, \boldsymbol{u}_\mu)} w_\mu(x) \boldsymbol{u}_\mu(y), & x < y, \\ \frac{1}{\mathrm{Wr}(w_\mu, \boldsymbol{u}_\mu)} w_\mu(y) \boldsymbol{u}_\mu(x), & x \geq y \end{cases}$$

and $\boldsymbol{u}_\lambda(\cdot)$ ($\lambda \in \mathbb{C} \setminus \mathbb{R}$) is a nontrivial solution of $\ell(u) = \lambda u$ satisfying (2.25) if b is regular or entrance and (2.26) if b is exit or natural; moreover, the $\mathcal{L}$-transform of the resolvent kernel is

$$\big(\mathcal{F}G(x, \cdot, \mu)\big)(\lambda) = \frac{w_\lambda(x)}{\lambda - \mu}.$$

(These facts follow from general results in Sturm-Liouville spectral theory, see [53, Lemma 10.6].) We have

$$\int_{\mathbb{R}_0^+} \frac{w_\lambda(x) w_\lambda(y)}{|\lambda - \mu|^2} \boldsymbol{\rho}_{\mathcal{L}}(d\lambda) = \int_a^b G(x, \xi, \mu) G(y, \xi, \mu)\, r(\xi) d\xi = \frac{1}{\mathrm{Im}(\mu)} \mathrm{Im}\big(G(x, y, \mu)\big),$$

where the first equality follows from the isometric property of $\mathcal{F}$ and the second equality is a consequence of the resolvent formula $(\mathcal{L}^{(2)} - \mu_1)^{-1} - (\mathcal{L}^{(2)} - \mu_2)^{-1} = (\mu_1 - \mu_2)(\mathcal{L}^{(2)} - \mu_1)^{-1}(\mathcal{L}^{(2)} - \mu_2)^{-1}$. Letting $\partial_\xi^{[1]} := p(\xi)\frac{\partial}{\partial \xi}$, it is easy to check that the functions $\mathrm{Im}\big(G(x, y, \mu)\big)$, $\partial_x^{[1]} \mathrm{Im}\big(G(x, y, \mu)\big)$ and $\partial_x^{[1]} \partial_y^{[1]} \mathrm{Im}\big(G(x, y, \mu)\big)$ are continuous in $a < x, y < b$. From this we can conclude, after a careful estimation of the differentiated integrals (see the proof of [144, §21.2, Corollary 3]), that

$$\int_{\mathbb{R}_0^+} \frac{(pw_\lambda')(x)\,(pw_\lambda')(y)}{|\lambda - \mu|^2} \boldsymbol{\rho}_{\mathcal{L}}(d\lambda) = \frac{1}{\mathrm{Im}(\mu)}\, \partial_x^{[1]} \partial_y^{[1]} \mathrm{Im}\big(G(x, y, \mu)\big)$$

and that the integrals (2.28) converge uniformly for x, y in compact subsets of J.

(b) By Theorem 2.5 and the classical theorem on differentiation under the integral sign for Riemann-Stieltjes integrals, to prove (2.29)–(2.30) it only remains to justify the absolute and uniform convergence of the integrals in the right-hand sides.

Recall from Theorem 2.5 that the condition $h \in \mathcal{D}(\mathcal{L}^{(2)})$ implies that $\mathcal{F}h \in L_2\big(\mathbb{R}_0^+, \boldsymbol{\rho}_{\mathcal{L}}\big)$ and also $\lambda\, (\mathcal{F}h)(\lambda) \in L_2\big(\mathbb{R}_0^+, \boldsymbol{\rho}_{\mathcal{L}}\big)$. As a consequence, we obtain

$$\begin{aligned}
&\int_{\mathbb{R}_0^+} \big|(\mathcal{F}h)(\lambda) w_\lambda(x)\big|\, \boldsymbol{\rho}_{\mathcal{L}}(d\lambda) \\
&\qquad \leq \int_{\mathbb{R}_0^+} \lambda\, \big|(\mathcal{F}h)(\lambda)\big| \left|\frac{w_\lambda(x)}{\lambda + i}\right| \boldsymbol{\rho}_{\mathcal{L}}(d\lambda) + \int_{\mathbb{R}_0^+} \big|(\mathcal{F}h)(\lambda)\big| \left|\frac{w_\lambda(x)}{\lambda + i}\right| \boldsymbol{\rho}_{\mathcal{L}}(d\lambda) \\
&\qquad \leq \big(\|\lambda\, (\mathcal{F}h)(\lambda)\|_\rho + \|(\mathcal{F}h)(\lambda)\|_\rho\big) \left\|\frac{w_\lambda(x)}{\lambda + i}\right\|_\rho \\
&\qquad < \infty,
\end{aligned}$$

where $\|\cdot\|_\rho$ denotes the norm of the space $L_2(\mathbb{R};\,\boldsymbol{\rho}_{\mathcal{L}})$, and similarly

$$\int_{\mathbb{R}_0^+} \bigl|(\mathcal{F}h)(\lambda)\,(pw_\lambda')(x)\bigr|\,\boldsymbol{\rho}_{\mathcal{L}}(d\lambda) \leq \bigl(\|\lambda\,(\mathcal{F}h)(\lambda)\|_\rho + \|(\mathcal{F}h)(\lambda)\|_\rho\bigr) \biggl\| \frac{(pw_\lambda')(x)}{\lambda + i} \biggr\|_\rho < \infty.$$

We know from part (a) that the integrals which define $\bigl\|\frac{w_\lambda(x)}{\lambda+i}\bigr\|_\rho$ and $\bigl\|\frac{(pw_\lambda')(x)}{\lambda+i}\bigr\|_\rho$ converge uniformly, hence the integrals in (2.29)–(2.30) converge absolutely and uniformly on compact subsets of J.

2.2.3 Diffusion Semigroups Generated by Sturm-Liouville Operators

Being a positive self-adjoint operator, the Neumann realization $(\mathcal{L}^{(2)}, \mathcal{D}(\mathcal{L}^{(2)}))$ of the Sturm-Liouville expression (2.10) is the (negative of the) infinitesimal generator of a strongly continuous semigroup $\{T_t^{(2)}\}_{t\geq 0}$ on $L^2(r)$. Since $T_t^{(2)} = e^{-t\mathcal{L}^{(2)}}$ (the latter being defined via the spectral calculus), the eigenfunction expansion of this semigroup is a by-product of Theorem 2.5:

Proposition 2.14 *The semigroup $\{T_t^{(2)}\}_{t\geq 0}$ generated by $(\mathcal{L}^{(2)}, \mathcal{D}(\mathcal{L}^{(2)}))$ is* sub-Markovian, *i.e. such that $\|T_t^{(2)} h\|_{L^\infty(r)} \leq \|h\|_{L^\infty(r)}$ for all $h \in L^2(r) \cap L^\infty(r)$ and $T_t^{(2)} h \geq 0$ whenever $h \geq 0$. Moreover, for $t > 0$ and $h \in L^2(r)$ this semigroup admits the representations*

$$(T_t^{(2)} h)(x) = \mathcal{F}^{-1}[e^{-t\cdot}(\mathcal{F}h)(\cdot)](x) = \int_{\mathbb{R}_0^+} e^{-t\lambda} w_\lambda(x)\,(\mathcal{F}h)(\lambda)\,\boldsymbol{\rho}_{\mathcal{L}}(d\lambda) \tag{2.31}$$

$$= \int_a^b h(y)\,p(t,x,y)\,r(y)dy, \tag{2.32}$$

where

$$p(t,x,y) := \mathcal{F}^{-1}[e^{-t\cdot} w_{(\cdot)}(x)](y) = \int_{\mathbb{R}_0^+} e^{-t\lambda} w_\lambda(x) w_\lambda(y)\,\boldsymbol{\rho}_{\mathcal{L}}(d\lambda) \qquad (x,y \in J) \tag{2.33}$$

(here J is defined as in Lemma 2.4) and the latter integral converges absolutely and uniformly on compact squares in $J \times J$ for each fixed $t > 0$.

Proof One can check (see [33, Section 1.1]) that $\mathcal{L}^{(2)}$ is the positive self-adjoint operator associated with the unbounded sesquilinear form $\mathcal{E}_{\mathcal{L}} : \mathcal{D}(\mathcal{E}_{\mathcal{L}}) \times \mathcal{D}(\mathcal{E}_{\mathcal{L}}) \longrightarrow \mathbb{C}$ defined by

$$\mathcal{D}(\mathcal{E}_{\mathcal{L}}) = \{u \in L^2(r) \cap L^2(q) \mid u \in \mathrm{AC}_{\mathrm{loc}}(\mathbb{R}^+),\ u' \in L^2(p)\},$$

$$\mathcal{E}_{\mathcal{L}}(u, v) = \int_0^\infty u'(x)\overline{v'(x)}\, p(x)dx + \int_0^\infty u(x)\overline{v(x)}\, q(x)dx,$$

where $L^2(p) = L^2\big((a,b); p(x)dx\big)$ and $L^2(q) = L^2\big((a,b); q(x)dx\big)$. According to [33, Section 2.3], $\big(\mathcal{E}_{\mathcal{L}}, \mathcal{D}(\mathcal{E}_{\mathcal{L}})\big)$ is closed and Markovian. (The closedness means that $\mathcal{D}(\mathcal{E}_{\mathcal{L}})$ is a Hilbert space with respect to the inner product $\mathcal{E}_{\mathcal{L}}(u, v) + \langle u, v\rangle_{L_2(r)}$, while the Markovianity means that if $u \in \mathcal{D}(\mathcal{E}_{\mathcal{L}})$ then $v := \max(\min(u, 1), 0) \in \mathcal{D}(\mathcal{E}_{\mathcal{L}})$ and $\mathcal{E}_{\mathcal{L}}(v, v) \leq \mathcal{E}_{\mathcal{L}}(u, u)$.) Using the well-known Beurling-Deny criterion (e.g. [33, Theorem 1.1.3]), it follows that the semigroup $\{T_t^{(2)}\}$ is sub-Markovian.

The representation (2.31) is a direct consequence of the spectral theorem for unbounded self-adjoint operators. It follows from Lemma 2.4(a) that the right hand side of (2.33) converges absolutely and uniformly in compact subsets of $J \times J$, hence for $x \in J$ the function $\lambda \mapsto e^{-t\lambda} w_\lambda(x)$ belongs to $L_2\big(\mathbb{R}; \boldsymbol{\rho}_{\mathcal{L}}\big)$. The representation (2.32) is therefore obtained by combining (2.31) with the isometric property of $\mathcal{F}$.

As noted in the beginning of this section, the (negative of the) Sturm-Liouville expression considered here is of the form (2.5). Assume that the endpoint b is not exit, and let $\bar{I} = [a, b)$ if b is natural and $\bar{I} = [a, b]$ if b is regular or entrance. By Theorem 2.3 and Proposition 2.7, there exists a diffusion process $\{X_t\}_{t\geq 0}$ on $\bar{I}$ which is a Feller process whose infinitesimal generator is the operator $(-\mathcal{L}^{(0)}, \mathcal{D}(\mathcal{L}^{(0)}))$, where

$$\mathcal{L}^{(0)} u = \ell(u),$$

$$\mathcal{D}(\mathcal{L}^{(0)}) = \left\{ u \in \mathrm{C}_0(\bar{I}) \;\middle|\; \begin{array}{l} u, u' \in \mathrm{AC}_{\mathrm{loc}}(a, b),\ \ell(u) \in \mathrm{C}_0(\bar{I}) \\ (pu')(a) = 0,\ (pu')(b) = 0 \text{ if } b \text{ is regular or entrance} \end{array} \right\}. \tag{2.34}$$

The Feller semigroup generated by $(\mathcal{L}^{(0)}, \mathcal{D}(\mathcal{L}^{(0)}))$ is the restriction to $\mathrm{C}_0(\bar{I})$ of the $L^\infty(r)$-extension of the semigroup $\{T_t^{(2)}\}$, cf. [33, Equation (1.1.9)]. The next corollary gives some consequences of the preceding remarks.

Corollary 2.2 *Assume that the endpoint b is not exit. Let $\{T_t\}_{t\geq 0}$ and $\{X_t\}_{t\geq 0}$ be, respectively, the Feller semigroup and diffusion generated by $\big(-\mathcal{L}^{(0)}, \mathcal{D}(\mathcal{L}^{(0)})\big)$. Then $\{T_t\}_{t\geq 0}$ is consistent with the strongly continuous contraction semigroup $\{T_t^{(2)}\}$ generated by $(\mathcal{L}^{(2)}, \mathcal{D}(\mathcal{L}^{(2)}))$, in the sense that $T_t h = T_t^{(2)} h$ if $h \in \mathrm{C}_0(\bar{I}) \cap L_2(r)$. The function $p(t, x, \cdot)$ defined in (2.33) is the density (with respect to $r(y)dy$) of the* *transition kernel of the Feller semigroup* $\{T_t\}_{t\geq 0}$*, i.e. we have*

$$(T_t h)(x) = \mathbb{E}_x[h(X_t)] = \int_a^b h(y)\, p(t, x, y)\, r(y) dy \qquad \bigl(t > 0,\ x \in J,\ h \in \mathrm{C}_0(\bar{I})\bigr).$$

If $q \equiv 0$*, then* $\{T_t\}_{t\geq 0}$ *is a conservative Feller semigroup and therefore* $p(t, x, \cdot)\, r(\cdot)$ *is, for each* $t > 0$ *and* $x \in J$*, the density of a probability measure on* $\bar{I}$*.*

2.2.4 Remarkable Particular Cases

The general family of Sturm-Liouville operators studied above includes many differential operators which are of *hypergeometric type* in the sense that the solutions of $\ell(u) = \lambda u$ can be written in terms of hypergeometric functions. In such cases, it is often possible to determine, using Propositions 2.11–2.13 and known identities from the theory of special functions, a closed-form expression for the spectral measure. As the examples below demonstrate, one can recover, in particular, the inversion theorem for many common integral transforms, as well as an explicit (spectral) representation for the transition probabilities of important diffusion processes. The following examples will be further reworked in the subsequent chapters of this book.

We start with an example which is nearly trivial, but quite instructive:

Example 2.1 The Sturm-Liouville operator

$$\ell = -\frac{d^2}{dx^2}, \qquad 0 < x < \infty$$

is obtained by setting $p = r = \mathbb{1}$ and $(a, b) = \mathbb{R}^+$. Since the solution of the Sturm-Liouville initial value problem (2.12) is $w_\lambda(x) = \cos(\tau x)$ (where $\lambda = \tau^2$), the $\mathcal{L}$-transform is simply the *cosine Fourier transform* $(\mathcal{F}h)(\tau) = \int_0^\infty h(x) \cos(\tau x) dx$. The function $\theta_\lambda(x) = \frac{1}{\tau} \sin(\tau x)$ satisfies the requirement of Proposition 2.13, and one can easily check that the Weyl-Titchmarsh m-function is given by $m(\lambda) = \frac{1}{i\sqrt{\lambda}}$ for $\lambda \in \mathbb{C}$ with $\operatorname{Im} \lambda > 0$. Using (2.27), we obtain that $\boldsymbol{\rho}_{\mathcal{L}}(d\lambda) = 0$ on $(-\infty, 0]$ and $\boldsymbol{\rho}_{\mathcal{L}}(d\lambda) = \frac{d\lambda}{\pi\sqrt{\lambda}} = \frac{2}{\pi} d\tau$ on $\mathbb{R}^+$. As one would expect, this result confirms the classical inversion formula $(\mathcal{F}^{-1}\varphi)(\tau) = \frac{2}{\pi} \int_0^\infty h(x) \cos(\tau x) dx$ for the cosine Fourier transform. The Feller process on $\mathbb{R}_0^+$ generated by ℓ is the *reflected Brownian motion*, whose transition density is given by

$$\begin{aligned} p(t, x, y) &= \frac{2}{\pi} \int_0^\infty e^{-t\tau^2} \cos(\tau x) \cos(\tau y) d\tau \\ &= \frac{1}{\sqrt{2\pi t}} \biggl(\exp\Bigl(-\frac{(x-y)^2}{2t}\Bigr) + \exp\Bigl(-\frac{(x+y)^2}{2t}\Bigr) \biggr) \end{aligned} \qquad (2.35)$$

(the second equality follows from integral 2.5.36.1 in [155]). The fact that the expression in the right-hand side of (2.35) is the transition density of the reflected Brownian motion is well-known, cf. e.g. [22, p. 250].

Our next case illustrates the fact that various classical expansions of functions as series of orthogonal polynomials are also a particular case of the general Sturm-Liouville spectral theory.

Example 2.2 Let $\alpha, \beta > -1$. The Jacobi differential operator

$$\ell = -(1-x^2)\frac{d^2}{dx^2} - (\beta - \alpha - (\alpha+\beta+1)x)\frac{d}{dx}, \qquad -1 < x < 1$$

is of the form (2.10) with $q \equiv 0$, $r(x) = (1-x)^\alpha(1+x)^\beta$ and $p(x) = (1-x)^{1+\alpha}(1+x)^{1+\beta}$. The endpoint 1 is regular if $-1 < \alpha < 0$ and entrance if $\alpha \geq 0$; similarly, the endpoint -1 is regular if $-1 < \beta < 0$ and entrance if $\beta \geq 0$. In all cases, the Neumann self-adjoint realization $(\mathcal{L}, \mathcal{D}(\mathcal{L}))$ has a purely discrete spectrum. The function

$$w_\lambda(x) = {}_2F_1\Bigl(\eta - \tau, \eta + \tau; \alpha + 1; \frac{1-x}{2}\Bigr) \qquad (\eta = \tfrac{1}{2}(\alpha+\beta+1),\ \lambda = \tau^2 - \eta^2)$$

is a solution of $\ell(u) = \lambda u$ such that $w_\lambda(1) = 1$ and $(pw_\lambda')(1) = 0$. Here ${}_2F_1$ denotes the hypergeometric function [145, Chapter 15]. One can verify that this is an eigenfunction of $(\mathcal{L}, \mathcal{D}(\mathcal{L}))$ if and only if $\lambda = n(2\eta + n)$ $(n \in \mathbb{N})$ [126]. Therefore, the eigenfunctions are the *Jacobi polynomials* $w_{k(2\eta+k)}(x) \equiv R_k^{(\alpha,\beta)}(x) := \frac{(-1)^k}{2^k(\alpha+1)_k} \frac{1}{(1-x)^\alpha(1+x)^\beta} \frac{d^k}{dx^k}[(1-x)^{k+\alpha}(1+x)^{k+\beta}]$. Since $\| R_k^{(\alpha,\beta)} \|_{L^2(r)}^2 = \frac{2^{2\eta-1} k!\, \Gamma(k+\alpha+1)\Gamma(k+\beta+1)}{[(1+\alpha)_k]^2 (k+\eta)\, \Gamma(k+2\eta)}$ [44, Section 15.2], the integral transform pair (2.21)–(2.22) is the Jacobi series expansion

$$h(x) = \sum_{k=0}^\infty \frac{[(\alpha+1)_k]^2 (k+\eta)\, \Gamma(k+2\eta)}{2^{2\eta-1} k!\, \Gamma(k+\alpha+1)\Gamma(k+\beta+1)} (\mathcal{F}h)(k)\, R_k^{(\alpha,\beta)}(x),$$

where $(\mathcal{F}h)(k) = \int_{-1}^1 h(x)\, R_k^{(\alpha,\beta)}(x)\, (1-x)^\alpha (1+x)^\beta dx$. Accordingly, the transition probability density of the diffusion process on $[-1, 1]$ generated by ℓ is

$$p(t,x,y) = \sum_{k=0}^\infty e^{-tk(2\eta+k)} \frac{[(\alpha+1)_k]^2 (k+\eta)\, \Gamma(k+2\eta)}{2^{2\eta-1} k!\, \Gamma(k+\alpha+1)\Gamma(k+\beta+1)} R_k^{(\alpha,\beta)}(x) R_k^{(\alpha,\beta)}(y).$$

In the literature, this stochastic process is known as the *Jacobi diffusion* [100, 122].

Next we present in some detail an example of how one can determine the spectral measure of a Sturm-Liouville operator whose spectrum is not discrete and whose fundamental solutions are nontrivial special functions of hypergeometric type.

Example 2.3 The Bessel process with drift $\mu > 0$ and index $\frac{\alpha-1}{2}$ $(\alpha > 0)$ is, according to [123], the diffusion generated by the differential operator

$$\ell = -\frac{d^2}{dx^2} - \Bigl(\frac{\alpha}{x} + 2\mu\Bigr)\frac{d}{dx}, \qquad 0 < x < \infty.$$

This Sturm-Liouville operator is obtained by choosing $q \equiv 0$ and $p(x) = r(x) = x^\alpha e^{2\mu x}$. One can check that the endpoint 0 is regular if $0 < \alpha < 1$ and entrance if $\alpha \geq 1$, while the endpoint $+\infty$ is natural.

The solution of the initial value problem (2.12) is

$$w_\lambda(x) = (2i\tau)^{-\frac{\alpha}{2}} e^{-\mu x} x^{-\frac{\alpha}{2}} M_{-\frac{\alpha\mu}{2i\tau}, \frac{\alpha-1}{2}}(2i\tau x), \tag{2.36}$$

where $\lambda = \tau^2 + \mu^2$ and $M_{\kappa,\nu}(z) := e^{-\frac{z}{2}} \sum_{n=0}^\infty \frac{(\frac{1}{2}+\nu-\kappa)_n}{(1+2\nu)_n n!} z^{\frac{1}{2}+\nu+n}$ is the Whittaker function of the first kind [145, §13.14]. (The fact that (2.36) is a solution of $\ell(u) = \lambda u$ follows from [153, Equation 2.1.2.108], and we can use the results of [145, §13.14(iii) and §13.15(ii)] to check that $w_\lambda(0) = 1$ and $(pw_\lambda')(0) = 0$.) If $\alpha \notin \mathbb{N}$, a suitable linearly independent solution of $\ell(u) = \lambda u$ is

$$\theta_\lambda(x) = (1-\alpha)^{-1}(2i\tau)^{\frac{\alpha}{2}-1} e^{-\mu x} x^{-\frac{\alpha}{2}} M_{-\frac{\alpha\mu}{2i\tau}, \frac{1-\alpha}{2}}(2i\tau x).$$

(Using [145, §13.2(i)] and [123, Remark 1], one verifies that $\theta_\lambda(x)$ is real entire; [145, Equation 13.2.33] yields that $\mathrm{Wr}(w_\lambda, \theta_\lambda) = 1$. The case $\alpha \in \mathbb{N}$ can be treated using [145, §13.2(v)].) Now, it follows from [145, Equation 13.14.21] that a solution of the Sturm-Liouville equation which is square-integrable with respect to $r(x)dx$ near infinity is

$$\psi_\lambda(x) = \frac{\Gamma(\frac{\alpha}{2}(1+\frac{\mu}{i\tau}))}{\Gamma(\alpha)} (2i\tau)^{\frac{\alpha}{2}-1} e^{-\mu x} x^{-\frac{\alpha}{2}} W_{-\frac{\alpha\mu}{2i\tau}, \frac{\alpha-1}{2}}(2i\tau x) \qquad (\lambda \in \mathbb{C} \setminus \mathbb{R}),$$

where $W_{\alpha,\nu}(x)$ is the Whittaker function of the second kind [145, §13.14]. By [145, Equation 13.14.33] this solution can be written as $\theta_\lambda(x) + m(\lambda) w_\lambda(x)$, where

$$m(\lambda) = -\Gamma(\alpha)^{-2}(2\tau)^{\alpha-1} \frac{\sin(\frac{\pi\alpha}{2}(1+\frac{\mu}{i\tau}))}{\sin(\pi\alpha)} \Gamma\bigl(\tfrac{\alpha}{2}(1+\tfrac{\mu}{i\tau})\bigr)\Gamma\bigl(\tfrac{\alpha}{2}(1-\tfrac{\mu}{i\tau})\bigr) \qquad (\lambda \in \mathbb{C} \setminus \mathbb{R}).$$

Taking the limit we obtain

$$\lim_{\varepsilon\downarrow 0} \frac{1}{\pi} m(\lambda + i\varepsilon) = \begin{cases} \frac{2^{\alpha-2}}{\pi\,\Gamma(\alpha)^2} \tau^{\alpha-1} \exp\bigl(-\frac{\pi\alpha\mu}{2\tau}\bigr) \bigl|\Gamma\bigl(\frac{\alpha}{2}(1+\frac{\mu}{i\tau})\bigr)\bigr|^2, & \lambda > \mu^2, \\ 0, & \lambda < \mu^2 \end{cases}$$

which, by Proposition 2.13, is the density of the (absolutely continuous) spectral measure $\rho_{\mathcal{L}}$.

Letting $\sigma(\tau) := \frac{2^{\alpha-1}}{\pi\Gamma(\alpha)^2}\tau^\alpha \exp(-\frac{\pi\alpha\mu}{2\tau})\big|\Gamma\big(\frac{\alpha}{2}(1+\frac{\mu}{i\tau})\big)\big|^2$, it follows that the pair of index transforms

$$(\mathcal{F}h)(\tau) = (2i\tau)^{-\frac{\alpha}{2}} \int_0^\infty h(x)\, M_{-\frac{\alpha\mu}{2i\tau},\frac{\alpha-1}{2}}(2i\tau x)\, e^{\mu x} x^{\frac{\alpha}{2}} dx, \tag{2.37}$$

$$(\mathcal{F}^{-1}\varphi)(x) = (2ix)^{-\frac{\alpha}{2}} e^{-\mu x} \int_0^\infty \varphi(\tau)\, M_{-\frac{\alpha\mu}{2i\tau},\frac{\alpha-1}{2}}(2i\tau x)\, \tau^{-\frac{\alpha}{2}}\sigma(\tau) d\tau \tag{2.38}$$

defines an isometry between the spaces $L^2(\mathbb{R}^+, x^\alpha e^{2\mu x} dx)$ and $L^2(\mathbb{R}^+, \sigma(\tau)d\tau)$. In the limit $\mu \to 0$, using [145, Equations 10.27.6 and 13.18.8] we recover the Hankel transform (2.47) whose kernel is the Bessel function of the first kind. From the above it also follows that the transition density of the Bessel process with drift is given by

$$p(t,x,y) = (-4xy)^{-\frac{\alpha}{2}} e^{-\mu(x+y)} \times$$
$$\times \int_0^\infty e^{-t(\tau^2+\mu^2)} M_{-\frac{\alpha\mu}{2i\tau},\frac{\alpha-1}{2}}(2i\tau x)\, M_{-\frac{\alpha\mu}{2i\tau},\frac{\alpha-1}{2}}(2i\tau y)\, \tau^{-\alpha}\sigma(\tau) d\tau.$$

This spectral representation for the law of the Bessel process was established by Linetsky in [123], based on the related results of Titchmarsh in [187, §4.17]. However, the pair of confluent hypergeometric type integral transforms (2.37)–(2.38) is apparently little known; in particular, it is not reported in reference monographs on integral transforms such as [158, 204, 207].

Example 2.4 Another integral transform related to the Whittaker functions is obtained by considering the operator

$$\ell = -x^2 \frac{d^2}{dx^2} - (1+2(1-\alpha)x)\frac{d}{dx}, \qquad 0 < x < \infty, \tag{2.39}$$

which is of the form (2.10) with $q \equiv 0$, $r(x) = x^{-2\alpha} e^{-1/x}$ and $p(x) = x^{2(1-\alpha)} e^{-1/x}$. The operator (2.39) is the generator of the Shiryaev process, which will be studied in detail in Chap. 3. Here the solution of the initial value problem (2.12) is a normalized Whittaker W function (Proposition 3.1). For $\alpha \leq \frac{1}{2}$, the corresponding spectral measure has density $\sigma(\tau) = \pi^{-2}\tau \sinh(2\pi\tau)\big|\Gamma\big(\frac{1}{2} - \alpha + i\tau\big)\big|^2$, where we write $\lambda = \tau^2 + (\frac{1}{2} - \alpha)^2$. The $\mathcal{L}$-transform specializes into

$$(\mathcal{F}h)(\tau) = \int_0^\infty h(x)\, W_{\alpha,i\tau}(\tfrac{1}{x})\, x^{-\alpha} e^{-\frac{1}{2x}} dx, \tag{2.40}$$

$$(\mathcal{F}^{-1}\varphi)(x) = \frac{1}{\pi^2} x^\alpha e^{\frac{1}{2x}} \int_0^\infty \varphi(\tau)\, W_{\alpha,i\tau}(\tfrac{1}{x})\, \tau \sinh(2\pi\tau)\big|\Gamma\big(\tfrac{1}{2} - \alpha + i\tau\big)\big|^2 d\tau. \tag{2.41}$$

Accordingly, (2.33) specializes into an explicit spectral representation for the transition density of the Shiryaev process.

The integral transform (2.40)–(2.41) is a modified form (cf. Remark 2.1) of the so-called *index Whittaker transform*, which was first introduced by Wimp [202] as a particular case of an integral transform having the Meijer-G function in the kernel. Its L^p theory was studied in [180]. The index Whittaker transform includes as a particular case the Kontorovich-Lebedev transform, which is one of the most well-known index transforms [204, 207] and has a wide range of applications in physics.

The spectral measure $\sigma(\tau)d\tau$ can be deduced using either the approach based on the Weyl-Titchmarsh m-function or the real variable approach (we refer to [177, Example 2] and [124] respectively).

Example 2.5 The coefficients $q \equiv 0$, $p(x) = r(x) = (\sinh x)^{2\alpha+1}(\cosh x)^{2\beta+1}$ (with $\beta \in \mathbb{R}, \alpha > -1, \alpha \pm \beta + 1 \geq 0$) give rise to the Jacobi operator

$$\ell = -\frac{d}{dx^2} - [(2\alpha+1)\coth x + (2\beta+1)\tanh x]\frac{d}{dx}, \qquad 0 < x < \infty. \tag{2.42}$$

The so-called Jacobi function

$$w_\lambda(x) = \phi_\tau^{(\alpha,\beta)}(x) := {}_2F_1\Bigl(\tfrac{1}{2}(\eta - i\tau), \tfrac{1}{2}(\eta + i\tau); \alpha+1; -(\sinh x)^2\Bigr)$$

(where $\eta = \alpha + \beta + 1$, $\lambda = \tau^2 + \eta^2$)) can be shown to be the unique solution of the Sturm-Liouville initial value problem (2.12). Using Proposition 2.13, one can show (cf. [105] and references therein, see also [177, Example 3]) that the spectral measure is absolutely continuous with density $\sigma(\tau) = \left|\frac{\Gamma(\frac{1}{2}(\eta+i\tau))\Gamma(\frac{1}{2}(\eta+i\tau)-\beta)}{\Gamma(\frac{1+i\tau}{2})\Gamma(\frac{i\tau}{2})\Gamma(\alpha+1)}\right|^2$, so that the $\mathcal{L}$-transform becomes

$$(\mathcal{F}h)(\tau) = \int_0^\infty h(x)\,\phi_\tau^{(\alpha,\beta)}(x)\,(\sinh x)^{2\alpha+1}(\cosh x)^{2\beta+1}dx,$$

$$(\mathcal{F}^{-1}\varphi)(x) = \int_0^\infty \varphi(\tau)\,\phi_\tau^{(\alpha,\beta)}(x)\left|\frac{\Gamma(\frac{1}{2}(\eta+i\tau))\Gamma(\frac{1}{2}(\eta+i\tau)-\beta)}{\Gamma(\frac{1+i\tau}{2})\Gamma(\frac{i\tau}{2})\Gamma(\alpha+1)}\right|^2 d\tau. \tag{2.43}$$

This is the so-called *(Fourier-)Jacobi transform*, which is closely related (via a suitable change of variables) to the Olevskii transform, the index hypergeometric transform or, in the case $\alpha = \beta$, the generalized Mehler-Fock transform [206].

If $\beta = -\frac{1}{2}$, the Feller process generated by the Neumann self-adjoint realization of (2.42) is known as the *hyperbolic Bessel process*; more generally, it is called a *hypergeometric diffusion* [21]. Like in the previous examples, the transition probabilities admit the explicit integral representation $p(t,x,y) = \int_0^\infty e^{-t(\tau^2+\eta^2)}\phi_\tau^{(\alpha,\beta)}(x)\,\phi_\tau^{(\alpha,\beta)}(y)\sigma(\tau)d\tau$.

For ease of presentation, in Examples 2.3–2.5 the range of the parameters α, β and μ was chosen so that the spectral measure is purely absolutely continuous. In general, the measure decomposes into a discrete and an absolutely continuous part, both of which can be determined using the results of Sect. 2.2.2 (for details, see [122]). For instance, if we let $\alpha > \frac{1}{2}$ in the Sturm-Liouville operator of Example 2.4 and let N_α be the integer part of $\alpha - \frac{1}{2}$, then the spectral measure becomes [115, 123]

$$\rho_{\mathcal{L}}(d\lambda) = \sum_{n=0}^{N_\alpha} \frac{2\alpha - 1 - 2n}{n!\,\Gamma(2\alpha - n)} \delta_{n(2\alpha-1-n)}(d\lambda)$$

$$+ \mathbb{1}_{[(\alpha-\frac{1}{2})^2,\infty)}(\lambda)\, \tau \sinh(2\pi\tau) \big|\Gamma\big(\tfrac{1}{2} - \alpha + i\tau\big)\big|^2 d\tau,$$

so that a finite sum must be added to the inversion formula (2.41) for the index Whittaker transform.

The examples presented above do not exhaust the class of Sturm-Liouville operators whose spectral measure is known in closed form. Additional examples can be found e.g. in [48, 67, 116, 179].

2.3 Generalized Convolutions and Hypergroups

Motivated by various considerations in analysis and in probability, various authors have introduced axiomatic notions of convolution-like structures and the closely related translation-like operators. These notions and the corresponding theories are broadly similar, but differ on significant aspects. Here we review some axiomatic definitions which have played an influential role in past studies on generalized harmonic analysis and give context to the materials in later chapters.

Generalized Translation Operators

The early historical development of the theory of abstract harmonic analysis is mostly due to the work of Levitan in the 1940s. Embracing the ideas laid out by Delsarte in [45], Levitan embarked on a comprehensive study of the so-called generalized translation operators, defined in [119] as follows (see also [121]):

Definition 2.1 Let E be a locally compact space endowed with a positive measure m. The linear operators $\mathcal{T}^y : L^p(E,m) \longrightarrow L^p(E,m)$ $(y \in E,\ 1 \leq p \leq \infty)$ are said to constitute a family of *generalized translation operators* if the following conditions hold:

T1. $\mathcal{T}_y \mathcal{T}^x = \mathcal{T}^x \mathcal{T}_y$, where we write $(\mathcal{T}_y f)(x) := (\mathcal{T}^x f)(y)$;

T2. There exists an element $\mathrm{e} \in E$ such that $\mathcal{T}^{\mathrm{e}} = \mathrm{Id}$;

T3. $\|\mathcal{T}^y f\|_{L^p(E,m)} \leq c_p(y)\, \|f\|_{L^p(E,m)}$, where the functions $c_p(\cdot)$ are positive and bounded on compact subsets of E;

T4. If $f \in L^p(E,m)$ and $\varepsilon > 0$, then for each $y \in E$ there exists $U_y \subset E$, a neighbourhood of y such that $\|\mathcal{T}^y f - \mathcal{T}^{y'} f\|_{L^p(E,m)} < \varepsilon$ for all $y' \in U_y$.

Generalized translations are naturally related to generalized convolutions by analogues of equality (2.44), below.

We refer to Levitan's monograph [121] for a comprehensive presentation of his results on Lie-type theorems with respect to generalized translation operators and other analogues of classical constructions in harmonic analysis.

Hypercomplex Systems
Influenced by Levitan's work, Berezansky and S. Krein initiated, in the early 1950s, a study of the so-called *hypercomplex systems*, whose definition reads as follows [13–15]:

Definition 2.2 Let E be a complete separable locally compact space. A measure $c(A, B, y)$ (A, B Borel subsets of E, $y \in E$) is called a *structure measure* if it is a positive regular Borel measure with respect to A (respectively B) for fixed (B, y) (resp. (A, y)), and the following properties hold:

S1. If A, B are relatively compact then $c(A, B, \cdot) \in \mathrm{C_c}(E)$;
S2. The identity $\int_E c(A, B, x)\, c(dx, C, y) = \int_E c(B, C, x)\, c(A, dx, y)$ holds for all $y \in E$ and A, B, C Borel subsets with A, B relatively compact;
S3. There exists a positive regular Borel measure m such that $m(A)m(B) = \int_E c(A, B, y)m(dy)$ for all relatively compact Borel sets A, B.

If $c(A, B, y)$ is a structure measure on E, then the space $L^1(E, m)$ with the convolution

$$(f * g)(x) = \int_E \int_E f(y)g(\xi)\, c(dy, d\xi, x)$$

is said to be a *hypercomplex system*.

A complete presentation of the extensive theory of harmonic analysis on hypercomplex systems developed by Berezansky and coauthors can be found on the monograph [15], which also provides an extensive discussion of examples related to orthogonal polynomials or Sturm-Liouville operators.

Urbanik Convolutions
The definition of Urbanik (generalized) convolutions on $\mathbb{R}_0^+$ was introduced by Urbanik in the 1960s and was thoroughly studied in a series of papers by the same author [190–194].

Definition 2.3 For $a > 0$, let $\Theta_a : \mathbb{R}_0^+ \longrightarrow \mathbb{R}_0^+$ be the multiplication map $x \mapsto \Theta_a(x) := ax$. A bilinear operator $\diamond$ on $\mathcal{M}_{\mathbb{C}}(\mathbb{R}_0^+)$ is said to be an *Urbanik convolution* if the following axioms hold:

U1. If $\mu, \nu \in \mathcal{P}(\mathbb{R}_0^+)$, then $\mu \diamond \nu \in \mathcal{P}(\mathbb{R}_0^+)$;
U2. $\mu \diamond \nu = \nu \diamond \mu$ and $\mu \diamond (\nu \diamond \pi) = (\mu \diamond \nu) \diamond \pi$ for all $\mu, \nu, \pi \in \mathcal{M}_{\mathbb{C}}(\mathbb{R}_0^+)$;
U3. If $\mu_n \overset{w}{\longrightarrow} \mu$, then $\mu_n \diamond \nu \overset{w}{\longrightarrow} \mu \diamond \nu$ for all $\nu \in \mathcal{M}_{\mathbb{C}}(\mathbb{R}_0^+)$;
U4. $\delta_0 \diamond \mu = \mu$ for all $\mu \in \mathcal{M}_{\mathbb{C}}(\mathbb{R}_0^+)$;

U5. $\Theta_a(\mu \diamond \nu) = (\Theta_a \mu) \diamond (\Theta_a \nu)$ for all $\mu, \nu \in \mathcal{P}(\mathbb{R}_0^+)$ and $a > 0$;

U6. There exists a sequence $\{c_n\}_{n\in\mathbb{N}} \subset \mathbb{R}^+$ such that $\Theta_{c_n} \delta_1^{\diamond n} \overset{w}{\longrightarrow} \mu$ for some measure $\mu \neq \delta_0$.

The topics studied in Urbanik's papers include infinite divisibility of probability measures with respect to any convolution satisfying axioms U1–U6. The following result, gives an analogous to the Fourier transform and a Lévy-Khintchine-type representation:

Proposition 2.15 *Let $\diamond$ be an Urbanik convolution, and assume there exists a mapping $\wp : \mathcal{P}(\mathbb{R}_0^+) \longrightarrow \mathbb{R}$ such that the identities*

$$\wp(c\mu_1 + (1-c)\mu_2) = c\wp(\mu_1) + (1-c)\wp(\mu_2), \qquad \wp(\mu_1 \diamond \mu_2) = \wp(\mu_1)\wp(\mu_2)$$

hold for all $0 \leq c \leq 1$, $\mu_1, \mu_2 \in \mathcal{P}(\mathbb{R}_0^+)$. Then:

(a) There exists a one-to-one correspondence $\mu \leftrightarrow \mathbb{H}\mu$ between measures $\mu \in \mathcal{P}(\mathbb{R}_0^+)$ and functions $\mathbb{H}\mu$ on $\mathbb{R}_0^+$ such that for $\mu, \mu_1, \mu_2 \in \mathcal{P}(\mathbb{R}_0^+)$, $a > 0$, $\lambda \geq 0$ we have

$$\big(\mathbb{H}(\mu_1 \diamond \mu_2)\big)(\lambda) = (\mathbb{H}\mu_1)(\lambda) \cdot (\mathbb{H}\mu_2)(\lambda), \qquad \big(\mathbb{H}(\Theta_a \mu)\big)(\lambda) = (\mathbb{H}\mu)(a\lambda).$$

The map $\mu \mapsto \mathbb{H}\mu$ can be written as an integral transform $\mathbb{H}\mu(\lambda) = \int_{\mathbb{R}_0^+} \Omega(\lambda x)\, \mu(dx)$ with kernel $\Omega : \mathbb{R}_0^+ \longrightarrow \mathbb{R}$.

(b) A function $h : \mathbb{R}_0^+ \longrightarrow \mathbb{R}$ can be written as $h = \mathbb{H}\mu$ for some $\diamond$-infinitely divisible measure μ (that is, a measure $\mu \in \mathcal{P}(\mathbb{R}_0^+)$ such that for each $n \in \mathbb{N}$ there exists $\nu_n \in \mathcal{P}(\mathbb{R}_0^+)$ for which $\mu = \nu_n^{\diamond n}$) *if and only if it has the representation*

$$h(\lambda) = \exp\!\left(-k\lambda^{\varkappa(\diamond)} + \int_{\mathbb{R}_0^+} \frac{\Omega(\lambda x) - 1}{\tau(x)} \nu(dx)\right), \qquad \lambda \geq 0,$$

where $k > 0$, $\varkappa(\diamond)$ is the so-called characteristic exponent *of the convolution $\diamond$ (see [23, 190]), ν is a finite Borel measure on $\mathbb{R}_0^+$, $\tau(x) := 1 - \Omega(x)$ for $0 \leq x \leq x_0$, $\tau(x) := 1 - \Omega(x_0)$ for $x \geq x_0$, and $x_0 > 0$ is chosen so that $\Omega(\cdot) < 1$ on $(0, x_0]$.*

The notions of $\diamond$-convolution-like semigroups and $\diamond$-Lévy-like processes with respect to an Urbanik convolution $\diamond$ can be defined exactly like the corresponding notions for the Kingman convolution, as explained in Sect. 2.4 below. The properties of such Lévy-like processes and their infinitesimal generators have been studied in [23, 195]. For instance, it has been shown in [23] that it is possible to define a stochastic integral with respect to a given $\diamond$-Lévy-like process and, moreover, the transition probabilities of this stochastic integral can be characterized using the integral transform $\mathbb{H}$ described above.

The so-called Kendall convolution is an example of an Urbanik convolution which has been the subject of numerous recent papers, due to its connection with the Williamson transform and the theory of Archimedean copulas [91–93]. We refer to [139–141] for further recent work on Urbanik convolutions and its particular cases.

Hypergroups

The following notion of hypergroup was introduced by Jewett in the mid-1970s [94] and (modulo slight changes in the axioms) independently also by Dunkl [49] and Spector [178]. The definition below, which singles out the more general concept of weak hypergroup, is taken from [82].

Definition 2.4 Let E be a locally compact space and $*$ a bilinear operator on $\mathcal{M}_\mathbb{C}(E)$. The pair $(E, *)$ is said to be a *weak hypergroup* if the following axioms are satisfied:

H1. If $\mu, \nu \in \mathcal{P}(E)$, then $\mu * \nu \in \mathcal{P}(E)$;
H2. $\mu * (\nu * \pi) = (\mu * \nu) * \pi$ for all $\mu, \nu, \pi \in \mathcal{M}_\mathbb{C}(E)$;
H3. The map $(\mu, \nu) \mapsto \mu*\nu$ is continuous (in the weak topology) from $\mathcal{M}_\mathbb{C}(E) \times \mathcal{M}_\mathbb{C}(E)$ to $\mathcal{M}_\mathbb{C}(E)$;
H4. There exists an element $\mathrm{e} \in E$ such that $\delta_\mathrm{e} * \mu = \mu * \delta_\mathrm{e} = \mu$ for all $\mu \in \mathcal{M}_\mathbb{C}(E)$;
H5. There exists a homeomorphism (called *involution*) $x \mapsto \check{x}$ of E onto itself such that $(\check{x})^\vee = x$ and $(\delta_x * \delta_y)^\vee = \delta_{\check{y}} * \delta_{\check{x}}$, where $(\delta_x * \delta_y)^\vee$ is defined via $\int f(\xi)\, (\delta_x * \delta_y)^\vee (d\xi) = \int f(\check{\xi})(\delta_x * \delta_y)(d\xi)$;
H6. $\mathrm{supp}(\delta_x * \delta_y)$ is compact for all $x, y \in E$.

A weak hypergroup $(E, *)$ is called a *hypergroup* if, in addition,

H7. $\mathrm{e} \in \mathrm{supp}(\delta_x * \delta_y)$ if and only if $y = \check{x}$;
H8. $(x, y) \mapsto \mathrm{supp}(\delta_x * \delta_y)$ is continuous from $E \times E$ into the space of compact subsets of E (endowed with the Michael topology, see [94]).

One should realize that the definitions of hypergroup and weak hypergroup include a compactness axiom on the support (axiom H6) which is comparable to axiom S1 in the definition of hypercomplex system (Definition 2.2). It is also important to note that other meanings for the word 'hypergroup' can be found on the literature on abstract harmonic analysis, particularly on older papers; we refer to the survey paper of Litvinov [127] for a thorough historical overview. However, the axioms of Definition 2.4 have been widely recognized as being appropriate for studying harmonic analysis, and are now well-established as the standard definition of hypergroup. In what follows, the word 'hypergroup' always refers to a structure satisfying axioms H1–H8.

The reference monograph on the theory of hypergroups is the book of Bloom and Heyer [19], which contains a complete bibliography of the theory developed up to the mid-1990s. A comprehensive treatment of wavelet theory on hypergroups is given in the book of Trimèche [188]. Recent work on hypergroup theory can be found on the book of Székelyhidi [184] and references therein.

The correspondence between the notions of (weak) hypergroup, generalized translation operator and hypercomplex system is described in the following proposition. As a preparation, we note that a positive Borel measure m on E is said to be *left invariant* if $\int_E f\, d(\delta_x * m) = \int_E f dm$ for all $x \in E$ and $f \in \mathrm{C_c}(E)$. Right invariant measures are defined similarly. It should be kept in mind that if the hypergroup $(E, *)$ is commutative (i.e. $\mu * \nu = \nu * \mu$ for all measures $\mu, \nu \in \mathcal{M}_{\mathbb{C}}(E)$), then there exists a left (and right) invariant measure on E [19, Theorem 1.3.15].

Proposition 2.16

(a) *If $(E, *)$ is a weak hypergroup endowed with a left invariant measure m, then the family of operators $\{\mathcal{T}^x\}_{x\in E}$ defined by*

$$(\mathcal{T}^x f)(y) := \int_E f(\xi)\,(\delta_x * \delta_y)(d\xi), \qquad f \text{ Borel measurable}$$

satisfies axioms T1 and T2 of Definition 2.1, as well as the following properties (which hold for $x, y \in E$):

(i) *$\mathcal{T}^x \mathbb{1} = \mathbb{1}$, and if $f \geq 0$ then $\mathcal{T}^x f \geq 0$;*
(ii) *$\|\mathcal{T}^x f\|_{L^2(E,m)} \leq \|f\|_{L^2(E,m)}$;*
(iii) *If $f \in \mathrm{C_c}(E)$, then the function $(x, y) \mapsto (\mathcal{T}^x f)(y)$ is continuous in each variable;*
(iv) *$(\mathcal{T}^x f)(y) = \overline{(\mathcal{T}^{\check{y}} \check{f})(\check{x})}$ for all $f \in L_2(E, m)$, where we set $\check{f}(x) = \overline{f(\check{x})}$;*
(v) *For every pair of relatively compact subsets $B_1, B_2 \subset E$ there exists a compact set $E_0 \subset E$ such that if $\operatorname{supp} f \cap E_0 = \emptyset$ then $(\mathcal{T}^x f)(y) = 0$ for m-a.e. $x \in B_1$, $y \in B_2$.*

Conversely, if $\{\mathcal{T}^x\}_{x\in E}$ is a family of linear operators satisfying the conditions above, then the convolution $*$ on $\mathcal{M}_{\mathbb{C}}(E)$ defined by

$$(\mu * \nu)(B) := \int_E \int_E (\mathcal{T}^x \mathbb{1}_B)(y)\, \mu(dx)\, \nu(dy) \qquad (B \text{ a Borel subset of } E) \tag{2.44}$$

endows E with a weak hypergroup structure.

(b) *If $(E, *)$ is a weak hypergroup endowed with a left invariant measure m, then the space $L^1(E, m)$ with the convolution*

$$(f * g)(x) := \int_E (\mathcal{T}^x f)(y)\, g(\check{y})\, \mathbf{m}(dy)$$

is a hypercomplex system with structure measure $\int_E (\mathcal{T}^x \mathbb{1}_A)(y)\, \mathbb{1}_B(\check{y}) dy$.

Proof See [15, pp. 60–62].

We note that a converse to the correspondence between weak hypergroups and hypercomplex systems stated in part (b) can be deduced as a corollary of part (a): *if* $(L^1(E,m),*)$ *is a hypercomplex system and the induced translation operators defined via* $\langle \mathcal{T}^x f, g\rangle := (f * \check{g})(x)$ *(where* $x \mapsto \check{x}$ *is an involution on* E*) satisfy the conditions listed in part (a) of the proposition, then the convolution defined in* (2.44) *endows* E *with a weak hypergroup structure.*

Concerning the relation between hypergroups and Urbanik convolutions, one should be aware that, in general, hypergroups on $\mathbb{R}_0^+$ do not satisfy the homogeneity axiom U5 of Urbanik convolutions, and Urbanik convolutions do not satisfy the compactness axiom H6 of hypergroups. We refer to [19, 190] for examples of both types. However, the Kingman convolution described in the next section (Definition 2.6) is an example both of a hypergroup structure on $\mathbb{R}_0^+$ and of an Urbanik convolution.

A *hypergroup homomorphism* between $(E_1, *_1)$ and $(E_2, *_2)$ is a map $\tau : \mathcal{M}_\mathbb{C}(E_1) \longrightarrow \mathcal{M}_\mathbb{C}(E_2)$ such that $\tau(\mu *_1 \nu) = \tau(\mu) *_2 \tau(\nu)$ for all $\mu, \nu \in \mathcal{M}_\mathbb{C}(E_1)$ and $\tau(\delta_x)$ is a Dirac measure for all $x \in E_1$. If τ is bijective, then it is said to be a *hypergroup isomorphism.* Given a hypergroup $(E_1, *_1)$ and a continuous bijection $\tau : E_1 \longrightarrow E_2$, one can define a convolution $*_2$ on E_2 by letting $\delta_x *_2 \delta_y = \tau(\delta_{\tau^{-1}(x)} *_1 \delta_{\tau^{-1}(y)})$ and $(\mu *_2 \nu)(\cdot) = \int_{E_2}\int_{E_2} (\delta_x *_2 \delta_y)(\cdot)\, \mu(dx)\nu(dy)$. (Here $\tau(\delta_{\tau^{-1}(x)} *_1 \delta_{\tau^{-1}(y)})$ stands for the pushforward of the measure $\delta_{\tau^{-1}(x)} *_1 \delta_{\tau^{-1}(y)}$ under the map $\xi \mapsto \tau(\xi)$.) It is then straightforward to check that the hypergroup axioms hold for $(E_2, *_2)$, so that $(E_1, *_1)$ and $(E_2, *_2)$ are isomorphic hypergroups.

Let $(E_1, *_1), \ldots, (E_n, *_n)$ be a finite family of hypergroups and write $E = E_1 \times \ldots \times E_n$. Define the convolution operator $* : \mathcal{M}_\mathbb{C}(E) \times \mathcal{M}_\mathbb{C}(E) \longrightarrow \mathcal{M}_\mathbb{C}(E)$ by

$$(\mu * \nu)(\cdot) = \int_E \int_E \bigl((\delta_{x_1} *_1 \delta_{y_1}) \otimes \ldots \otimes (\delta_{x_n} *_n \delta_{y_n})\bigr)(\cdot)\, \mu(dx)\nu(dy).$$

One can easily verify that this operator satisfies all the hypergroup axioms. The hypergroup $(E, *)$ is called the *product* of the hypergroups $(E_1, *_1), \ldots, (E_n, *_n)$.

As noted above, the structural properties captured in the hypergroup axioms allow for the development of several analogues of standard theorems in classical harmonic analysis [19, 184, 188]. Here we only highlight the Lévy-Khintchine type theorem stated below. (We say that a measure $\mu \in \mathcal{M}_\mathbb{C}(E)$ is *symmetric* if $\mu(B) = \mu(\check{B})$ for all Borel subsets $B \subset E$. The notion of infinitely divisible distribution is defined as in Proposition 2.15(b).)

Proposition 2.17 *Let* $(E, *)$ *be a commutative hypergroup. Any symmetric infinitely divisible measure* $\mu \in \mathcal{P}(E)$ *can be represented as*

$$\mu = \gamma * \mathbf{e}(\theta),$$

where:

- $\theta = \lim_{n\to\infty} (n \cdot \mu_n)|_{E\setminus\{e\}}$ *is a σ-finite positive measure;*
- $\mathbf{e}(\theta)$ *is the so-called* ∗-Poisson measure *associated with θ, defined as* $\mathbf{e}(\theta) := e^{-\|\theta\|} \sum_{n=0}^{\infty} \frac{\theta^{*n}}{n!}$;
- γ *is a* ∗-Gaussian measure, *i.e. an infinitely divisible measure such that* $\lim_{n\to\infty} n \cdot \gamma_n(E \setminus V) = 0$ *for all open sets V containing* e.

(Here μ_n and γ_n are the measures such that $\mu = \mu_n^{*n}$ and $\gamma = \gamma_n^{*n}$.) The representation is unique, i.e. if $\mu = \widetilde{\gamma} * \mathbf{e}(\widetilde{\theta})$ for a σ-finite positive measure $\widetilde{\theta}$ and a ∗-Gaussian measure $\widetilde{\gamma}$, then $\theta = \widetilde{\theta}$ and $\gamma = \widetilde{\gamma}$.

Proof See [162, Theorems 4.4 and 4.7].

Stochastic Convolutions

Unlike the definitions of convolution-like structures presented above, the so-called stochastic convolutions, which were introduced and studied by Volkovich in [196, 198], include the existence of a compatible (generalized) characteristic function in their defining axioms:

Definition 2.5 Let E be a locally compact space. A bilinear operator $\circ$ on $\mathcal{M}_\mathbb{C}(E)$ is said to be a *stochastic convolution* (in the sense of Volkovich) if it has the following properties:

V1. If $\mu, \nu \in \mathcal{P}(E)$, then $\mu \circ \nu \in \mathcal{P}(E)$;

V2. There exists a separable complete metric space S and a bounded real continuous function $\omega(x, \sigma)$ on $E \times S$ such that the ∘*-characteristic function*

$$\Phi_\mu(\sigma) := \int_E \omega(x, \sigma)\mu(dx) \qquad (\mu \in \mathcal{P}(E),\ \sigma \in S)$$

determines uniquely the probability measure μ, and no proper closed subset of S has the same property;

V3. There exists $\mathrm{e} \in E$ such that $\omega(\mathrm{e}, \sigma) = 1$ for all $\sigma \in S$;

V4. $\mu_n \overset{w}{\longrightarrow} \mu$ if and only if $\Phi_{\mu_n}(\sigma) \to \Phi_\mu(\sigma)$ for all $\sigma \in S$;

V5. $\mu_3 = \mu_1 \circ \mu_2$ if and only if $\Phi_{\mu_3}(\sigma) = \Phi_{\mu_1}(\sigma)\Phi_{\mu_2}(\sigma)$ for all $\sigma \in S$;

V6. Let $\mathfrak{P} \subset \mathcal{P}(E)$. The set $\mathrm{D}(\mathfrak{P})$ of all divisors (with respect to the convolution $\circ$) of measures $\nu \in \mathfrak{P}$ is relatively compact if and only if $\mathfrak{P}$ is relatively compact.

The following Lévy-Khintchine type theorem of Volkovich provides us with a characterization of the family of infinitely divisible distributions with respect to a given stochastic convolution. (The reader should compare it with the corresponding results for Urbanik convolutions and hypergroups stated in Propositions 2.15(b) and 2.17 respectively.)

Proposition 2.18 *Let $\circ$ be a stochastic convolution on E. A function $\Phi : E \longrightarrow \mathbb{R}$ is a $\circ$-characteristic function of an infinitely divisible measure $\mu \in \mathcal{P}(E)$ if and only it can be represented in the form*

$$\Phi(\sigma) = \Phi_\gamma(\sigma) \exp\biggl(\int_{E \setminus \{\mathrm{e}\}} \bigl(\omega(x,\sigma) - 1\bigr)\nu(dx)\biggr), \tag{2.45}$$

where:

- *ν is a σ-finite measure on $E \setminus \{\mathrm{e}\}$ which is finite on the complement of any neighbourhood of* e *and such that*

$$\int_E \bigl(1 - \omega(x,\sigma)\bigr)\nu(dx) < \infty;$$

- *$\Phi_\gamma(\cdot)$ is the $\circ$-characteristic function of an infinitely divisible measure γ such that*

$$\gamma = \mathbf{e}(a\nu) \circ \eta \quad \bigl(a > 0,\ \nu, \eta \in \mathcal{P}(E),\ \eta \textit{ infinitely divisible}\bigr) \quad \Longrightarrow \quad \nu = \delta_{\mathrm{e}}.$$

Proof See [196, pp. 465–466].

Any Urbanik convolution is a stochastic convolution [198], as well as all known examples of hypergroups on $\mathbb{R}_0^+$ (cf. Sect. 4.1 below). Additional examples are listed in [198]. We can therefore interpret the defining properties V1–V6 of stochastic convolutions as a less restrictive set of axioms which still enable one to study infinite divisibility of measures on the convolution algebra and establish an analogue of the usual Lévy-Khintchine representation. The additional structure provided by the stronger axioms of hypergroups or of Urbanik convolutions gives rise to many other analogues of fundamental results of harmonic analysis, such as laws of large numbers and characterizations of Lévy processes (these results can be found in the literature cited above).

Generalized Integral Convolutions
This notion, which is complementary to that of generalized translation operator and of hypercomplex system, is not based on probabilistic motivations; rather, the motivation is to construct analogues of the ordinary convolution associated with prescribed integral transforms.

Let $\mathcal{K} : \mathfrak{C} \longrightarrow \widetilde{\mathfrak{C}}$ be an integral transformation between function spaces $\mathfrak{C}$ and $\widetilde{\mathfrak{C}}$. An operator $*$ such that $\mathcal{K}(f * g) = \mathcal{K}f \cdot \mathcal{K}g$ is said to be a *generalized integral convolution* associated with the integral transform $\mathcal{K}$ [73, 97]. (Note the analogy with the corresponding property between the Fourier transform and the ordinary convolution.) This notion has been generalized by allowing for the presence of a weight function $\beta \in \widetilde{\mathfrak{C}}$ on the right-hand side, so that $\mathcal{K}(f * g) = \beta \cdot \mathcal{K}f \cdot \mathcal{K}g$ [98], or by considering the more general property $\mathcal{K}_3(f * g) = \mathcal{K}_1 f \cdot \mathcal{K}_2 g$, where $\mathcal{K}_1, \mathcal{K}_2, \mathcal{K}_3$ are different integral transforms [186, 204, 207].

It is noted in [46] that if $*$ is a generalized integral convolution associated with an integral transform $\mathcal{K}$ and L is a linear operator on $\mathfrak{C}$ such that $\mathcal{K}\mathrm{L} = \mathrm{M}\mathcal{K}$, where M is a multiplication operator on $\widetilde{\mathfrak{C}}$, then the linear operator $*$ commutes with the convolution, i.e. the identity $\mathrm{L}(f * g) = (\mathrm{L}f) * g$ holds. As mentioned in the Introduction, this gives rise to applications in differential equations, because many integral transforms determine isomorphisms between differential operators and multiplication operators.

Many generalized integral convolutions have been constructed for one-dimensional integral transforms of the form $(\mathcal{K}f)(x) = \int_I k(x, y) f(y) dy$, where I is an interval of the real line and $k : I \times I \longrightarrow \mathbb{R}$. We refer to [73] for an extensive list of examples where the convolution has been constructed in closed form, and to [186, 207] for an approach based on double Mellin-Barnes integrals which is applicable to a general family of integral transforms with Meijer-G and Fox-H functions in the kernel.

2.4 Harmonic Analysis with Respect to the Kingman Convolution

We saw in Sect. 2.1 that the drifted Brownian motion $\{B_t\}$ has the convolution semigroup property, namely it satisfies $P_x[B_t \in \cdot] = \mu_t * \delta_x$ for some convolution semigroup $\{\mu_t\}$. In Sect. 2.3, we reviewed various notions of generalized convolutions. Now, we conclude the chapter with an example of a generalized convolution and corresponding convolution semigroups.

The Kingman convolution is the seminal example of a binary operator $\circ$ on the space of probability measures which allows us to obtain the following analogue of the convolution semigroup property of drifted Brownian motion: for a diffusion process $\{X_t\}$ other than the Brownian motion (in the case of the Kingman convolution, the Bessel process), we have $P_x[X_t \in \cdot] = \mu_t \circ \delta_x$, where $\{\mu_t\}$ is a family of measures satisfying $\mu_{t+s} = \mu_t \circ \mu_s$. In this section we briefly present the construction of the Kingman convolution and some properties which mirror well-known facts in classical harmonic analysis. This construction should be kept in mind throughout the subsequent chapters, as it serves as a benchmark for our later work in developing structures of generalized harmonic analysis associated with other diffusion processes.

Let $\{X_t\}_{t\geq 0}$ be the *Bessel process* with index $\eta > -\frac{1}{2}$ (started at $x_0 \geq 0$), defined as $X_t = \sqrt{Z_t}$ where $\{Z_t\}_{t\geq 0}$ is the unique strong solution of the SDE

$$dZ_t = 2(\eta + 1)dt + 2\sqrt{Z_t}dW_t, \qquad Z_0 = x_0^2$$

(cf. [164, §XI.1]). The process $\{X_t\}$ is a one-dimensional diffusion with infinitesimal generator $\mathcal{G} = \frac{1}{2}\frac{d^2}{dx^2} + \frac{\eta+\frac{1}{2}}{x}\frac{d}{dx}$. In the case $0 < \eta < 1$, the boundary $x = 0$ is instantaneously reflecting, while in the case $\eta \geq 1$ the endpoint $x = 0$ is never reached by $\{X_t\}$. The transition probabilities of the Bessel process are given by the closed-form expression

$$
\begin{aligned}
p_{t,x}(dy) &:= P[X_t \in dy | X_0 = x] \\
&= \begin{cases} t^{-1} x^{-\eta}\, y^{\eta+1} \exp\!\big(-\frac{x^2+y^2}{2t}\big) I_\eta\big(\frac{\sqrt{xy}}{t}\big) dy, & \text{if } x, t > 0, \\ \frac{2^{-\eta} t^{-\eta-1}}{\Gamma(\eta+1)} y^{2\eta+1} \exp\!\big(-\frac{y^2}{2t}\big) dy, & \text{if } x = 0,\ t > 0, \\ \delta_x(dy), & \text{if } t = 0, \end{cases}
\end{aligned}
\tag{2.46}
$$

where $\Gamma(\cdot)$ is the Gamma function and $I_\eta(z) := \sum_{k=0}^{\infty} \frac{(z/2)^{\eta+2k}}{k!\Gamma(\eta+k+1)}$ is the *modified Bessel function of the first kind* with index η [145, §10.25].

The infinitesimal generator $\frac{1}{2}\frac{d^2}{dx^2} + \frac{\eta+\frac{1}{2}}{x}\frac{d}{dx}$ is associated with the invertible integral transform $\mathcal{H}: L^2(\mathbb{R}^+; x^{2\eta+1}dx) \longrightarrow L^2(\mathbb{R}^+; \tau^{2\eta+1}d\tau)$ defined by

$$
\begin{aligned}
(\mathcal{H}f)(\tau) &= \int_0^\infty f(x)\, \boldsymbol{J}_\eta(\tau x)\, x^{2\eta+1} dx, \\
(\mathcal{H}^{-1}\varphi)(x) &= \frac{2^{-2\eta}}{\Gamma(\eta+1)^2} \int_0^\infty \varphi(\tau)\, \boldsymbol{J}_\eta(\tau x)\, \tau^{2\eta+1} d\tau,
\end{aligned}
\tag{2.47}
$$

where $\boldsymbol{J}_\eta(z) := 2^\eta \Gamma(\eta+1) z^{-\eta} J_\eta(z)$ and $J_\eta(z) := \sum_{k=0}^{\infty} \frac{(-1)^k (z/2)^{\eta+2k}}{k!\Gamma(\eta+k+1)}$ is the *Bessel function of the first kind* [145, §10.2]. The operator $\mathcal{H}$, which is known as the *Hankel transform* [84], is a particular case of the general Sturm-Liouville integral transform (2.21)–(2.22), introduced in Sect. 2.2. To see this, notice that the function $x \mapsto \boldsymbol{J}_\eta(\tau x)$ is the unique solution of the boundary value problem $-\frac{1}{2}\frac{d^2}{dx^2} - \frac{\eta+\frac{1}{2}}{x}\frac{d}{dx}u = \tau^2 u$, $u(0) = 1$, $\lim_{x\downarrow 0} x^{2\eta+1}u'(x) = 0$ (see [145]), which is a problem of type (2.1) with $r(x) = x^{2\eta+1}$, $p(x) = \frac{x^{2\eta+1}}{2}$ and $q = 0$.

Proposition 2.19 *Define the extension of the Hankel transform to finite complex measures by*

$$
(\mathcal{H}\mu)(\tau) := \int_{\mathbb{R}_0^+} \boldsymbol{J}_\eta(\tau x)\mu(dx), \qquad (\mu \in \mathcal{M}_\mathbb{C}(\mathbb{R}_0^+),\ \tau \geq 0). \tag{2.48}
$$

Then $(\mathcal{H}\mu)(\tau)$ *is, for each* $\mu \in \mathcal{M}_\mathbb{C}(\mathbb{R}_0^+)$, *a continuous function of* $\tau \geq 0$ *which determines uniquely the measure* μ. *Moreover, the Hankel transform of the transition probabilities* (2.46) *equals*

$$
(\mathcal{H}p_{t,x})(\tau) = e^{-t\tau^2}\, \boldsymbol{J}_\eta(\tau x) \qquad (t > 0,\ x \geq 0).
$$

Proof Since $|\boldsymbol{J}_\eta(y)| \leq 1$ for $y \geq 0$ [84, Theorem 2a], dominated convergence yields that $\tau \mapsto (\mathcal{H}\mu)(\tau)$ is continuous. By [101, Lemma 2], $(\mathcal{H}\mu)(\tau)$ determines uniquely the measure μ. The fact that $(\mathcal{H}p_{t,x})(\tau) = e^{-t\tau^2} \boldsymbol{J}_\eta(\tau x)$ can be verified using [156, Equation 2.12.39.3].

We will say that $\circ : \mathcal{M}_\mathbb{C}(\mathbb{R}_0^+) \times \mathcal{M}_\mathbb{C}(\mathbb{R}_0^+) \longrightarrow \mathcal{M}_\mathbb{C}(\mathbb{R}_0^+)$ is a *generalized convolution for the Bessel process* if the transition probabilities are such that

$$p_{t,x} = \mu_t \circ \delta_x, \tag{2.49}$$

where $\{\mu_t\}_{t\geq 0} \subset \mathcal{P}(\mathbb{R}_0^+)$ is such that $\mu_{t+s} = \mu_t \circ \mu_s$ for every $t, s \geq 0$.

It follows from Proposition 2.19 that if $\circ$ is such that $\mathcal{H}(\mu \circ \nu) \equiv (\mathcal{H}\mu)\cdot(\mathcal{H}\nu)$ for all $\mu, \nu \in \mathcal{P}(\mathbb{R}_0^+)$, then $\circ$ is a generalized convolution for the Bessel process. This suggests that a crucial requirement for the generalized convolution $\circ$ is that it should satisfy the *product formula* $\big(\mathcal{H}(\delta_x \circ \delta_y)\big)(\tau) \equiv (\mathcal{H}\delta_x)(\tau)\cdot(\mathcal{H}\delta_y)(\tau)$ or, equivalently, $\boldsymbol{J}_\eta(\tau x)\, \boldsymbol{J}_\eta(\tau y) = \int_{\mathbb{R}_0^+} \boldsymbol{J}_\eta(\tau\xi)(\delta_x \circ \delta_y)(d\xi)$, where the measure $\delta_x \circ \delta_y$ should not depend on τ. It turns out that such a product formula indeed exists, and we will see below that it gives rise to a convolution for which the desired (generalized) convolution semigroup property (2.49) holds.

Theorem 2.6 (Product Formula for the Bessel Function of the First Kind) *The following identity holds for all $x, y > 0$, $\tau \geq 0$ and $\eta > 0$:*

$$\begin{aligned} \boldsymbol{J}_\eta(\tau x)\, \boldsymbol{J}_\eta(\tau y) =& \frac{2^{1-2\eta}\Gamma(\eta+1)}{\sqrt{\pi}\,\Gamma(\eta+\frac{1}{2})}(xy)^{-2\eta} \\ & \int_{|x-y|}^{x+y} \boldsymbol{J}_\eta(\tau\xi)\big[(\xi^2-(x-y)^2)((x+y)^2-\xi^2)\big]^{\eta-1/2}\, \xi\, d\xi. \end{aligned} \tag{2.50}$$

Proof This follows from a classical integration formula for the Bessel function [200, p. 411], [84].

Definition 2.6 The operator $\circ : \mathcal{M}_\mathbb{C}(\mathbb{R}_0^+) \times \mathcal{M}_\mathbb{C}(\mathbb{R}_0^+) \longrightarrow \mathcal{M}_\mathbb{C}(\mathbb{R}_0^+)$ defined by

$$(\mu \circ \nu)(B) := \int_{\mathbb{R}_0^+}\int_{\mathbb{R}_0^+} \gamma_{x,y}(B)\, \mu(dx)\nu(dy) \qquad (\mu, \nu \in \mathcal{M}_\mathbb{C}(\mathbb{R}_0^+)),$$

where B is an arbitrary Borel subset of $\mathbb{R}_0^+$, $\gamma_{x,0} = \gamma_{0,x} = \delta_x$ and $\gamma_{x,y}(d\xi) = \boldsymbol{k}(x,y,\xi)\xi^{2\eta+1}d\xi$, with

$$\begin{aligned} \boldsymbol{k}(x,y,\xi) = & \frac{2^{1-2\eta}\Gamma(\eta+1)}{\sqrt{\pi}\,\Gamma(\eta+\frac{1}{2})}(xy\xi)^{-2\eta} \\ & \big[(\xi^2-(x-y)^2)((x+y)^2-\xi^2)\big]^{\eta-1/2}\mathbb{1}_{[|x-y|,x+y]}(\xi), \qquad x, y, \xi > 0 \end{aligned}$$

is called the *Kingman convolution* (of order η) [101, 194].

One can easily verify that the Kingman convolution preserves the space $\mathcal{P}(\mathbb{R}_0^+)$ (i.e. the Kingman convolution of two probability measures is indeed a probability measure) and, moreover, that it is trivialized by the Hankel transform of measures:

Proposition 2.20 *Let $\pi, \mu, \nu \in \mathcal{M}_{\mathbb{C}}(\mathbb{R}_0^+)$. We have $\pi = \mu \circ \nu$ if and only if $(\mathcal{H}\pi)(\tau) = (\mathcal{H}\mu)(\tau) \cdot (\mathcal{H}\nu)(\tau)$ for all $\tau \geq 0$.*

We observe that the theorem, definition and proposition above are counterparts of the following facts from classical harmonic analysis: *the kernel $e^{iz\cdot x}$ of the Fourier transform satisfies the trivial product formula $e^{iz\cdot x}e^{iz\cdot y} = \int_{\mathbb{R}^d} e^{iz\cdot\xi}\,\delta_{x+y}(d\xi)$; the ordinary convolution is computed as $(\mu * \nu)(B) = \int_{\mathbb{R}^d}\int_{\mathbb{R}^d} \gamma_{x,y}(B)\,\mu(dx)\nu(dy)$, where $\gamma_{x,y} = \delta_{x+y}$ is the measure of the product formula; we have $\pi = \mu * \nu$ if and only if $(\mathfrak{F}\pi)(z) = (\mathfrak{F}\mu)(z) \cdot (\mathfrak{F}\nu)(z)$ for all $z \in \mathbb{R}^d$, where $\mathfrak{F}$ is the Fourier transform.* Proposition 2.20 should also be compared with statement ***(a)*** of Proposition 2.15. Indeed, it can be checked that the Kingman convolution is also a Urbanik convolution, whose $\circ$-characteristic function $\mu \mapsto \mathbb{H}\mu$ has the kernel $\Omega(x) \equiv \boldsymbol{J}_\eta(\tau x)$.

Using Proposition 2.20, one can verify (see Chap. 4 below or [19]) that the Kingman convolution satisfies axioms H1–H8 of Definition 2.4 and, therefore, induces a hypergroup structure in $\mathbb{R}_0^+$, which is known as the *Bessel-Kingman hypergroup*. When $\eta = n/2 - 1$ ($n \in \mathbb{N}$), this hypergroup can be constructed as the projection on $\mathbb{R}_0^+$ of the space of all radial measures on $\mathbb{R}^n$, cf. [163, Section 7].

From Definition 2.6, the notions of Kingman convolution semigroups and Kingman Lévy processes can be constructed as follows.

Definition 2.7 A family $\{\mu_t\}_{t\geq 0} \subset \mathcal{P}(\mathbb{R}_0^+)$ is said to be a *Kingman convolution semigroup* if

$$\mu_s \circ \mu_t = \mu_{s+t} \text{ for all } s,t \geq 0, \qquad \mu_0 = \delta_0 \qquad \text{and} \quad \mu_t \xrightarrow{w} \delta_0 \text{ as } t \downarrow 0.$$

Convolution semigroups are similarly defined for a general Urbanik convolution or a general hypergroup, see [23, Section 4] or [19, Section 5.2] respectively.

Corollary 2.3 *Let $\mu_t = p_{t,0}$, where $\{p_{t,x}\}_{t,x\geq 0}$ are the transition probabilities* (2.46) *of the Bessel process started at zero. Then $\{\mu_t\}_{t\geq 0}$ is a Kingman convolution semigroup. Moreover, we have $p_{t,x} = \mu_t \circ \delta_x$ for all $t, x \geq 0$ (i.e. $\circ$ is a generalized convolution for the Bessel process).*

Proof See the comments before Theorem 2.6 and observe that the weak continuity $\mu_t \xrightarrow{w} \delta_0$ as $t \downarrow 0$ follows from the fact that the Bessel process is a Feller process (Proposition 2.7).

The next two results show that (an analogue of) two important properties of ordinary convolution semigroups—the fact that a convolution semigroup determines a Feller semigroup on $\mathbb{R}$ (cf. Proposition 2.8), and the Lévy-Khintchine representation (cf. Theorem 2.4)—can also be established for Kingman convolution semigroups.

Proposition 2.21 *Let* $\{\mu_t\}_{t\geq 0} \subset \mathcal{P}(\mathbb{R}_0^+)$ *be a Kingman convolution semigroup. Then the family* $\{T_t\}_{t\geq 0}$ *defined by*

$$T_t : \mathrm{C_b}(\mathbb{R}_0^+) \longrightarrow \mathrm{C_b}(\mathbb{R}_0^+),$$

$$T_t f = \mathcal{T}^{\mu_t} f, \quad \textit{where} \quad (\mathcal{T}^{\mu_t} f)(x) := \int_{\mathbb{R}_0^+} f\, d(\delta_x \circ \mu_t)$$

is a conservative Feller semigroup.

Proof See [163, Proposition 2.1].

Theorem 2.7 (Lévy-Khintchine Type Representation) *If* $\{\mu_t\}_{t\geq 0} \subset \mathcal{P}(\mathbb{R}_0^+)$ *is a Kingman convolution semigroup, then*

$$(\mathcal{H}\mu_t)(\tau) = e^{-t\psi(\tau)} \tag{2.51}$$

for some function $\psi(\cdot)$ *of the form*

$$\psi(\tau) = c\tau^2 + \int_{\mathbb{R}^+} \bigl(1 - \boldsymbol{J}_\eta(\tau x)\bigr)\nu(dx), \tag{2.52}$$

where $c \geq 0$ *and* ν *is a measure on* $\mathbb{R}^+$ *which is finite on the complement of any neighbourhood of* 0 *and such that for* $\tau > 0$ *we have*

$$\int_{\mathbb{R}^+} \bigl(1 - \boldsymbol{J}_\eta(\tau x)\bigr)\nu(dx) < \infty.$$

Conversely, for each function of the form (2.52) *there exists a Kingman convolution semigroup* $\{\mu_t\}$ *such that* $(\mathcal{H}\mu_t)(\tau) = e^{-t\psi(\tau)}$ *for all* $\tau \geq 0$.

In particular, the functions $\psi_\beta(\tau) := \tau^\beta$ $(0 < \beta < 2)$ *belong to the set of admissible functions of the form* (2.52).

Proof See [190, Theorem 13] and [194, Theorem 2].

Unlike the Lévy symbol (2.9) in the Lévy-Khintchine representation for ordinary convolution semigroups, the symbol $\psi(\cdot)$ in the Lévy-Khintchine type formula (2.51)–(2.52) for the Kingman convolution has no imaginary terms. This is unsurprising, because the Hankel transform of a probability measure on $\mathbb{R}_0^+$ is real-valued, while the Fourier transform of probability measures on $\mathbb{R}^d$ is complex-valued. The resemblance between the two formulas becomes yet more evident when $\{\mu_t\} \subset \mathcal{P}(\mathbb{R}^d)$ is an ordinary convolution semigroup of symmetric measures: an ordinary convolution semigroup is symmetric if and only if $\alpha = 0$ and the measure ν is symmetric, so that $\phi(z) = z \cdot Qz + \int_{\mathbb{R}^d \setminus \{0\}} \bigl(1 - \cos(z \cdot y)\bigr)\nu(dy)$. (Since $\boldsymbol{J}_{-\frac{1}{2}}(\xi) = \cos\xi$ [145, §10.16], this right-hand side is, for $d = 1$, the limiting form of the representation (2.52) when $\eta \downarrow -\frac{1}{2}$.)

Theorem 2.7 can also be compared with the Lévy-Khintchine representation for Urbanik convolutions in statement ***(b)*** of Proposition 2.15. An analogous representation in the context of Sturm-Liouville hypergroups is given in [30, Theorem 7].

A *Kingman Lévy process* is a Feller process associated with a Kingman convolution semigroup. By the above Lévy-Khintchine type representation, the class of Kingman Lévy processes generalizes the Bessel processes in an analogous way as the class of (ordinary) Lévy processes generalizes the Brownian motion. We note, in particular, that the class of Kingman Lévy processes includes many processes which do not admit continuous versions (cf. [163, Theorem 2.2]). For further properties of the Kingman convolution and the associated Lévy processes, we refer to [19, 23, 101, 163, 190].

The results stated thus far refer to the probabilistic properties of the Kingman convolution (seen as a binary operator on the space of probability measures). There is also an extensive literature on the *Hankel convolution* of functions, which is defined by

$$(f \circ g)(x) := \int_0^\infty \int_0^\infty f(y)\boldsymbol{k}(x, y, \xi) y^{2\eta+1} dy\, g(\xi)\, \xi^{2\eta+1} d\xi. \tag{2.53}$$

In other words, the Hankel convolution $f \circ g$ is defined as the density of the Kingman convolution of the measures $\mu_f(dx) = f(x)x^{2\eta+1}dx$ and $\mu_g(dx) = g(x)x^{2\eta+1}dx$. It is thus clear that the Kingman convolution is the generalized integral convolution associated with the Hankel transform (2.47).

It is clear that $\mathcal{H}(f \circ g) = (\mathcal{H}f) \cdot (\mathcal{H}g)$ for $f, g \in L^1(\mathbb{R}^+; x^{2\eta+1}dx)$; this result is the Hankel counterpart of the usual convolution theorem $\mathfrak{F}(f * g) = (\mathfrak{F}f) \cdot (\mathfrak{F}g)$, where $(f * g)(x) = \int_{\mathbb{R}^d} f(x - y)g(y)dy$ and $\mathfrak{F}$ is the Fourier transform on $\mathbb{R}^d$. The Hankel convolution has many other properties which are parallel to those of the ordinary convolution of functions, such as a Young-type inequality. Let us recall that the classical Young convolution inequality [62, Proposition 8.9] states that *if $f \in L^{p_1}(\mathbb{R}^d)$, $g \in L^{p_2}(\mathbb{R}^d)$ $(p_1, p_2 \in [1, \infty])$ and $s \in [1, \infty]$ is defined by $\frac{1}{s} = \frac{1}{p_1} + \frac{1}{p_2} - 1$, then the integral defining $(f * g)(x)$ converges for a.e. x and $\|f * g\|_{L^s(\mathbb{R}^d)} \leq \|f\|_{L^{p_1}(\mathbb{R}^d)} \cdot \|g\|_{L^{p_2}(\mathbb{R}^d)}$.* The following analogue holds for the Hankel convolution. (We write $L_\eta^p := L^p(\mathbb{R}^+; x^{2\eta+1}dx)$.)

Proposition 2.22 *Let $f \in L_\eta^{p_1}$, $g \in L_\eta^{p_2}$ $(p_1, p_2 \in [1, \infty])$ and let $s \in [1, \infty]$ be defined by $\frac{1}{s} = \frac{1}{p_1} + \frac{1}{p_2} - 1$. Then $(f \circ g)(x)$ converges for a.e. $x > 0$ and*

$$\|f \circ g\|_{L_\eta^s} \leq \|f\|_{L_\eta^{p_1}} \cdot \|g\|_{L_\eta^{p_2}}.$$

If $f \in L_\eta^p$ and $g \in L_\eta^q$ with $\frac{1}{p} + \frac{1}{q} = 1$, then $f \circ g \in \mathrm{C_b}(\mathbb{R}^+)$.

Proof See [84, Theorem 2b].

Additional examples of analogues of classical properties which have been established for the Hankel convolution include: an analogue of the Marcinkiewicz multiplier theorem [78], a characterization of variation diminishing convolution kernels similar to that for the ordinary convolution [84], a parallel theory for Hankel convolution equations [35], among others.

Chapter 3
The Whittaker Convolution

The goal of this chapter is to construct a generalized convolution for the one-dimensional diffusion process known as the Shiryaev process. The properties of this convolution-like operator will allow us to interpret the Shiryaev process as a Lévy-like process, thereby providing a positive answer to the general question formulated in the Introduction.

The *Shiryaev process* (started at $y_0 \geq 0$) is defined in [149] as the unique strong solution $\{Y_t\}_{t\geq 0}$ of the SDE

$$dY_t = (1 + \mu Y_t)dt + \sigma Y_t \, dW_t, \qquad Y_0 = y_0 \tag{3.1}$$

where $\mu \in \mathbb{R}$, $\sigma > 0$ and $\{W_t\}_{t\geq 0}$ is a standard Brownian motion. The infinitesimal generator of the Shiryaev process is the differential operator $\mathcal{A}$ defined as $\mathcal{A}u(y) = \frac{\sigma^2}{2} y^2 u''(y) + (1 + \mu y)u'(y)$.

The Shiryaev process, whose defining SDE (3.1) was first derived by Shiryaev in the context of quickest detection problems [168], has various applications in mathematical finance; in particular, it plays a fundamental role in the problem of Asian option pricing under the famous Black-Scholes model [47, 124]. See [152] for a survey of other applications in physics and finance.

We will restrict our attention to the *standardized* Shiryaev process with parameters $\sigma = \sqrt{2}$ and $\mu = 2(1 - \alpha) \in \mathbb{R}$, i.e. the one-dimensional diffusion generated by the operator

$$\mathcal{A}_\alpha u(y) = y^2 u''(y) + (1 + 2(1-\alpha)y)u'(y) = \frac{1}{r_\alpha(y)} \big(p_\alpha u'\big)'(y), \tag{3.2}$$

where $r_\alpha(\xi) := \xi^{-2\alpha} e^{-1/\xi}$ and $p_\alpha(\xi) := \xi^{2(1-\alpha)} e^{-1/\xi}$. This restriction does not introduce any loss of generality, because we know (cf. [21, Section II.8]) that if $\{X_t\}_{t\geq 0}$ is a one-dimensional diffusion generated by $\mathcal{A}_{\alpha,\gamma,c} := \gamma x^2 \frac{d^2}{dx^2} + \gamma(c +$

R. Sousa et al., *Convolution-like Structures, Differential Operators and Diffusion Processes*, Lecture Notes in Mathematics 2315,
https://doi.org/10.1007/978-3-031-05296-5_3

$2(1-\alpha)x)\frac{d}{dx}$ $(c, \gamma > 0)$, then the process $\{Y_t = \frac{1}{c}X_{\gamma t}\}_{t\geq 0}$ is a one-dimensional diffusion generated by (3.2). In fact, once we have constructed the convolution structure for the standardized Shiryaev process, the convolution structure for $\mathcal{A}_{\alpha,\gamma,c}$ is simply a by-product which is obtained via elementary changes of variable (cf. Remark 3.11).

From Sect. 3.3 onwards we will mostly assume that $\alpha \leq \frac{1}{2}$; this is a necessary and sufficient condition for the underlying product formula to have the positivity and conservativeness property which is required for the induced convolution to be a binary operator on the space of probability measures.

3.1 A Special Case: The Kontorovich–Lebedev Convolution

Before advancing to our general discussion of convolutions associated with the standardized Shiryaev process (3.2), we take a look into a motivating particular case: the so-called Kontorovich–Lebedev convolution, which was introduced by Kakichev in 1967 [97] and whose properties and applications have been studied in numerous works [85, 86, 154, 204, 205, 207].

Definition 3.1

(a) The *Kontorovich–Lebedev transform* $(\mathcal{F}_{\mathrm{KL}}f)(\cdot)$ of a function $f \in L^2(\mathbb{R}^+, dx)$ is defined by

$$(\mathcal{F}_{\mathrm{KL}}f)(\tau) = \int_0^\infty f(x)K_{i\tau}(x)\,dx, \tag{3.3}$$

where $K_\nu(x)$ is the modified Bessel function of the second kind [145, §10.25].

(b) Let $f, g : \mathbb{R}^+ \longrightarrow \mathbb{C}$. If the double integral

$$(f \underset{KL}{*} g)(x) := \frac{1}{2x}\int_0^\infty\int_0^\infty \exp\biggl(-\frac{xy}{2\xi} - \frac{x\xi}{2y} - \frac{y\xi}{2x}\biggr) f(y)g(\xi)dyd\xi \tag{3.4}$$

exists for almost every $x \in \mathbb{R}^+$, then we call it the *Kontorovich–Lebedev convolution* of f and g.

The connection between the Kontorovich–Lebedev transform and convolution is analogous to that between the Hankel transform and convolution, discussed in Sect. 2.4. Indeed, like in (2.50), (2.53), the definition of the Kontorovich–Lebedev convolution is based on the product formula for the kernel of the Kontorovich–Lebedev transform, which is given by

$$K_\nu(x)K_\nu(y) = \frac{1}{2}\int_0^\infty K_\nu(\xi)\exp\biggl(-\frac{xy}{2\xi} - \frac{x\xi}{2y} - \frac{y\xi}{2x}\biggr)\frac{d\xi}{\xi} \qquad (x, y > 0,\ \nu \in \mathbb{C}). \tag{3.5}$$

This identity is a classical result known as the *Macdonald formula*, which can be found in standard texts on special functions, cf. [56, §7.7.6] or [204, Equation (1.103)].

Since the modified Bessel function of the second kind and the Whittaker W function are related by $K_\nu(x) = \pi^{\frac{1}{2}}(2x)^{-\frac{1}{2}} W_{0,\nu}(2x)$ (cf. Remark 3.1 below), the definition (3.3) can be rewritten as $(\mathcal{F}_{\mathrm{KL}} f)(\tau) = \pi^{\frac{1}{2}} \int_0^\infty f(x)\, W_{0,i\tau}(2x)\,(2x)^{-\frac{1}{2}} dx$. The Kontorovich–Lebedev transform is therefore (up to a change of variables) the particular case $\alpha = 0$ of the index Whittaker transform (2.40). Accordingly, a straightforward corollary of (3.5) is that the product formula

$$\boldsymbol{W}_{0,\nu}(x)\,\boldsymbol{W}_{0,\nu}(y) = \int_0^\infty \boldsymbol{W}_{0,\nu}(\xi)\,k_0(x,y,\xi)\,e^{-1/\xi} d\xi, \tag{3.6}$$

where

$$k_0(x,y,\xi) := \frac{1}{2(\pi x y \xi)^{1/2}} \exp\left(\frac{1}{2x} + \frac{1}{2y} + \frac{1}{2\xi} - \frac{x}{4y\xi} - \frac{y}{4x\xi} - \frac{\xi}{4xy}\right)$$

holds for the functions $\boldsymbol{W}_{0,\nu}(y) := e^{\frac{1}{2y}} W_{0,\nu}(\frac{1}{y})$, i.e., for the generalized eigenfunctions of the generator $\mathcal{A}_0$ of the standardized Shiryaev process with parameter $\alpha = 0$ (cf. Proposition 3.1 below).

One can check (using the identity (3.17) below with $\nu = -\frac{1}{2}$) that the measures $k_0(x,y,\xi)\,e^{-1/\xi} d\xi$ constitute a family of probability measures: $\{k_0(x,y,\xi)\,e^{-1/\xi} d\xi\}_{x,y>0} \subset \mathcal{P}(\mathbb{R}_0^+)$. As pointed out in the Introduction, this probabilistic property of the product formula means that the normalized Kontorovich–Lebedev convolution

$$(f \underset{KL}{\diamond} g)(x) := \int_0^\infty \int_0^\infty k_0(x,y,\xi) f(y) g(\xi) e^{-1/y} dy e^{-1/\xi} d\xi, \tag{3.7}$$

together with its extension to probability measures, are especially suitable for developing generalized harmonic analysis. We note that, despite this, the normalized form (3.7) has rarely been adopted in the literature on the Kontorovich–Lebedev convolution.

3.2 The Product Formula for the Whittaker Function

For our purposes, the Kontorovich–Lebedev transform is merely a particular case of the index Whittaker transform; this makes it natural to conjecture that a probabilistic product formula similar to (3.6) might hold for the generalized eigenfunctions of $\mathcal{A}_\alpha$ with $\alpha \neq 0$. The goal of this section is to prove that this conjecture is true.

It should be noted that no counterpart of the product formula (3.6) for the operator $\mathcal{A}_\alpha$ $(\alpha \neq 0)$ can be found on standard texts on special functions; in order to establish such counterpart, we will rely on integration formulas and other known results from the theory of special functions.

To prepare our work, we begin by summarizing some relevant facts on the Whittaker W functions and their role as eigenfunctions of the Sturm–Liouville operator $\mathcal{A}_\alpha$.

Proposition 3.1 *The unique solution of the boundary value problem*

$$-\mathcal{A}_\alpha u = \lambda u \quad (y \in \mathbb{R}^+,\ \lambda \in \mathbb{C}), \qquad u(0) = 1, \qquad (p_\alpha u')(0) = 0. \tag{3.8}$$

is given by

$$\boldsymbol{W}_{\alpha,\Delta_\lambda}(y) := y^\alpha e^{\frac{1}{2y}} W_{\alpha,\Delta_\lambda}(\tfrac{1}{y}), \tag{3.9}$$

where $\Delta_\lambda = \sqrt{(\frac{1}{2} - \alpha)^2 - \lambda}$ *and* $W_{\alpha,\nu}(x)$ *is the Whittaker function of the second kind.*

Throughout this chapter, the function $\boldsymbol{W}_{\alpha,\nu}(y) = y^\alpha e^{\frac{1}{2y}} W_{\alpha,\nu}(\frac{1}{y})$ will be called the *normalized Whittaker W function*. To prove Proposition 3.1, one just needs to check, using the basic properties of the Whittaker function stated below, that (3.9) is a solution of $-\mathcal{A}_\alpha u = \lambda u$ which satisfies the given boundary conditions.

Remark 3.1 (Some Basics on the Whittaker Function of the Second Kind) Let $\alpha, \nu \in \mathbb{C}$. The Whittaker function $W_{\alpha,\nu}(x)$ is, by definition, the solution of Whittaker's differential equation $\frac{d^2u}{dx^2} + \bigl(-\frac{1}{4} + \frac{\alpha}{x} + \frac{1/4-\nu^2}{x^2}\bigr)u = 0$ which is determined uniquely by the property

$$W_{\alpha,\nu}(x) \sim x^\alpha e^{-\frac{x}{2}}, \qquad |x| \to \infty,\ \operatorname{Re} x > 0. \tag{3.10}$$

The Whittaker W function is an analytic function of x on the half-plane $\operatorname{Re} x > 0$, and for fixed x it is an entire function of the first and the second parameter [145, §13.14(ii)]. For $\operatorname{Re} x > 0$ and $\operatorname{Re}\alpha < \frac{1}{2} + \operatorname{Re}\nu$, it admits the integral representation (cf. [155], integral 2.3.6.9)

$$W_{\alpha,\nu}(x) = \frac{e^{-\frac{x}{2}} x^\alpha}{\Gamma(\frac{1}{2} - \alpha + \nu)} \int_0^\infty e^{-s} s^{-\frac{1}{2}-\alpha+\nu} \Bigl(1 + \frac{s}{x}\Bigr)^{-\frac{1}{2}+\alpha+\nu} ds. \tag{3.11}$$

The Whittaker W function is an even function of the parameter ν [145, Equation 13.14.31]. For $\alpha \neq \frac{1}{2} \pm \nu, \frac{3}{2} \pm \nu, \ldots$, its asymptotic behaviour near the origin is, cf. [145, §13.14(iii)]

$$\begin{aligned} W_{\alpha,\nu}(x) &= O\bigl(x^{\frac{1}{2}-\operatorname{Re}\nu}\bigr) \qquad (\operatorname{Re}\nu \geq 0,\ \nu \neq 0), \\ W_{\alpha,0}(x) &= O\bigl(-x^{\frac{1}{2}}\log x\bigr), \end{aligned} \qquad x \to 0. \tag{3.12}$$

The asymptotic expansion for $W_{\alpha,\nu}(x)$ as $|x| \to \infty$ is given by [145, Equation 13.19.3]

$$W_{\alpha,\nu}(x) \sim e^{-\frac{x}{2}} x^{\alpha} \sum_{k=0}^{\infty} \frac{(\frac{1}{2}-\alpha+\nu)_k (\frac{1}{2}-\alpha-\nu)_k}{k!} (-x)^{-k}, \qquad |x| \to \infty, \ \operatorname{Re} x > 0. \tag{3.13}$$

The Whittaker function satisfies the recurrence relation and the differentiation formula [145, Equations 13.15.13 and 13.15.23]

$$x^{\frac{1}{2}} W_{\alpha+\frac{1}{2},\nu+\frac{1}{2}}(x) = (x+2\nu) W_{\alpha,\nu}(x) + (\tfrac{1}{2}-\alpha-\nu)\, x^{\frac{1}{2}} W_{\alpha-\frac{1}{2},\nu-\frac{1}{2}}(x), \tag{3.14}$$

$$\Bigl(x\frac{d}{dx}x\Bigr)^{n}\Bigl(e^{x/2}x^{-\alpha-1}W_{\alpha,\nu}(x)\Bigr) = \bigl(\tfrac{1}{2}+\nu-\alpha\bigr)_n \bigl(\tfrac{1}{2}-\nu-\alpha\bigr)_n\, e^{x/2} x^{n-\alpha-1} W_{\alpha-n,\nu}(x), \tag{3.15}$$

where $n \in \mathbb{N}$ and $(a)_n = \prod_{j=0}^{n-1}(a+j)$ is the Pochhammer symbol. When the parameter α is equal to zero (resp., equal to $\frac{1}{2}+\nu$), the Whittaker function reduces to the modified Bessel function of the second kind (resp., to an elementary function) [145, §13.18(i), (iii)],

$$W_{0,\nu}(2x) = \pi^{-\frac{1}{2}} (2x)^{\frac{1}{2}} K_\nu(x), \tag{3.16}$$

$$W_{\frac{1}{2}+\nu,\nu}(x) = x^{\frac{1}{2}+\nu} e^{-x/2}. \tag{3.17}$$

By [204, Theorem 1.11], for $\alpha \in \mathbb{R}$ the asymptotic expansion of the Whittaker function with imaginary parameter $\nu = i\tau$ as $\tau \to \infty$ is

$$W_{\alpha,i\tau}(x) = (2x)^{\frac{1}{2}} \tau^{\alpha-\frac{1}{2}} e^{-\pi\tau/2} \cos\biggl(\tau \log\Bigl(\frac{x}{4\tau}\Bigr) + \frac{\pi}{2}\Bigl(\frac{1}{2}-\alpha\Bigr) + \tau\biggr)\bigl[1 + O(\tau^{-1})\bigr], \tag{3.18}$$

the expansion being uniform in $0 < x \le M$ $(M > 0)$.

The desired generalization of the product formula (3.5) to the Whittaker function $W_{\alpha,\nu}(x)$ with general parameter $\alpha \in \mathbb{C}$ is stated next:

Theorem 3.1 *The product $W_{\alpha,\nu}(x) W_{\alpha,\nu}(y)$ of two Whittaker functions of the second kind with different arguments admits the integral representation*

$$W_{\alpha,\nu}(x) W_{\alpha,\nu}(y) = \int_0^\infty W_{\alpha,\nu}(\xi)\, k_\alpha(x,y,\xi)\, \frac{d\xi}{\xi^2} \qquad (x, y > 0,\ \alpha, \nu \in \mathbb{C}), \tag{3.19}$$

where

$$\kappa_\alpha(x, y, \xi) := 2^{-1-\alpha}\pi^{-\frac{1}{2}}(xy\xi)^{\frac{1}{2}} \exp\left(\frac{x}{2}+\frac{y}{2}+\frac{\xi}{2}-\frac{(xy+x\xi+y\xi)^2}{8xy\xi}\right) D_{2\alpha}\left(\frac{xy+x\xi+y\xi}{(2xy\xi)^{1/2}}\right)$$

being $D_\mu(z)$ the parabolic cylinder function [56, Section 8.2].

Remark 3.2 Before the proof, let us collect some facts on the parabolic cylinder function $D_\mu(z)$ which will be needed in the sequel.

The parabolic cylinder function is given in terms of the Whittaker function by

$$D_\mu(z) = 2^{\frac{\mu}{2}+\frac{1}{4}} z^{-\frac{1}{2}} W_{\frac{\mu}{2}+\frac{1}{4},\frac{1}{4}}\left(\tfrac{z^2}{2}\right).$$

This function is a solution of the differential equation $\frac{d^2u}{dz^2} + (\mu + \frac{1}{2} - \frac{z^2}{4})u = 0$, and it is an entire function of the parameter μ. An integral representation for the parabolic cylinder function is [145, Equation 12.5.3]

$$D_\mu(z) = \frac{z^\mu\, e^{-\frac{z^2}{4}}}{\Gamma\big(\frac{1}{2}(1-\mu)\big)} \int_0^\infty e^{-s} s^{-\frac{1}{2}(1+\mu)} \left(1+\frac{2s}{z^2}\right)^{\frac{\mu}{2}} ds \qquad (\operatorname{Re} z > 0,\ \operatorname{Re}\mu < 1). \tag{3.20}$$

The asymptotic form of $D_\mu(z)$ for large z is [56, Equation 8.4(1)]

$$D_\mu(z) \sim z^\mu e^{-\frac{z^2}{4}}, \qquad z \to \infty. \tag{3.21}$$

The recurrence relation and differentiation formula for $D_\mu(z)$ are [56, Equations 8.2(14) and 8.2(16)]

$$D_{\mu+1}(z) = zD_\mu(z) - \mu D_{\mu-1}(z), \tag{3.22}$$

$$\frac{d^n}{dz^n}\left[e^{-\frac{z^2}{4}} D_\mu(z)\right] = (-1)^n e^{-\frac{z^2}{4}} D_{\mu+n}(z) \qquad (n \in \mathbb{N}), \tag{3.23}$$

and the parabolic cylinder function reduces to an exponential function when its parameter equals zero [56, Equation 8.2(9)],

$$D_0(z) = e^{-\frac{z^2}{4}}. \tag{3.24}$$

We will prove Theorem 3.1 through a sequence of lemmas, where we shall assume that α is a negative real number and ν is purely imaginary. In the final step of the proof, an analytic continuation argument will be used to remove this restriction.

Our first lemma gives an alternative product formula which is less useful than (3.19) because its kernel also depends on the second parameter of the Whittaker function.

Lemma 3.1 *If $\alpha \in (-\infty, 0)$ and $\tau \in \mathbb{R}$, then the integral representation*

$$W_{\alpha,i\tau}(x)W_{\alpha,i\tau}(p) = \frac{(xp)^{\alpha}e^{-\frac{x}{2}-\frac{p}{2}}}{|\Gamma(\frac{1}{2}-\alpha+i\tau)|^2}\times$$
$$\times \int_0^\infty \xi^{-1-\alpha}e^{-\frac{\xi}{2}}W_{\alpha,i\tau}(\xi)\int_0^\infty w^{-2\alpha}\exp\biggl(-w-\Bigl(\frac{1}{x}+\frac{1}{p}+\frac{w}{xp}\Bigr)w\xi\biggr)dw\,d\xi \tag{3.25}$$

is valid for $x, p > 0$.

Proof From relation 2.21.2.17 in [157] it follows that

$$\begin{aligned}
&W_{\alpha,i\tau}(x)W_{\alpha,i\tau}(p)\\
&= (xp)^{\frac{1}{2}-i\tau}e^{-\frac{x}{2}-\frac{p}{2}}\Psi\Bigl(\frac{1}{2}-\alpha-i\tau, 1-2i\tau; x\Bigr)\Psi\Bigl(\frac{1}{2}-\alpha-i\tau, 1-2i\tau; p\Bigr)\\
&= \frac{(xp)^{\frac{1}{2}-i\tau}e^{-\frac{x}{2}-\frac{p}{2}}}{\Gamma(1-2\alpha)}\int_0^\infty e^{-w}\,w^{-2\alpha}\bigl[(w+x)(w+p)\bigr]^{-\frac{1}{2}+\alpha+i\tau}\times\\
&\qquad\times {}_2F_1\Bigl(\frac{1}{2}-\alpha-i\tau, \frac{1}{2}-\alpha-i\tau; 1-2\alpha; 1-\frac{xp}{(w+x)(w+p)}\Bigr)dw\\
&= \frac{(xp)^{\alpha}e^{-\frac{x}{2}-\frac{p}{2}}}{\Gamma(1-2\alpha)}\int_0^\infty e^{-w}\,w^{-2\alpha}\times \qquad (3.26)\\
&\qquad\times {}_2F_1\Bigl(\frac{1}{2}-\alpha-i\tau, \frac{1}{2}-\alpha+i\tau; 1-2\alpha; -\Bigl(\frac{1}{x}+\frac{1}{p}\Bigr)w-\frac{w^2}{px}\Bigr)dw.
\end{aligned}$$

Here $\Psi(a, b; x) := e^{x/2}x^{-b/2}W_{\frac{b}{2}-a,\frac{b}{2}-\frac{1}{2}}(x)$ is the confluent hypergeometric function of the second kind [55, Chapter VI] (also known as the Tricomi function or the Kummer function of the second kind); in the last step we used the transformation formula ${}_2F_1(a, b; c; z) = (1-z)^{-a}{}_2F_1\bigl(a, c-b; c; \frac{z}{z-1}\bigr)$ for the Gauss hypergeometric function, cf. [145, Equation 15.8.1].

Next, according to integral 2.19.3.5 in [157], the Gauss hypergeometric function in (3.26) admits the integral representation

$$\begin{aligned}
&{}_2F_1\Bigl(\frac{1}{2}-\alpha-i\tau, \frac{1}{2}-\alpha+i\tau; 1-2\alpha; -\Bigl(\frac{1}{x}+\frac{1}{p}\Bigr)w-\frac{w^2}{px}\Bigr)\\
&\qquad= \frac{\Gamma(1-2\alpha)}{|\Gamma(\frac{1}{2}-\alpha+i\tau)|^2}\int_0^\infty \xi^{-1-\alpha}\exp\biggl(-\frac{\xi}{2}-\Bigl(\frac{1}{x}+\frac{1}{p}+\frac{w}{xp}\Bigr)w\xi\biggr)W_{\alpha,i\tau}(\xi)d\xi,
\end{aligned}$$

and thus we have

$$W_{\alpha,i\tau}(x)W_{\alpha,i\tau}(p)=\frac{(xp)^{\alpha}e^{-\frac{x}{2}-\frac{p}{2}}}{|\Gamma(\frac{1}{2}-\alpha+i\tau)|^2}$$

$$\int_0^\infty e^{-w}w^{-2\alpha}\int_0^\infty \xi^{-1-\alpha}\exp\biggl(-\frac{\xi}{2}-\Bigl(\frac{1}{x}+\frac{1}{p}+\frac{w}{xp}\Bigr)w\xi\biggr)W_{\alpha,i\tau}(\xi)\,d\xi\,dw. \tag{3.27}$$

Using the assumption $\operatorname{Re}\alpha < 0$ and the limiting forms (3.10), (3.12) of the Whittaker function, we see that the integrals $\int_0^\infty e^{-w}w^{-2\alpha}\,dw$ and $\int_0^\infty \xi^{-1-\alpha}e^{-\frac{\xi}{2}}W_{\alpha,i\tau}(\xi)d\xi$ converge absolutely. Therefore, we can use Fubini's theorem to reverse the order of integration in (3.27); doing so, we obtain (3.25).

The previous lemma gives an integral representation for $|\Gamma(\frac{1}{2}-\alpha+i\tau)|^2W_{\alpha,i\tau}(x)W_{\alpha,i\tau}(p)$ whose kernel does not depend on τ. Integral representations for $|\Gamma(\frac{1}{2}-\alpha+i\tau)|^2W_{\alpha,i\tau}(x)$ which share the same property are also known. In the next two lemmas we take advantage of these integral representations and of the uniqueness theorem for Laplace transforms in order to deduce that the product formula (3.19) holds when α is a negative real number and $\nu=i\tau\in i\mathbb{R}$.

Lemma 3.2 *The identity*

$$2^{2\alpha}x^{-\alpha}W_{\alpha,i\tau}(x)\int_0^\infty e^{-\frac{s}{2y}-\frac{y}{2}}y^{\alpha-2}\,W_{\alpha,i\tau}(y)dy$$

$$=\int_0^\infty\Bigl(1+\frac{2s}{x\xi}\Bigr)^{-\frac{1}{2}}\biggl(\Bigl(1+\frac{2s}{x\xi}\Bigr)^{1/2}+1\biggr)^{2\alpha}\exp\biggl[-\Bigl(\frac{x}{2}+\frac{\xi}{2}\Bigr)\Bigl(1+\frac{2s}{x\xi}\Bigr)^{1/2}\biggr]W_{\alpha,i\tau}(\xi)\,\xi^{\alpha-2}d\xi \tag{3.28}$$

holds for $\alpha\in(-\infty,0)$, $\tau\in\mathbb{R}$ *and* $x,s>0$.

Proof Using the change of variable $s=2w\xi(1+\frac{w}{x})$, we rewrite (3.25) as

$$|\Gamma(\tfrac{1}{2}-\alpha+i\tau)|^2W_{\alpha,i\tau}(x)W_{\alpha,i\tau}(p)$$

$$=\frac{1}{2}(xp)^{\alpha}e^{-\frac{x}{2}-\frac{p}{2}}\int_0^\infty e^{-\frac{\xi}{2}}\xi^{\alpha-2}W_{\alpha,i\tau}(\xi)\int_0^\infty e^{-\frac{s}{2p}}s^{-2\alpha}\Bigl(1+\frac{2s}{x\xi}\Bigr)^{-\frac{1}{2}}\times$$

$$\times\biggl(\Bigl(1+\frac{2s}{x\xi}\Bigr)^{\frac{1}{2}}+1\biggr)^{2\alpha}\exp\biggl[\Bigl(\frac{x}{2}+\frac{\xi}{2}\Bigr)\Bigl(1-\Bigl(1+\frac{2s}{x\xi}\Bigr)^{\frac{1}{2}}\Bigr)\biggr]ds\,d\xi$$

$$=\frac{1}{2}(xp)^{\alpha}e^{-\frac{p}{2}}\int_0^\infty e^{-\frac{s}{2p}}s^{-2\alpha}\int_0^\infty\Bigl(1+\frac{2s}{x\xi}\Bigr)^{-\frac{1}{2}}\biggl(\Bigl(1+\frac{2s}{x\xi}\Bigr)^{\frac{1}{2}}+1\biggr)^{2\alpha}\times$$

$$\times\exp\biggl[-\Bigl(\frac{x}{2}+\frac{\xi}{2}\Bigr)\Bigl(1+\frac{2s}{x\xi}\Bigr)^{\frac{1}{2}}\biggr]W_{\alpha,i\tau}(\xi)\,\xi^{\alpha-2}d\xi\,ds, \tag{3.29}$$

where the absolute convergence of the iterated integral (see the proof of the previous lemma) justifies the change of order of integration.

On the other hand, by relation 2.19.5.18 in [157] we have

$$\begin{aligned}
&|\Gamma(\tfrac{1}{2}-\alpha+i\tau)|^2 W_{\alpha,i\tau}(p)\\
&= 2^{2\alpha-1}\Gamma(1-2\alpha)p^\alpha e^{-\frac{p}{2}}\int_0^\infty \Big(\frac{1}{2y}+\frac{1}{2p}\Big)^{-1+2\alpha} e^{-\frac{y}{2}} y^{\alpha-2}\, W_{\alpha,i\tau}(y)\,dy\\
&= 2^{2\alpha-1}p^\alpha e^{-\frac{p}{2}}\int_0^\infty\!\!\int_0^\infty e^{-\frac{s}{2y}-\frac{s}{2p}} s^{-2\alpha} ds\, e^{-\frac{y}{2}} y^{\alpha-2}\, W_{\alpha,i\tau}(y)\,dy\\
&= 2^{2\alpha-1}p^\alpha e^{-\frac{p}{2}}\int_0^\infty e^{-\frac{s}{2p}} s^{-2\alpha}\int_0^\infty e^{-\frac{s}{2y}-\frac{y}{2}} y^{\alpha-2}\, W_{\alpha,i\tau}(y)\,dy\,ds.
\end{aligned} \tag{3.30}$$

Comparing (3.29) and (3.30), and recalling the injectivity of Laplace transform, we deduce that (3.28) holds.

Lemma 3.3 *The product formula* (3.19) *holds for* $\alpha < 0$, $\tau \in \mathbb{R}$ *and* $x, y > 0$.

Proof We begin by deriving the following representation for the function of s appearing in the right-hand side of (3.28):

$$\begin{aligned}
&\Big(1+\frac{2s}{x\xi}\Big)^{-\frac{1}{2}}\bigg(\Big(1+\frac{2s}{x\xi}\Big)^{\frac{1}{2}}+1\bigg)^{2\alpha}\exp\bigg[-\Big(\frac{x}{2}+\frac{\xi}{2}\Big)\Big(1+\frac{2s}{x\xi}\Big)^{\frac{1}{2}}\bigg]\\
&= \frac{1}{\Gamma(-2\alpha)}\exp\bigg[-\Big(\frac{x}{2}+\frac{\xi}{2}\Big)\Big(1+\frac{2s}{x\xi}\Big)^{\frac{1}{2}}\bigg]\int_0^\infty \exp\bigg(-u\Big(1+\frac{2s}{x\xi}\Big)^{\frac{1}{2}}\bigg)\gamma(-2\alpha,u)\,du\\
&= \frac{(\pi x\xi)^{-\frac{1}{2}}}{\Gamma(-2\alpha)}\int_0^\infty \Big(u+\frac{x}{2}+\frac{\xi}{2}\Big)\gamma(-2\alpha,u)\times\\
&\qquad\times\int_0^\infty y^{-\frac{1}{2}}\exp\bigg[-(2s+x\xi)\frac{1}{4y}-\Big(u+\frac{x}{2}+\frac{\xi}{2}\Big)^2\frac{y}{x\xi}\bigg]dy\,du\\
&= \frac{(\pi x\xi)^{-\frac{1}{2}}}{\Gamma(-2\alpha)}\int_0^\infty e^{-\frac{s}{2y}}\exp\Big(-\frac{x\xi}{4y}\Big)y^{-\frac{1}{2}}\times\\
&\qquad\times\int_0^\infty\Big(u+\frac{x}{2}+\frac{\xi}{2}\Big)\exp\bigg(-\Big(u+\frac{x}{2}+\frac{\xi}{2}\Big)^2\frac{y}{x\xi}\bigg)\gamma(-2\alpha,u)du\,dy,
\end{aligned}$$

where $\gamma(\cdot,\cdot)$ is the incomplete Gamma function [56, Chapter IX]. In the first two equalities we have used integral 8.14.1 in [145] and integral 2.3.16.3 in [155], respectively, and the positivity of the integrand allows us to change the order of integration. Substituting in (3.28), we find that

$$\begin{aligned}
&\Gamma(-2\alpha)2^{\frac{1}{2}+2\alpha}\pi^{-\frac{1}{2}}x^{\frac{1}{2}-\alpha}W_{\alpha,i\tau}(x)\int_0^\infty e^{-\frac{s}{2y}-\frac{y}{2}}y^{\alpha-2}W_{\alpha,i\tau}(y)dy \\
&\quad=\int_0^\infty \xi^{-\frac{5}{2}+\alpha}W_{\alpha,i\tau}(\xi)\int_0^\infty e^{-\frac{s}{2y}}\exp\Bigl(-\frac{x\xi}{4y}\Bigr)y^{-\frac{1}{2}} \\
&\qquad\quad\times\int_0^\infty\Bigl(u+\frac{x}{2}+\frac{\xi}{2}\Bigr)\exp\biggl(-\Bigl(u+\frac{x}{2}+\frac{\xi}{2}\Bigr)^2\frac{y}{x\xi}\biggr)\gamma(-2\alpha,u)du\,dy\,d\xi \\
&\quad=\int_0^\infty e^{-\frac{s}{2y}}y^{-\frac{1}{2}}\int_0^\infty \xi^{-\frac{5}{2}+\alpha}\exp\Bigl(-\frac{x\xi}{4y}\Bigr)W_{\alpha,i\tau}(\xi) \\
&\qquad\quad\times\int_0^\infty\Bigl(u+\frac{x}{2}+\frac{\xi}{2}\Bigr)\exp\biggl(-\Bigl(u+\frac{x}{2}+\frac{\xi}{2}\Bigr)^2\frac{y}{x\xi}\biggr)\gamma(-2\alpha,u)du\,d\xi\,dy,
\end{aligned} \tag{3.31}$$

where the order of integration can be interchanged because of the absolute convergence of the triple integral, which follows from the inequality $\gamma(-2\alpha,u)\leq\Gamma(-2\alpha)$ and the equalities

$$\begin{aligned}
&\int_0^\infty \xi^{-\frac{5}{2}+\alpha}\bigl|W_{\alpha,i\tau}(\xi)\bigr|\int_0^\infty e^{-\frac{s}{2y}}y^{-\frac{1}{2}}\exp\Bigl(-\frac{x\xi}{2y}\Bigr)\times \\
&\qquad\qquad\times\int_0^\infty\Bigl(u+\frac{x}{2}+\frac{\xi}{2}\Bigr)\exp\biggl(-\Bigl(u+\frac{x}{2}+\frac{\xi}{2}\Bigr)^2\frac{y}{x\xi}\biggr)du\,dy\,d\xi \\
&\quad=\frac{x}{2}\int_0^\infty \xi^{-\frac{5}{2}+\alpha}\bigl|W_{\alpha,i\tau}(\xi)\bigr|\int_0^\infty \exp\biggl(-\frac{s}{2y}-\frac{x\xi}{4y}-\frac{y}{2}-\frac{xy}{4\xi}-\frac{\xi y}{4x}\biggr)y^{-\frac{3}{2}}dy\,d\xi \\
&\quad=2^{-\frac{1}{2}}(\pi x)^{\frac{1}{2}}\int_0^\infty \xi^{-3+\alpha}\Bigl(1+\frac{2s}{x\xi}\Bigr)^{-\frac{1}{2}}\exp\biggl(-\Bigl(\frac{x}{2}+\frac{\xi}{2}\Bigr)\Bigl(1+\frac{2s}{x\xi}\Bigr)^{\frac{1}{2}}\biggr)\bigl|W_{\alpha,i\tau}(\xi)\bigr|\,d\xi \\
&\quad<\infty
\end{aligned}$$

(which follow from integral 2.3.16.3 in [155] and straighforward calculations; the convergence of the latter integral can be verified using the limiting forms (3.10), (3.12) of the Whittaker function).

Using, as in the previous proof, the injectivity of Laplace transform, from (3.31) it follows that

$$W_{\alpha,i\tau}(x)W_{\alpha,i\tau}(y) = \frac{2^{-2\alpha}\pi^{-\frac{1}{2}}}{\Gamma(-2\alpha)}x^{-\frac{1}{2}+\alpha}y^{\frac{3}{2}-\alpha}e^{\frac{y}{2}}\int_0^\infty \xi^{-\frac{5}{2}+\alpha}\exp\Bigl(-\frac{x\xi}{4y}\Bigr)W_{\alpha,i\tau}(\xi)$$
$$\times \int_0^\infty \Bigl(u+\frac{x}{2}+\frac{\xi}{2}\Bigr)\exp\biggl(-\Bigl(u+\frac{x}{2}+\frac{\xi}{2}\Bigr)^2\frac{y}{x\xi}\biggr)\gamma(-2\alpha,u)du\,d\xi. \tag{3.32}$$

Let us compute the inner integral. Since $\frac{d}{du}\gamma(-2\alpha,u)=u^{-1-2\alpha}e^{-u}$ and

$$\int\Bigl(u+\frac{x}{2}+\frac{\xi}{2}\Bigr)\exp\biggl(-\Bigl(u+\frac{x}{2}+\frac{\xi}{2}\Bigr)^2\frac{y}{x\xi}\biggr)du = -\frac{x\xi}{2y}\exp\biggl(-\Bigl(u+\frac{x}{2}+\frac{\xi}{2}\Bigr)^2\frac{y}{x\xi}\biggr),$$

we obtain, using integration by parts,

$$\begin{aligned}
&\int_0^\infty\Bigl(u+\frac{x}{2}+\frac{\xi}{2}\Bigr)\exp\biggl(-\Bigl(u+\frac{x}{2}+\frac{\xi}{2}\Bigr)^2\frac{y}{x\xi}\biggr)\gamma(-2\alpha,u)du\\
&\quad=\frac{x\xi}{2y}\int_0^\infty u^{-1-2\alpha}e^{-u}\exp\biggl(-\Bigl(u+\frac{x}{2}+\frac{\xi}{2}\Bigr)^2\frac{y}{x\xi}\biggr)du\\
&\quad=\Gamma(-2\alpha)\Bigl(\frac{x\xi}{2y}\Bigr)^{1-\alpha}\exp\Bigl(\frac{x}{4}+\frac{\xi}{4}-\frac{y}{4}+\frac{x\xi}{8y}-\frac{xy}{8\xi}-\frac{y\xi}{8x}\Bigr)D_{2\alpha}\biggl(\frac{xy+x\xi+y\xi}{(2xy\xi)^{1/2}}\biggr),
\end{aligned} \tag{3.33}$$

where we applied relation 2.3.15.3 in [155]. Substituting this in (3.32), we conclude that (3.19) holds for all $\alpha<0$ and $\nu=i\tau\in i\mathbb{R}$.

Proof of Theorem 3.1 To simplify the notation, throughout the proof we write $\boldsymbol{v}_{\alpha,\nu}(t):=t^{-\alpha}W_{\alpha,\nu}(t)$. We use an analytic continuation argument to extend the identity (3.19) to all $\alpha,\nu\in\mathbb{C}$. To that end, let us prove that the right-hand side of (3.19) is an entire function of each of the variables α and ν. Let $M>0$ and suppose that $\frac{1}{M}\le\frac{1}{2}-\operatorname{Re}\alpha\le M$ and $0\le\operatorname{Re}\nu\le M$. Then for $t>0$ we have

$$\begin{aligned}
\bigl|\boldsymbol{v}_{\alpha,\nu}(t)\bigr| = \bigl|\boldsymbol{v}_{\alpha,-\nu}(t)\bigr| &= \frac{e^{-\frac{t}{2}}}{|\Gamma(\frac{1}{2}-\alpha+\nu)|}\biggl|\int_0^\infty e^{-s}s^{-\frac{1}{2}-\alpha+\nu}\Bigl(1+\frac{s}{t}\Bigr)^{-\frac{1}{2}+\alpha+\nu}ds\biggr|\\
&\le \frac{e^{-\frac{t}{2}}}{|\Gamma(\frac{1}{2}-\alpha+\nu)|}\int_0^\infty e^{-s}s^{-1}(s^{1/M}+s^{2M})\Bigl(1+\frac{s}{t}\Bigr)^M ds\\
&= \frac{1}{|\Gamma(\frac{1}{2}-\alpha+\nu)|}\Bigl[\Gamma\bigl(\tfrac{1}{M}\bigr)\boldsymbol{v}_{\frac{1}{2}(M-\frac{1}{M}+1),\frac{1}{2}(M+\frac{1}{M})}(t)+\Gamma(2M)\boldsymbol{v}_{\frac{1}{2}(1-M),\frac{3M}{2}}(t)\Bigr],
\end{aligned}$$

where we have used the integral representation (3.11). Moreover, letting $n \in \mathbb{N}$, a repeated application of the recurrence relation (3.14) shows that

$$\boldsymbol{v}_{\alpha+\frac{n}{2},\nu+\frac{n}{2}}(t) = Q^{(1)}_{n,\alpha,\nu}\Big(\frac{1}{t}\Big)\boldsymbol{v}_{\alpha,\nu}(t) + Q^{(2)}_{n,\alpha,\nu}\Big(\frac{1}{t}\Big)\boldsymbol{v}_{\alpha-\frac{1}{2},\nu-\frac{1}{2}}(t),$$

where the $Q^{(i)}_{n,\alpha,\nu}(\cdot)$ are polynomials of degree at most n whose coefficients depend on α and ν. Therefore, for $\frac{1}{M} \leq \frac{1}{2} - \operatorname{Re}\alpha \leq M - \frac{1}{2}$ and $-M + \frac{1}{2} \leq \operatorname{Re}\nu \leq M$ we have

$$\begin{aligned} \big|\boldsymbol{v}_{\alpha+\frac{n}{2},\nu+\frac{n}{2}}(t)\big| &\leq \big|Q^{(1)}_{n,\alpha,\nu}\big(\tfrac{1}{t}\big)\boldsymbol{v}_{\alpha,\nu}(t)\big| + \big|Q^{(2)}_{n,\alpha,\nu}\big(\tfrac{1}{t}\big)\boldsymbol{v}_{\alpha-\frac{1}{2},\nu-\frac{1}{2}}(t)\big| \\ &\leq \Big(\big|Q^{(1)}_{n,\alpha,\nu}\big(\tfrac{1}{t}\big)\big| + \big|Q^{(2)}_{n,\alpha,\nu}\big(\tfrac{1}{t}\big)\big|\Big)\, G(\alpha,\nu) \times \\ &\qquad \times \Big[\Gamma\big(\tfrac{1}{M}\big)\boldsymbol{v}_{\frac{1}{2}(M-\frac{1}{M}+1),\frac{1}{2}(M+\frac{1}{M})}(t) + \Gamma(2M)\boldsymbol{v}_{\frac{1}{2}(1-M),\frac{3M}{2}}(t)\Big], \end{aligned} \tag{3.34}$$

where

$$G(\alpha,\nu) = \begin{cases} 2|\Gamma(\frac{1}{2}-\alpha+\nu)|^{-1}, & \frac{1}{2} \leq \operatorname{Re}\nu \leq M, \\ |\Gamma(\frac{1}{2}-\alpha+\nu)|^{-1} + |\Gamma(\frac{3}{2}-\alpha-\nu)|^{-1}, & 0 \leq \operatorname{Re}\nu < \frac{1}{2}, \\ |\Gamma(\frac{1}{2}-\alpha-\nu)|^{-1} + |\Gamma(\frac{3}{2}-\alpha-\nu)|^{-1}, & -M+\frac{1}{2} \leq \operatorname{Re}\nu < 0. \end{cases}$$

Similarly, for $\frac{1}{M} \leq \frac{1}{2} - \operatorname{Re}\alpha \leq M$ the integral representation (3.20) gives

$$\begin{aligned} |D_{2\alpha}(t)| &= \frac{e^{-\frac{t^2}{4}}}{|\Gamma(\frac{1}{2}-\alpha)|} t^{2\operatorname{Re}\alpha} \left| \int_0^\infty e^{-s} s^{-\frac{1}{2}-\alpha}\Big(1+\frac{2s}{t^2}\Big)^{\alpha} ds \right| \\ &\leq \frac{e^{-\frac{t^2}{4}} t}{|\Gamma(\frac{1}{2}-\alpha)|} (t^{-2M} + t^{-\frac{2}{M}}) \int_0^\infty e^{-s} s^{-1} (s^{\frac{1}{M}} + s^M)\Big(1+\frac{2s}{t^2}\Big)^{\frac{1}{2}-\frac{1}{M}} ds \\ &= \frac{t}{|\Gamma(\frac{1}{2}-\alpha)|} (t^{-2M} + t^{-\frac{2}{M}}) \Big[\Gamma\big(\tfrac{1}{M}\big)\boldsymbol{v}_{\frac{3}{4}-\frac{1}{M},\frac{1}{4}}\big(\tfrac{t^2}{2}\big) + \Gamma(M)\boldsymbol{v}_{\frac{3}{4}-\frac{1}{2}(M+\frac{1}{M}),\frac{1}{4}+\frac{1}{2}(M+\frac{1}{M})}\big(\tfrac{t^2}{2}\big)\Big] \end{aligned}$$

and, by (3.22), for each $n \in \mathbb{N}$ we have $D_{2\alpha+n}(t) = Q^{(3)}_{n,\alpha}(t)D_{2\alpha}(t) + Q^{(4)}_{n,\alpha}(t)D_{2\alpha-1}(t)$, being $Q^{(j)}_{n,\alpha}(\cdot)$ polynomials of degree at most n with coefficients depending on α, hence

$$\begin{aligned} \big|D_{2\alpha+n}(t)\big| &\leq \big(|\Gamma(\tfrac{1}{2}-\alpha)|^{-1} + |\Gamma(1-\alpha)|^{-1}\big)\big(|Q^{(3)}_{n,\alpha,\nu}(t)| + |Q^{(4)}_{n,\alpha}(t)|\big)\, t\, (t^{-2M} + t^{-\frac{2}{M}}) \\ &\qquad \times \Big[\Gamma\big(\tfrac{1}{M}\big)\boldsymbol{v}_{\frac{3}{4}-\frac{1}{M},\frac{1}{4}}\big(\tfrac{t^2}{2}\big) + \Gamma(M)\boldsymbol{v}_{\frac{3}{4}-\frac{1}{2}(M+\frac{1}{M}),\frac{1}{4}+\frac{1}{2}(M+\frac{1}{M})}\big(\tfrac{t^2}{2}\big)\Big]. \end{aligned} \tag{3.35}$$

Using the inequalities (3.34), (3.35) and the limiting forms (3.10), (3.12) for the Whittaker function, one can verify without difficulty that

$$\sup_{(\alpha,\nu)\in\mathcal{R}_M} \int_0^\infty \Bigl| W_{\alpha+\frac{n}{2},\nu+\frac{n}{2}}(\xi)\, \kappa_{\alpha+\frac{n}{2}}(x,y,\xi)\Bigr| \frac{d\xi}{\xi^2} < \infty,$$

where $\mathcal{R}_M = \bigl\{(\alpha,\nu) \mid \frac{1}{M} \leq \frac{1}{2} - \operatorname{Re}\alpha \leq M - \frac{1}{2},\ -M + \frac{1}{2} \leq \operatorname{Re}\nu \leq M\bigr\}$. Since M and n are arbitrary, the known results on the analyticity of parameter-dependent integrals (e.g. [135]) yield that $\int_0^\infty W_{\alpha,\nu}(\xi)\, \kappa_\alpha(x,y,\xi) \frac{d\xi}{\xi^2}$ is an entire function of the parameter α and the parameter ν. As the left-hand side of (3.19) is also an entire function of α and ν, by analytic continuation we conclude that the product formula (3.19) extends to all $\alpha, \nu \in \mathbb{C}$, as we wanted to show.

Remark 3.3

(a) The product formula (3.19) can be equivalently written in terms of the normalized Whittaker W function as

$$\mathbf{W}_{\alpha,\nu}(x)\, \mathbf{W}_{\alpha,\nu}(y) = \int_0^\infty \mathbf{W}_{\alpha,\nu}(\xi)\, k_\alpha(x,y,\xi)\, \xi^{-2\alpha} e^{-1/\xi} d\xi, \tag{3.36}$$

where

$$\begin{aligned} k_\alpha(x,y,\xi) &:= (xy\xi)^\alpha e^{\frac{1}{2x}+\frac{1}{2y}+\frac{1}{2\xi}} \kappa_\alpha\bigl(\tfrac{1}{x},\tfrac{1}{y},\tfrac{1}{\xi}\bigr) \\ &= 2^{-1-\alpha}\pi^{-\frac{1}{2}}(xy\xi)^{-\frac{1}{2}+\alpha} \exp\biggl(\frac{1}{x}+\frac{1}{y}+\frac{1}{\xi}-\frac{(x+y+\xi)^2}{8xy\xi}\biggr) D_{2\alpha}\biggl(\frac{x+y+\xi}{(2xy\xi)^{1/2}}\biggr). \end{aligned} \tag{3.37}$$

(b) It follows from (3.16) and (3.24) that in the particular case $\alpha = 0$, (3.19) specializes into

$$K_\nu(x)K_\nu(y) = \frac{1}{2}\int_0^\infty K_\nu(\xi) \exp\biggl(-\frac{xy}{2\xi} - \frac{x\xi}{2y} - \frac{y\xi}{2x}\biggr)\frac{d\xi}{\xi},$$

which is the product formula for the modified Bessel function stated in (3.5).

(c) Since the parabolic cylinder function $D_\nu(t)$ is a positive function of $t > 0$ whenever $\nu \in (-\infty, 1]$ (as can be seen e.g. from the representation (3.20)), we have

$$k_\alpha(x,y,\xi) > 0 \qquad \text{for all } \alpha \leq \tfrac{1}{2} \text{ and } x,y,\xi > 0. \tag{3.38}$$

This positivity property means that the convolution operator induced by the product formula (3.36) (cf. Sect. 3.5) is positivity-preserving.

(d) A useful upper bound for the kernel of the product formula (3.36), valid for $\alpha \in \mathbb{R}$, is the following:

$$\bigl|k_\alpha(x,y,\xi)\bigr| \leq A(y)\,(xy\xi)^{-\frac{1}{2}}(x+y+\xi)^{2\alpha}\exp\biggl(\frac{1}{\xi}-\frac{(x+y-\xi)^2}{4xy\xi}\biggr) \qquad (x,y,\xi>0), \tag{3.39}$$

where

$$A(y) = 2^{-1-\alpha}\pi^{-\frac{1}{2}}\cdot\Bigl(\max_{t\geq y^{-1/2}} t^{-2\alpha}e^{\frac{t^2}{4}}D_\alpha(t)\Bigr) < \infty \qquad (y>0).$$

This upper bound follows from the inequality $\frac{(x+y+\xi)^2}{2xy\xi} \geq \frac{1}{y}$ and the fact that, by (3.21), the function $t^{-2\alpha}e^{t^2/4}D_{2\alpha}(t)$ is bounded on the interval $[y^{-1/2},\infty)$.

The normalized Whittaker W function includes, as a particular case, the *(generalized) Bessel polynomial* $B_n(x;\alpha)$ ($\alpha \in \mathbb{R}\backslash\{1,\frac{3}{2},2,\ldots\}$, $n \in \mathbb{N}_0$) introduced in [109] as the polynomial of degree n, with constant term equal to 1, which is a solution of the Sturm–Liouville equation $y^2u''(y)+(1+2(1-\alpha)y)u'(y) = n(n+1-2\alpha)u(y)$. By [145, Equation 13.14.9] and [109, §15], these polynomials are given by

$$B_n(x;\alpha) = \mathbf{W}_{\alpha,\frac{1}{2}-\alpha+n}(x) = x^{2\alpha}e^{\frac{1}{x}}\frac{d^n}{dx^n}[x^{2(n-\alpha)}e^{-\frac{1}{x}}] \qquad (n \in \mathbb{N}_0,\ \alpha \neq 1,\tfrac{3}{2},2,\ldots).$$

The Bessel polynomials are one of the four canonical families of classical orthogonal polynomials (see [132, 133]). We refer to the book [77] for a detailed exposition on the properties and applications of the Bessel polynomials. As an immediate corollary of (3.36), the following product formula holds for the Bessel polynomials:

Corollary 3.1 *The product $B_n(x;\alpha)B_n(y;\alpha)$ of Bessel polynomials admits the integral representation*

$$B_n(x;\alpha)B_n(y;\alpha) = \int_0^\infty B_n(\xi;\alpha)\,k_\alpha(x,y,\xi)\,\xi^{-2\alpha}e^{-1/\xi}d\xi,$$

where $k_\alpha(x,y,\xi)$ is defined by (3.37).

3.3 Whittaker Translation

We now define the generalized translation operator induced by the product formula (3.36) for the normalized Whittaker function:

Definition 3.2 Let $1 \le p \le \infty$ and $\alpha \le \frac{1}{2}$. The linear operator

$$(\mathcal{T}_\alpha^y f)(x) = \int_0^\infty f(\xi) k_\alpha(x, y, \xi)\, r_\alpha(\xi) d\xi \qquad \bigl(f \in L^p(\mathbb{R}^+; r_\alpha(x)dx),\ x, y > 0\bigr), \tag{3.40}$$

where $k_\alpha(x, y, \xi)$ is defined by (3.37) and r_α is as in (3.2), will be called the *Whittaker translation operator (of order α).*

The operator $\mathcal{T}_\alpha^y$ is called a translation operator because it is obtained from the ordinary translation operator $(T^y f)(x) = f(x + y) \equiv \int f(\xi) d\delta_{x+y}(d\xi)$ by replacing the measure δ_{x+y} of the product formula for the Fourier kernel by the measure $k_\alpha(x, y, \xi)$ of the product formula for the Whittaker function. Accordingly, many of the properties given in Proposition 3.2 below resemble the properties of the ordinary translation operator. We first establish the following lemma which gives the closed-form expression for the Whittaker translation of the power function $\vartheta(x) = x^\beta$.

Lemma 3.4 *For $\alpha, \beta \in \mathbb{C}$, we have*

$$\int_0^\infty \xi^\beta k_\alpha(x, y, \xi)\, r_\alpha(\xi) d\xi = (x + y)^\beta\, \mathbf{W}_{\alpha,\alpha-\frac{1}{2}-\beta}\Bigl(\frac{xy}{x + y}\Bigr) \qquad (x, y > 0). \tag{3.41}$$

In particular, $\int_0^\infty k_\alpha(x, y, \xi)\, r_\alpha(\xi) d\xi = 1$ for $\alpha \in \mathbb{C}$ and $x, y > 0$.

Proof Fix $x, y > 0$, and suppose that $\alpha < 0$ and $\beta \in \mathbb{R}$. Using the definition (3.37) and the integral representation of $D_{2\alpha}\bigl(\frac{x+y+\xi}{(2xy\xi)^{1/2}}\bigr)$ obtained by exchanging the variables (x, y, ξ) by $(\frac{1}{x}, \frac{1}{\xi}, \frac{1}{y})$ in (3.33), we find that for each $\alpha < 0$ we have

$$k_\alpha(x, y, \xi)\, \xi^{-2\alpha} e^{-1/\xi} = \frac{2^{-1-2\alpha}\pi^{-\frac{1}{2}}}{\Gamma(-2\alpha)} (xy\xi)^{-\frac{1}{2}} \times$$
$$\times \exp\Bigl(\frac{1}{2x} + \frac{1}{2y} - \frac{\xi}{4xy}\Bigr) \int_0^\infty u^{-1-2\alpha} \exp\biggl(-u - \Bigl(u + \frac{1}{2x} + \frac{1}{2y}\Bigr)^2 \frac{xy}{\xi}\biggr) du.$$

Consequently, we may compute

$$\int_0^\infty \xi^\beta k_\alpha(x, y, \xi)\, r_\alpha(\xi) d\xi$$
$$= \frac{2^{-1-2\alpha}\pi^{-\frac{1}{2}}}{\Gamma(-2\alpha)} (xy)^{-\frac{1}{2}} \exp\Bigl(\frac{1}{2x} + \frac{1}{2y}\Bigr) \times$$
$$\times \int_0^\infty u^{-1-2\alpha} e^{-u} \int_0^\infty \xi^{\beta-\frac{1}{2}} \exp\biggl(-\Bigl(u + \frac{1}{2x} + \frac{1}{2y}\Bigr)^2 \frac{xy}{\xi} - \frac{\xi}{4xy}\biggr) d\xi du$$

$$= \frac{2^{\beta+\frac{1}{2}-2\alpha}}{\Gamma(-2\alpha)}\pi^{-\frac{1}{2}}(xy)^{\beta}\exp\Big(\frac{1}{2x}+\frac{1}{2y}\Big)\times$$
$$\times\int_0^\infty u^{-1-2\alpha}\Big(u+\frac{1}{2x}+\frac{1}{2y}\Big)^{\beta+\frac{1}{2}}e^{-u}K_{\beta+\frac{1}{2}}\Big(u+\frac{1}{2x}+\frac{1}{2y}\Big)du$$
$$= \frac{2^{\beta+\frac{1}{2}-2\alpha}}{\Gamma(-2\alpha)}\pi^{-\frac{1}{2}}(xy)^{\beta}\exp\Big(\frac{1}{x}+\frac{1}{y}\Big)\int_{\frac{1}{2x}+\frac{1}{2y}}^\infty t^{\beta+\frac{1}{2}}\Big(t-\frac{1}{2x}-\frac{1}{2y}\Big)^{-1-2\alpha}e^{-t}K_{\beta+\frac{1}{2}}(t)\,dt$$
$$= (x+y)^{\beta}\,\mathbf{W}_{\alpha,\alpha-\frac{1}{2}-\beta}\Big(\frac{xy}{x+y}\Big),$$

where the first equality is obtained by changing the order of integration (note the positivity of the integrand), the second equality follows from integral 2.3.16.1 in [155] and a few simplifications, the third equality results from the change of variables $u = t - \frac{1}{2x} - \frac{1}{2y}$, and the last equality uses relation 2.16.7.5 in [156]. This proves that (3.41) holds in the case $\alpha < 0$ and $\beta \in \mathbb{R}$.

To extend the result to all $\alpha, \beta \in \mathbb{C}$, we can use an analytic continuation argument similar to that of the proof of Theorem 3.1. Indeed, using (3.35) and the elementary inequality $|\xi^{\beta}| \leq \xi^{-M} + \xi^{M}$ ($\xi > 0$, $\beta \in [-M, M]$) one can verify, as in the previous proof, that

$$\sup_{(\alpha,\beta)\in\bar{\mathcal{R}}_M}\int_0^\infty \big|\,\xi^{\beta}k_{\alpha+\frac{n}{2}}(x,y,\xi)\,\xi^{-2\alpha-n}\big|\,e^{-1/\xi}d\xi < \infty,$$

where $\bar{\mathcal{R}}_M = \big\{(\alpha,\beta) \;\big|\; \frac{1}{M} \leq \frac{1}{2} - \operatorname{Re}\alpha \leq M - \frac{1}{2},\ -M \leq \operatorname{Re}\beta \leq M\big\}$, being $M > 0$ and $n \in \mathbb{N}$ arbitrary. Both sides of (3.41) are therefore entire functions of the parameter a and the parameter β; consequently, the principle of analytic continuation gives (3.41) in the general case. By (3.17), the right-hand side of (3.41) equals 1 when $\beta = 0$.

The next proposition gives the basic continuity and L^p properties of the Whittaker translation operator. We consider the weighted L^p spaces

$$L^p(r_\alpha) := L^p\big(\mathbb{R}^+; r_\alpha(x)dx\big) \qquad (1 \leq p \leq \infty,\ -\infty < \alpha \leq \tfrac{1}{2}) \tag{3.42}$$

with the usual norms

$$\|f\|_{p,\alpha} = \Big(\int_0^\infty |f(x)|^p r_\alpha(x)dx\Big)^{1/p} \qquad (1 \leq p < \infty),$$
$$\|f\|_\infty \equiv \|f\|_{\infty,\alpha} = \operatorname*{ess\,sup}_{0<x<\infty}|f(x)|.$$

Proposition 3.2 *Fix $\alpha \leq \frac{1}{2}$ and $y > 0$. Then:*

(a) If $f \in L^\infty(r_\alpha)$ is such that $0 \leq f \leq 1$, then $0 \leq \mathcal{T}_\alpha^y f \leq 1$;
(b) For each $1 \leq p \leq \infty$, we have

$$\|\mathcal{T}_\alpha^y f\|_{p,\alpha} \leq \|f\|_{p,\alpha} \qquad \textit{for all } f \in L^p(r_\alpha)$$

(in particular, $\mathcal{T}_\alpha^y\big(L^p(r_\alpha)\big) \subset L^p(r_\alpha)$);
(c) If $f \in L^p(r_\alpha)$ where $1 < p \leq \infty$, then $\mathcal{T}_\alpha^y f \in \mathrm{C}(\mathbb{R}^+)$, and for $1 < p < \infty$ we also have

$$\lim_{h\to 0} \|\mathcal{T}_\alpha^{y+h} f - \mathcal{T}_\alpha^y f\|_{p,\alpha} = 0;$$

(d) If $f \in \mathrm{C_b}(\mathbb{R}^+)$, then $(\mathcal{T}_\alpha^x f)(y) \to f(y)$ as $x \to 0$;
(e) If $f \in L^\infty(r_\alpha)$ is such that $\lim_{x\to\infty} f(x) = 0$, then $\lim_{x\to\infty}(\mathcal{T}_\alpha^y f)(x) = 0$.

Proof Throughout this proof the letter C stands for a constant whose exact value may change from line to line.

(a) By Lemma 3.4, if $f = 1$ then $\mathcal{T}_\alpha^y f \equiv 1$. Moreover, Remark 3.3(c) means that $\mathcal{T}_\alpha^y f$ is nonnegative whenever f is nonnegative. Recalling that $\mathcal{T}_\alpha^y$ is a linear operator, we see that we have $0 \leq \mathcal{T}_\alpha^y f \leq 1$ whenever $0 \leq f \leq 1$.

(b) The case $p = \infty$ was proved in part (a). Now, for $1 \leq p < \infty$ and $f \in L^p(r_\alpha)$ we have

$$\begin{aligned}
\|\mathcal{T}_\alpha^y f\|_{p,\alpha}^p &= \int_0^\infty \left| \int_0^\infty f(\xi) k_\alpha(x,y,\xi)\, r_\alpha(\xi) d\xi \right|^p r_\alpha(x) dx \\
&\leq \int_0^\infty \int_0^\infty |f(\xi)|^p k_\alpha(x,y,\xi)\, r_\alpha(\xi) d\xi\, r_\alpha(x) dx \\
&= \int_0^\infty \int_0^\infty k_\alpha(x,y,\xi)\, r_\alpha(x) dx\, |f(\xi)|^p r_\alpha(\xi) d\xi = \|f\|_{p,\alpha}^p,
\end{aligned}$$

where we have used the final statement in Lemma 3.4, the fact that $k_\alpha(x, y, \xi)$ is positive and symmetric, and Hölder's inequality.

(c) For $f \in L^p(r_\alpha)$ $(1 < p < \infty)$, by Young's inequality we have

$$\int_0^\infty |f(\xi)| k_\alpha(x,y,\xi)\, r_\alpha(\xi) d\xi \leq \frac{1}{p}\|f\|_{p,\alpha}^p + \frac{1}{q}\int_0^\infty |k_\alpha(x,y,\xi)|^q r_\alpha(\xi) d\xi$$

and therefore the continuity of $\mathcal{T}_\alpha^y f$ will be proved if we show that, for each $1 \leq q < \infty$, the integral $\int_0^\infty |k_\alpha(x, y, \xi)|^q \, r_\alpha(\xi) d\xi$ converges absolutely and locally uniformly. In fact, let us fix $M > 0$; then,

$$k_\alpha(x, y, \xi) \leq A_1(y) \, \xi^{-\frac{1}{2}} (1+\xi) \exp\biggl(-\frac{\xi}{4My}\biggr), \qquad \tfrac{1}{M} \leq x \leq M, \; \xi > 0, \tag{3.43}$$

where $A_1(y) = 2A(y) y^{2\alpha - \frac{3}{2}} (1+y) M^{\frac{1}{2}} \exp(\frac{M}{2} + \frac{1}{2y})$; this estimate is obtained using (3.39), together with the inequalities $x+y+\xi \leq (1+x)(1+y)(1+\xi)$ and $(x+y+\xi)^{2\alpha-1} \leq y^{2\alpha-1}$. Clearly, (3.43) implies that $\int_0^\infty |k_\alpha(x, y, \xi)|^q r_\alpha(\xi) d\xi$ converges absolutely and uniformly in $x \in [\frac{1}{M}, M]$, and it follows that $\mathcal{T}_\alpha^y f \in \mathrm{C}(\mathbb{R}^+)$.

To prove the L^p-continuity of the translation, let $f \in \mathrm{C_c}(\mathbb{R}^+)$ and $1 < p < \infty$. Fix $M > 0$ such that the support of f is contained in $[\frac{1}{M}, M]$. Interchanging the role of x and ξ in the estimate (3.43), we easily see that

$$\begin{aligned} |\mathcal{T}_\alpha^{y+h} f(x)| &\leq \|f\|_\infty \int_{\frac{1}{M}}^M k_\alpha(x, y+h, \xi) \, r_\alpha(\xi) d\xi \\ &\leq \|f\|_\infty \, A_2(y+h) \, x^{-\frac{1}{2}} (1+x) \exp\biggl(-\frac{x}{4M(y+h)}\biggr), \end{aligned} \tag{3.44}$$

where $A_2(y) = A_1(y) \int_{1/M}^M r_\alpha(\xi) d\xi$. It is easy to check that the function $A_2(y)$ is locally bounded on $\mathbb{R}^+$, so it follows from (3.44) that there exists $g \in L^p(r_\alpha)$ such that $|\mathcal{T}_\alpha^{y+h} f(x)| \leq g(x)$ for all $0 < x < \infty$ and all $|h| < \delta$ (where $\delta > 0$ is sufficiently small). We have already proved that $(\mathcal{T}_\alpha^y f)(x) \equiv (\mathcal{T}_\alpha^x f)(y)$ is continuous in y, hence by L^p-dominated convergence we conclude that $\|\mathcal{T}_\alpha^{y+h} f - \mathcal{T}_\alpha^y f\|_{p,\alpha} \to 0$ as $h \to 0$. As in the proof of the L^p-continuity of the ordinary translation, for general $f \in L^p(r_\alpha)$ the result is proved by taking a sequence of functions $f_n \in \mathrm{C_c}(\mathbb{R}^+)$ which tend to f in the norm $\|\cdot\|_{p,\alpha}$.

(d) We start by studying the behaviour of the integral $\int_{E_\delta} k_\alpha(x, y, \xi) \, r_\alpha(\xi) d\xi$ as $x \to 0$, where $E_\delta = \bigl\{\xi \in \mathbb{R}^+ \bigm| |y - \xi| > \delta\bigr\}$ and $\delta \in (0, y)$ is some fixed constant. We have

$$k_\alpha(x, y, \xi) e^{-1/\xi} \leq C \, \frac{x+y+\xi}{|x+y-\xi|} \exp\biggl(-\frac{(x+y-\xi)^2}{8xy\xi}\biggr), \qquad x, \xi > 0 \tag{3.45}$$

(where $C < \infty$ is independent of x and ξ). This follows by combining (3.39) with the boundedness of the function $|t|e^{-t^2}$ and the inequality $(x + y + \xi)^{2\alpha-1} \le y^{2\alpha-1}$. Furthermore, if $x \le \frac{\delta}{2}$ and $\xi \in E_\delta$, the inequalities

$$\frac{x+y+\xi}{|x+y-\xi|} = \left|1 + \frac{2\xi}{x+y-\xi}\right| \le 1 + \frac{4\xi}{\delta}$$

$$\exp\left(-\frac{(x+y-\xi)^2}{8xy\xi}\right) \le \exp\left(\frac{1}{4y} - \frac{1}{4\xi} - \frac{(y-\xi)^2}{4\delta y\xi}\right)$$

lead us to

$$k_\alpha(x,y,\xi)\xi^{-2\alpha}e^{-1/\xi} \le C\,\xi^{-2\alpha}(1+\xi)\exp\left(-\frac{1}{4\xi} - \frac{(y-\xi)^2}{4\delta y\xi}\right), \qquad x \le \tfrac{\delta}{2},\ \xi \in E_\delta. \tag{3.46}$$

Since the right-hand side of (3.46) clearly belongs to $L^1(E_\delta)$, the dominated convergence theorem is applicable, and letting $x \to 0$ in (3.45) we find that

$$\lim_{x\downarrow 0}\int_{E_\delta} k_\alpha(x,y,\xi)\, r_\alpha(\xi)d\xi = \int_{E_\delta}\Bigl(\lim_{x\downarrow 0} k_\alpha(x,y,\xi)\Bigr) r_\alpha(\xi)d\xi = 0. \tag{3.47}$$

Let us now fix $\varepsilon > 0$, and write $V_\delta = \mathbb{R}^+ \setminus E_\delta$. Since f is continuous, we can choose $\delta > 0$ such that $|f(\xi) - f(y)| < \varepsilon$ for all $\xi \in V_\delta$. By this choice of δ and the positivity of $k_\alpha(x,y,\xi)$, we find

$$\begin{aligned}
\bigl|(\mathcal{T}_\alpha^y f)(x) - f(y)\bigr| &= \left|\int_0^\infty k_\alpha(x,y,\xi)\bigl(f(\xi)-f(y)\bigr)r_\alpha(\xi)d\xi\right| \\
&\le \left|\int_{E_\delta} k_\alpha(x,y,\xi)\bigl(f(\xi)-f(y)\bigr)r_\alpha(\xi)d\xi\right| + \varepsilon\int_{V_\delta} k_\alpha(x,y,\xi)r_\alpha(\xi)d\xi \\
&\le 2\|f\|_\infty\int_{E_\delta} k_\alpha(x,y,\xi)\, r_\alpha(\xi)d\xi + \varepsilon.
\end{aligned}$$

By (3.47), it follows that $\limsup_{x\downarrow 0}\bigl|(\mathcal{T}_\alpha^y f)(x) - f(y)\bigr| \le \varepsilon$. Since ε is arbitrary, the proof of part (d) is finished.

(e) We begin by claiming that for each $M > 0$ we have $\int_0^M k_\alpha(x,y,\xi)\, r_\alpha(\xi)d\xi \to 0$ as $x \to \infty$. Indeed, if $x > 2M$ and $\xi \le M$, combining (3.45) with the inequalities

$$\frac{x+y+\xi}{|x+y-\xi|} = 1 + \frac{2\xi}{x+y-\xi} \le 1 + \frac{2M}{y+M}, \tag{3.48}$$

$$\exp\left(-\frac{(x+y-\xi)^2}{8xy\xi}\right) \le \exp\left(\frac{1}{4y} - \frac{1}{4\xi} - \frac{M}{4y\xi}\right), \tag{3.49}$$

we see that

$$k_\alpha(x,y,\xi)\xi^{-2\alpha}e^{-1/\xi} \le C\,\xi^{-2\alpha}\exp\biggl(-\frac{1}{4\xi}-\frac{M}{4y\xi}\biggr), \qquad x \ge 2M,\ \xi \le M,$$

where the right-hand side is integrable on the interval $(0, M]$; hence, if we let $x \to \infty$ in (3.45), by dominated convergence we obtain

$$\lim_{x\to\infty}\int_0^M k_\alpha(x,y,\xi)\,r_\alpha(\xi)d\xi = \int_0^M \Bigl(\lim_{x\to\infty} k_\alpha(x,y,\xi)\Bigr)r_\alpha(\xi)d\xi = 0. \tag{3.50}$$

Let $f \in \mathrm{B_b}(\mathbb{R}^+)$ be such that $\lim_{x\to\infty} f(x) = 0$, and let $\varepsilon > 0$. Choose M such that $|f(x)| < \varepsilon$ for all $x \ge M$. Then

$$\begin{aligned}|(\mathcal{T}_\alpha^y f)(x)| &\le \|f\|_\infty \int_0^M k_\alpha(x,y,\xi)\,r_\alpha(\xi)d\xi + \varepsilon\int_M^\infty k_\alpha(x,y,\xi)\,r_\alpha(\xi)d\xi\\ &\le \|f\|_\infty \int_0^M k_\alpha(x,y,\xi)\,r_\alpha(\xi)d\xi + \varepsilon,\end{aligned}$$

so that (3.50) yields $\limsup_{x\to\infty} |(\mathcal{T}_\alpha^y f)(x)| \le \varepsilon$, where ε is arbitrary.

We observe that, as a consequence of Proposition 3.2, the Whittaker translation (3.40) with the convention that $(\mathcal{T}_\alpha^x f)(0) = (\mathcal{T}_\alpha^0 f)(x) = f(x)$ for all x, satisfies the properties

$$\mathcal{T}_\alpha^y\bigl(\mathrm{C_b}(\mathbb{R}_0^+)\bigr) \subset \mathrm{C_b}(\mathbb{R}_0^+) \qquad \text{and} \qquad \mathcal{T}_\alpha^y\bigl(\mathrm{C_0}(\mathbb{R}_0^+)\bigr) \subset \mathrm{C_0}(\mathbb{R}_0^+) \qquad (y \ge 0), \tag{3.51}$$

as well as the obvious symmetry property

$$(\mathcal{T}_\alpha^y f)(x) = (\mathcal{T}_\alpha^x f)(y) \qquad (x, y \ge 0).$$

It is also easy to check that the Whittaker translation is symmetric with respect to the measure $r_\alpha(x)dx$, in the sense that for $f, g \in \mathrm{C_b}(\mathbb{R}_0^+) \cap L^1(r_\alpha)$ we have

$$\int_0^\infty (\mathcal{T}_\alpha^y f)(x)g(x)\,r_\alpha(x)dx = \int_0^\infty f(x)(\mathcal{T}_\alpha^y g)(x)\,r_\alpha(x)dx. \tag{3.52}$$

3.4 Index Whittaker Transforms

The integral transform determined by the generator (3.2) of the Shiryaev process, which we will call the *index Whittaker transform (of order α)*, is defined by

$$(\mathcal{W}_\alpha f)(\tau) := \int_0^\infty f(y)\, W_{\alpha,i\tau}(y)\, r_\alpha(y) dy, \qquad \tau \geq 0. \tag{3.53}$$

(This is a modified form of the index Whittaker transform defined in [180].) This integral transform is a fundamental tool for studying the Whittaker convolution (defined in the next section), since it is the object which will play a role similar to that of the Hankel transform in the construction of the Kingman convolution.

As noted in Example 2.4, the spectral expansion of the differential operator $\mathcal{A}_\alpha$ yields the following theorem:

Proposition 3.3 *For $\alpha < \frac{1}{2}$, the index Whittaker transform* (3.53) *defines an isometric isomorphism*

$$\mathcal{W}_\alpha : L^2(r_\alpha) \longrightarrow L^2\big(\mathbb{R}^+;\, \rho_\alpha(\tau)d\tau\big),$$

where $\rho_\alpha(\tau) := \pi^{-2}\tau \sinh(2\pi\tau)\big|\Gamma\big(\frac{1}{2} - \alpha + i\tau\big)\big|^2$, whose inverse is given by

$$(\mathcal{W}_\alpha^{-1} \varphi)(x) = \int_0^\infty \varphi(\tau)\, W_{\alpha,i\tau}(x)\, \rho_\alpha(\tau) d\tau, \tag{3.54}$$

the convergence of the integrals (3.53) *and* (3.54) *being understood with respect to the norm of the spaces $L^2\big(\mathbb{R}^+;\, \rho_\alpha(\tau)d\tau\big)$ and $L^2(r_\alpha)$, respectively. Moreover, the differential operator* (3.2) *is connected with the index Whittaker transform via the identity*

$$\big[\mathcal{W}_\alpha(-\mathcal{A}_\alpha f)\big](\tau) = \big(\tau^2 + (\tfrac{1}{2} - \alpha)^2\big) \cdot \big(\mathcal{W}_\alpha f\big)(\tau), \qquad f \in \mathcal{D}_\alpha^{(2)},$$

where

$$\begin{aligned} \mathcal{D}_\alpha^{(2)} &:= \Big\{ u \in L^2(r_\alpha) \;\Big|\; u, u' \in \mathrm{AC_{loc}}(\mathbb{R}^+),\ \mathcal{A}_\alpha u \in L^2(r_\alpha),\ (p_\alpha u')(0) = 0 \Big\} \\ &= \Big\{ u \in L^2(r_\alpha) \;\Big|\; \big(\tau^2 + (\tfrac{1}{2} - \alpha)^2\big) \cdot \big(\mathcal{W}_\alpha f\big)(\tau) \in L^2\big(\mathbb{R}^+;\, \rho_\alpha(\tau)d\tau\big) \Big\}. \end{aligned}$$

We note that, for $\alpha < \frac{1}{2}$ and $\nu = i\tau$, the product formula (3.36) can be written as

$$W_{\alpha,i\tau}(x)\, W_{\alpha,i\tau}(y) = [\mathcal{W}_\alpha k_\alpha(x, y, \cdot)](\tau), \qquad (x, y > 0,\ \alpha < \tfrac{1}{2},\ \tau \geq 0).$$

Applying the inverse Whittaker transform (3.54), we find that for $x, y, \xi > 0$ and $\alpha < \frac{1}{2}$ we have

$$k_\alpha(x, y, \xi) = \int_0^\infty \boldsymbol{W}_{\alpha,i\tau}(x)\, \boldsymbol{W}_{\alpha,i\tau}(y)\, \boldsymbol{W}_{\alpha,i\tau}(\xi)\, \rho_\alpha(\tau) d\tau, \tag{3.55}$$

where the integral on the right-hand side converges absolutely, as can be verified using the asymptotic forms (3.18) and $\bigl|\Gamma(\frac{1}{2} - \alpha + i\tau)\bigr| \sim (2\pi)^{\frac{1}{2}} \tau^{-\alpha} \exp(-\frac{\pi\tau}{2})$, $\tau \to +\infty$ (cf. [145, Equation 5.11.9]).

The product formula (3.36) ensures that for each fixed $\alpha < \frac{1}{2}$ the normalized Whittaker functions $\boldsymbol{W}_{\alpha,\nu}(\cdot)$ $(\nu \in \mathbb{C})$ are solutions of the functional equation

$$\omega(x)\omega(y) = \int_0^\infty \omega(\xi)\, k_\alpha(x, y, \xi)\, r_\alpha(\xi) d\xi \qquad (x, y > 0). \tag{3.56}$$

Using the representation (3.55) for the kernel of the product formula, one can prove a lemma which rules out the existence of other nontrivial solutions for this functional equation:

Lemma 3.5 *Let $\alpha < \frac{1}{2}$ and $\nu \geq 0$. Suppose that the function $\omega(x)$ is such that there exists $C > 0$ for which*

$$\bigl|\omega(x)\bigr| \leq C\, \boldsymbol{W}_{\alpha,\nu}(x) \qquad \textit{for a.e. } x > 0, \tag{3.57}$$

and that $\omega(x)$ is a nontrivial solution of the functional equation (3.56). *Then $\omega(x) = \boldsymbol{W}_{\alpha,\rho}(x)$ for some $\rho \in \mathbb{C}$ with $|\mathrm{Re}\, \rho| \leq \nu$.*

Proof We begin by noting that

$$\begin{aligned} \mathcal{A}_{\alpha,x}\, k_\alpha(x, y, \xi) &= \mathcal{A}_{\alpha,y}\, k_\alpha(x, y, \xi) \\ &= -\int_0^\infty \boldsymbol{W}_{\alpha,i\tau}(x)\, \boldsymbol{W}_{\alpha,i\tau}(y)\, \boldsymbol{W}_{\alpha,i\tau}(\xi)\, \bigl(\tau^2 + (\tfrac{1}{2} - \alpha)^2\bigr) \rho_\alpha(\tau) d\tau, \end{aligned} \tag{3.58}$$

where $\mathcal{A}_{\alpha,x}$ and $\mathcal{A}_{\alpha,y}$ denote the differential operator (3.2) acting on the variable x and y respectively. The identity (3.58) is obtained via differentiation of (3.55) under the integral sign, which is admissible because the differentiated integrals converge absolutely and locally uniformly, as can be verified in a straightforward way using the identity

$$\frac{d}{dy} \boldsymbol{W}_{\alpha,\nu}(y) = \bigl(\nu^2 - (\tfrac{1}{2} - \alpha)^2\bigr) \boldsymbol{W}_{\alpha-1,\nu}(y) \tag{3.59}$$

(which follows from (3.15)) and the asymptotic expansion (3.18). (Recall also that, by Proposition 3.1, the function $\boldsymbol{W}_{\alpha,\nu}(\cdot)$ satisfies the differential equation $\mathcal{A}_\alpha u = (\nu^2 - (\frac{1}{2}-\alpha)^2)u$.)

Now, assuming that the right-hand side of the functional equation (3.56) can also be differentiated under the integral sign, it follows from (3.58) that

$$\big(\mathcal{A}_{\alpha,x}\,\omega(x)\big)\,\omega(y) = \big(\mathcal{A}_{\alpha,y}\,\omega(y)\big)\,\omega(x) \qquad (x, y > 0). \tag{3.60}$$

Here the possibility of interchanging derivative and integral follows again from the locally uniform convergence of the differentiated integrals, which can be straightforwardly checked using (3.57), the identity

$$\frac{\partial k_\alpha(x,y,\xi)}{\partial x} = \frac{y+\xi-x}{2x^2 y\xi}\,k_{\alpha+\frac{1}{2}}(x,y,\xi) - \big(x^{-2} + (1-2\alpha)x^{-1}\big)\,k_\alpha(x,y,\xi) \tag{3.61}$$

(which is a consequence of (3.23)) and the upper bound (3.39) for the function $k_\alpha(x, y, \xi)$.

Notice that (3.60) holds for arbitrary values of x and y. Therefore, we must have

$$\frac{\mathcal{A}_{\alpha,x}\,\omega(x)}{\omega(x)} = \frac{\mathcal{A}_{\alpha,y}\,\omega(y)}{\omega(y)} = \lambda$$

for some $\lambda \in \mathbb{C}$, meaning that $\omega(x)$ is a solution of the Sturm–Liouville equation

$$\mathcal{A}_\alpha \omega(x) = \big(\rho^2 - (\tfrac{1}{2}-\alpha)^2\big)\omega(x),$$

where ρ is the principal square root of $\lambda + (\frac{1}{2}-\alpha)^2$. Consequently, the function $\omega(x)$ is a linear combination of the functions $\boldsymbol{W}_{\alpha,\rho}(x)$ and

$$\boldsymbol{M}_{\alpha,\rho}(x) := \sum_{k=0}^{\infty} \frac{(\frac{1}{2}-\alpha+\rho)_k}{\Gamma(1+2\rho+k)k!}\,x^{-(\frac{1}{2}-\alpha+\rho+k)} \equiv \frac{1}{\Gamma(1+2\rho)}\,x^\alpha e^{\frac{1}{2x}}\,M_{\alpha,\rho}(\tfrac{1}{x}),$$

where $M_{\alpha,\rho}(x)$ is the Whittaker function of the first kind [145, §13.14]. (Here we are using the well-known fact that the Whittaker functions $W_{\alpha,\rho}(z)$ and $\frac{1}{\Gamma(1+2\rho)}M_{\alpha,\rho}(z)$ are, for $\frac{1}{2}-\alpha+\rho \neq 0, -1, -2, \ldots$, two linearly independent solutions of $\frac{d^2u}{dz^2} + \big(-\frac{1}{4} + \frac{\alpha}{z} + \frac{1/4-\rho^2}{z^2}\big)u = 0$. Recall also that the vector space of solutions of a Sturm–Liouville equation is two-dimensional.) However, it follows from the limiting forms for the Whittaker M function [145, Equation 13.14.20] that $\boldsymbol{M}_{\alpha,\rho}(x)$ is, for all $\rho \in \mathbb{C}$, unbounded as x goes to zero, and this violates (3.57). In addition, the limiting forms (3.10), (3.12) for the Whittaker function show that $\big|\boldsymbol{W}_{\alpha,\rho}(x)\big| \leq C\,\boldsymbol{W}_{\alpha,\nu}(x)$ holds if and only if $|\mathrm{Re}\,\rho| \leq \nu$. Therefore, we must have $\omega(x) = \boldsymbol{W}_{\alpha,\rho}(x)$ for ρ belonging to the strip $|\mathrm{Re}\,\rho| \leq \nu$.

We proceed with the definition of the index Whittaker transform of finite complex measures, which will allow us to interpret (3.53) as the index Whittaker transform of an absolutely continuous measure with density $f(\cdot)r_\alpha(\cdot)$:

Definition 3.3 Let $\mu \in \mathcal{M}_\mathbb{C}(\mathbb{R}_0^+)$. The *index Whittaker transform* of the measure μ is the function defined by the integral

$$\widehat{\mu}(\lambda) \equiv \widehat{\mu}(\lambda; \alpha) = \int_{\mathbb{R}_0^+} \boldsymbol{W}_{\alpha,\Delta_\lambda}(y)\, \mu(dy), \qquad \lambda \geq 0. \tag{3.62}$$

For convenience this transformation is regarded as a function of $\lambda = \tau^2 + (\frac{1}{2} - \alpha)^2$ (recall that $\Delta_\lambda = \sqrt{(\frac{1}{2} - \alpha)^2 - \lambda}$).

Before we state the basic properties of the index Whittaker transform of finite measures, we prove an auxiliary result of independent interest: an integral representation for the normalized Whittaker function $\boldsymbol{W}_{\alpha,\nu}(y)$ which we call the *Laplace-type representation* for $\boldsymbol{W}_{\alpha,\nu}(y)$ because it is of the same form as the Laplace representation for the characters of Sturm–Liouville hypergroups, cf. [210, (4.7)–(4.8)]. The integral representation below cannot be found in standard references on integration formulas for special functions such as [155–157].

Theorem 3.2 *The normalized Whittaker W function admits the integral representation*

$$\boldsymbol{W}_{\alpha,\nu}(y) = \int_{-\infty}^{\infty} e^{\nu s} \eta_{\alpha,y}(s) ds = 2 \int_0^{\infty} \cosh(\nu s) \eta_{\alpha,y}(s) ds \qquad (\alpha, \nu \in \mathbb{C},\ y > 0), \tag{3.63}$$

where $\eta_{\alpha,y}$ is the function defined by

$$\eta_{\alpha,y}(s) := 2^{-1-\alpha} \pi^{-\frac{1}{2}} y^{-\frac{1}{2}+\alpha} \exp\left(\frac{1}{y} - \frac{1}{2y} \cosh^2\!\left(\frac{s}{2}\right)\right) D_{2\alpha}\left(2^{\frac{1}{2}} y^{-\frac{1}{2}} \cosh\!\left(\frac{s}{2}\right)\right)$$

and $D_\mu(z)$ is the parabolic cylinder function.

Proof Only the first equality in (3.63) needs proof. Let us temporarily assume that $\nu \geq 0$ and $-\infty < \alpha < \frac{1}{2}$, and let $\xi > 0$. We begin by noting the identity

$$\begin{aligned} \xi^{1-2\alpha} \int_0^\infty & \exp\left(-\frac{\xi^2 y}{4} - \frac{1}{y}\right) y^{-2\alpha} \boldsymbol{W}_{\alpha,\nu}(y)\, dy \\ &= 2^{2-2\alpha} K_{2\nu}(\xi) \\ &= 2^{-2\alpha} \int_{-\infty}^{\infty} e^{\nu s} \exp\!\left(-\xi \cosh\!\left(\frac{s}{2}\right)\right) ds, \end{aligned} \tag{3.64}$$

which is a consequence of integrals 2.4.18.12 in [155] and 2.19.4.7 in [157]. To deduce the theorem from this identity, we will use the injectivity property of the Laplace transform, after rewriting the right-hand side as an iterated integral. To that end, we point out that, according to integral 2.11.4.4 in [156], for $s, \xi > 0$ we have

$$\begin{aligned}
&\xi^{2\alpha-1}\exp\Bigl(-\xi\cosh\Bigl(\frac{s}{2}\Bigr)\Bigr)\\
&= 2^{\alpha-1}\pi^{-\frac{1}{2}}\int_0^\infty \exp\biggl(-\frac{\xi^2 y}{4}-\frac{1}{2y}\cosh^2\Bigl(\frac{s}{2}\Bigr)\biggr)y^{-\frac{1}{2}-\alpha}D_{2\alpha}\Bigl(2^{\frac{1}{2}}y^{-\frac{1}{2}}\cosh\Bigl(\frac{s}{2}\Bigr)\Bigr)dy.
\end{aligned}$$

Substituting in (3.64) and interchanging the order of integration (which is valid because, as noted in Remark 3.3(c), we have $D_\mu(y) > 0$ for $y > 0$ and $\mu < 1$, and therefore the iterated integral has positive integrand), we find that

$$\begin{aligned}
&\int_0^\infty \exp\biggl(-\frac{\xi^2 y}{4}-\frac{1}{y}\biggr)y^{-2\alpha}\boldsymbol{W}_{\alpha,\nu}(y)\,dy =\\
&\quad = 2^{-1-\alpha}\pi^{-\frac{1}{2}}\int_0^\infty \exp\biggl(-\frac{\xi^2 y}{4}\biggr)y^{-\frac{1}{2}-\alpha}\times\\
&\qquad\times\int_{-\infty}^\infty \exp\biggl(\nu s-\frac{1}{2y}\cosh^2\Bigl(\frac{s}{2}\Bigr)\biggr)D_{2\alpha}\Bigl(2^{\frac{1}{2}}y^{-\frac{1}{2}}\cosh\Bigl(\frac{s}{2}\Bigr)\Bigr)ds\,dy.
\end{aligned}$$

Given that the Laplace transform is one-to-one, this identity yields

$$e^{-\frac{1}{y}}\boldsymbol{W}_{\alpha,\nu}(y) = 2^{-1-\alpha}\pi^{-\frac{1}{2}}y^{-\frac{1}{2}+\alpha}\int_{-\infty}^\infty \exp\biggl(\nu s-\frac{1}{2y}\cosh^2\Bigl(\frac{s}{2}\Bigr)\biggr)D_{2\alpha}\Bigl(2^{\frac{1}{2}}y^{-\frac{1}{2}}\cosh\Bigl(\frac{s}{2}\Bigr)\Bigr)ds,$$

finishing the proof for the case $-\infty < \alpha < \frac{1}{2}$, $\nu \in \mathbb{R}$.

To extend (3.63) to all $\alpha, \nu \in \mathbb{C}$, it is enough to show that $\int_{-\infty}^\infty e^{\nu s}\eta_{\alpha,x}(s)ds$ is an entire function of the parameter α and the parameter ν (so that the usual analytic continuation argument can be applied). For $t > 0$ and $\alpha \in \mathbb{C}$ with $\operatorname{Re}\alpha \leq 0$, the integral representation (3.20) gives

$$\begin{aligned}
\bigl|D_{2\alpha}(t)\bigr| &= \frac{e^{-\frac{t^2}{4}}\,t^{2\operatorname{Re}\alpha}}{|\Gamma(\frac{1}{2}-\alpha)|}\biggl|\int_0^\infty e^{-s}s^{-\frac{1}{2}-\alpha}\Bigl(1+\frac{2s}{t^2}\Bigr)^\alpha ds\biggr|\\
&\leq \frac{e^{-\frac{t^2}{4}}\,t^{2\operatorname{Re}\alpha}}{|\Gamma(\frac{1}{2}-\alpha)|}\int_0^\infty e^{-s}s^{-\frac{1}{2}-\operatorname{Re}\alpha}ds\\
&= \frac{\Gamma(\frac{1}{2}-\operatorname{Re}\alpha)}{|\Gamma(\frac{1}{2}-\alpha)|}e^{-\frac{t^2}{4}}\,t^{2\operatorname{Re}\alpha}.
\end{aligned}$$

Furthermore, for each $n \in \mathbb{N}_0$ we have $D_{2\alpha+n}(t) = Q^{(3)}_{n,\alpha}(t) D_{2\alpha}(t) + Q^{(4)}_{n,\alpha}(t) D_{2\alpha-1}(t)$ (cf. proof of Theorem 3.1), being $Q^{(j)}_{n,\alpha}(\cdot)$ polynomials of degree at most n whose coefficients are continuous functions of α. It is easy to see that $|Q^{(j)}_{n,\alpha}(t)| \leq C_n(\alpha)\,(1+t^n)$ for some function $C_n(\alpha)$ that depends continuously on $\alpha \in \mathbb{C}$ and, consequently,

$$\begin{aligned}\big|D_{2\alpha+n}(t)\big| &\leq C_n(\alpha)\,(1+t^n)\big[D_{2\alpha}(t) + D_{2\alpha-1}(t)\big] \\ &\leq C_n(\alpha)\, e^{-\frac{t^2}{4}}\, t^{2\operatorname{Re}\alpha}(1+t^n)\bigg[\frac{\Gamma(\frac{1}{2} - \operatorname{Re}\alpha)}{|\Gamma(\frac{1}{2}-\alpha)|} + \frac{\Gamma(1-\operatorname{Re}\alpha)}{|\Gamma(1-\alpha)|} t^{-1}\bigg],\end{aligned}$$

$$\sup_{\substack{|\alpha|\leq M \\ \operatorname{Re}\alpha \leq 0}} \big|D_{2\alpha+n}(t)\big| \leq C_{M,n}\, e^{-\frac{t^2}{4}}(t^{-2M-1} + t^n), \tag{3.65}$$

where $M > 0$ and $n \in \mathbb{N}_0$ are arbitrary and the constant $C_{M,n}$ depends on M and n. Using (3.65), we see that

$$\sup_{(\alpha,\nu)\in\mathcal{R}_M} \int_{-\infty}^{\infty} \bigg| \exp\Big(\nu s - \frac{1}{2y}\cosh^2\Big(\frac{s}{2}\Big)\Big) D_{2\alpha+n}\Big(2^{\frac{1}{2}} y^{-\frac{1}{2}} \cosh\Big(\frac{s}{2}\Big)\Big)\bigg|\, ds < \infty,$$

where $\mathcal{R}_M = \big\{(\alpha,\nu) \;\big|\; |\alpha| \leq M,\ \operatorname{Re}\alpha \leq 0,\ |\nu| \leq M\big\}$. Applying the standard results on the analyticity of parameter-dependent integrals (e.g. [135]), we obtain the entireness in α and in ν of $\int_{-\infty}^{\infty} e^{\nu s} \eta_{\alpha,x}(s) ds$, completing the proof.

It is worth observing that

$$\eta_{\alpha,y}(s) \geq 0 \qquad \text{for all } \alpha \leq \tfrac{1}{2},\ y > 0$$

and so it follows from Theorem 3.2 that

$$|\boldsymbol{W}_{\alpha,\nu}(y)| \leq \boldsymbol{W}_{\alpha,\nu_0}(y) \qquad \text{whenever } \alpha \leq \tfrac{1}{2},\ |\operatorname{Re}\nu| \leq \nu_0 \quad (\nu_0 \geq 0). \tag{3.66}$$

Together with the identity $\boldsymbol{W}_{\alpha,\frac{1}{2}-\alpha}(y) = 1$ (see (3.17)), this implies that

$$|\boldsymbol{W}_{\alpha,\nu}(y)| \leq 1 \qquad \text{for all } y > 0,\ \alpha \leq \tfrac{1}{2},\ \nu \text{ in the strip } |\operatorname{Re}\nu| \leq \tfrac{1}{2} - \alpha. \tag{3.67}$$

We are now ready to establish some important facts on the index Whittaker transform (3.62):

Proposition 3.4 *For $\alpha < \frac{1}{2}$, the index Whittaker transform $\widehat{\mu} \equiv \widehat{\mu}(\cdot;\alpha)$ of $\mu \in \mathcal{M}_{\mathbb{C}}(\mathbb{R}_0^+)$ has the following properties:*

(i) *$\widehat{\mu}$ is uniformly continuous on $\mathbb{R}_0^+$. Moreover, if a family of measures $\{\mu_j\} \subseteq \mathcal{M}_{\mathbb{C}}(\mathbb{R}_0^+)$ is such that the family of restricted measures $\{\mu_j|_{\mathbb{R}^+}\}$ is tight and uniformly bounded, then $\{\widehat{\mu_j}\}$ is uniformly equicontinuous on $\mathbb{R}_0^+$.*

(ii) *Each measure $\mu \in \mathcal{M}_{\mathbb{C}}(\mathbb{R}_0^+)$ is uniquely determined by $\widehat{\mu}|_{[(\frac{1}{2}-\alpha)^2,\infty)}$.*

(iii) *If $\{\mu_n\}$ is a sequence of measures belonging to $\mathcal{M}_+(\mathbb{R}_0^+)$, $\mu \in \mathcal{M}_+(\mathbb{R}_0^+)$, and $\mu_n \xrightarrow{w} \mu$, then*

$$\widehat{\mu_n} \xrightarrow[n\to\infty]{} \widehat{\mu} \qquad \textit{uniformly on compact sets.}$$

(iv) *If $\{\mu_n\}$ is a sequence of measures belonging to $\mathcal{M}_+(\mathbb{R}_0^+)$ whose index Whittaker transforms are such that*

$$\widehat{\mu_n}(\lambda) \xrightarrow[n\to\infty]{} f(\lambda) \qquad \textit{pointwise in } \lambda \geq 0 \tag{3.68}$$

for some real-valued function f which is continuous at a neighborhood of zero, then $\mu_n \xrightarrow{w} \mu$ for some measure $\mu \in \mathcal{M}_+(\mathbb{R}_0^+)$ such that $\widehat{\mu} = f$.

Proof

(i) Let us prove the second statement, which implies the first. Fix $\varepsilon > 0$. By the tightness assumption, we can choose $M > 0$ such that $\mu_j\big((0,\frac{1}{M})\cup(M,\infty)\big) < \varepsilon$. Moreover, noting that $|\mathrm{Re}\,\Delta_\lambda| \leq \frac{1}{2} - \alpha$, it is easily seen that $|\exp(\Delta_{\lambda_1}s) - \exp(\Delta_{\lambda_2}s)| \leq |\Delta_{\lambda_1} - \Delta_{\lambda_2}|s\, e^{(\frac{1}{2}-\alpha)s}$ for all $s, \lambda_1, \lambda_2 \geq 0$ and, consequently, from Theorem 3.2 we get

$$\big|\mathbf{W}_{\alpha,\Delta_{\lambda_1}}(y) - \mathbf{W}_{\alpha,\Delta_{\lambda_2}}(y)\big| \leq |\Delta_{\lambda_1} - \Delta_{\lambda_2}| \int_{-\infty}^{\infty} s\, e^{(\frac{1}{2}-\alpha)s}\, \eta_{\alpha,y}(s)\, ds, \tag{3.69}$$

where the integral on the right-hand side converges uniformly with respect to y in compact subsets of $\mathbb{R}^+$ and is therefore a continuous function of $y > 0$. By continuity of $\lambda \mapsto \Delta_\lambda$, we can choose $\delta > 0$ such that

$$|\Delta_{\lambda_1} - \Delta_{\lambda_2}| < \frac{\varepsilon}{C_M} \quad \text{whenever } |\lambda_1 - \lambda_2| < \delta \quad (\lambda_1, \lambda_2 \geq 0), \tag{3.70}$$

where $C_M = \max_{y \in [\frac{1}{M}, M]} \int_{-\infty}^{\infty} s\, e^{(\frac{1}{2}-\alpha)s} \eta_{\alpha,y}(s)\, ds < \infty$. Set $S = \sup_j \|\mu_j\|$. Combining (3.69)–(3.70), (3.67) and the fact that $\boldsymbol{W}_{\alpha,\Delta_\lambda}(0) \equiv 1$, we deduce that

$$\begin{aligned}\big|\widehat{\mu_j}(\lambda_1) - \widehat{\mu_j}(\lambda_2)\big| &= \bigg| \int_{\mathbb{R}^+} \big(\boldsymbol{W}_{\alpha,\Delta_{\lambda_1}}(y) - \boldsymbol{W}_{\alpha,\Delta_{\lambda_2}}(y)\big)\mu_j(dy) \bigg| \\ &\leq \int_{(0,\frac{1}{M})\cup(M,\infty)} \big|\boldsymbol{W}_{\alpha,\Delta_{\lambda_1}}(y) - \boldsymbol{W}_{\alpha,\Delta_{\lambda_2}}(y)\big|\mu_j(dy) \\ &\quad + \int_{[\frac{1}{M},M]} \big|\boldsymbol{W}_{\alpha,\Delta_{\lambda_1}}(y) - \boldsymbol{W}_{\alpha,\Delta_{\lambda_2}}(y)\big|\mu_j(dy) \\ &\leq 2\varepsilon + S\varepsilon = (S+2)\varepsilon\end{aligned}$$

for all j, provided that $|\lambda_1 - \lambda_2| < \delta$, which means that $\{\widehat{\mu_j}\}$ is uniformly equicontinuous.

(ii) Writing $\lambda = \tau^2 + (\frac{1}{2}-\alpha)^2$ with $\tau \geq 0$, the index Whittaker transform $\widehat{\mu}(\tau^2 + (\frac{1}{2}-\alpha)^2)$ can be written as

$$\begin{aligned}\widehat{\mu}(\tau^2 + (\tfrac{1}{2}-\alpha)^2) &= \frac{2^{1+2\alpha}}{|\Gamma(\frac{1}{2}-\alpha+i\tau)|^2} \int_{\mathbb{R}_0^+} \int_0^\infty \exp\big(-\tfrac{yt^2}{4}\big) t^{-2\alpha} K_{2i\tau}(t)\, dt\, \mu(dy) \\ &= \frac{2^{1+2\alpha}}{|\Gamma(\frac{1}{2}-\alpha+i\tau)|^2} \int_0^\infty K_{2i\tau}(t)\, t^{-2\alpha} \int_{\mathbb{R}_0^+} \exp\big(-\tfrac{yt^2}{4}\big)\mu(dy)\, dt,\end{aligned} \tag{3.71}$$

where we have applied integral 2.16.8.4 in [156], and the change of order of integration is easily justified. Suppose that $\widehat{\mu_1}(\lambda) = \widehat{\mu_2}(\lambda)$ for all $\lambda \geq (\frac{1}{2}-\alpha)^2$. Then (3.71), together with the injectivity of the Kontorovich–Lebedev transform (see [207, Theorem 6.5]), imply that

$$\int_0^\infty \exp\big(-\tfrac{yt^2}{4}\big)\mu_1(dy) = \int_0^\infty \exp\big(-\tfrac{yt^2}{4}\big)\mu_2(dy) \qquad \text{for almost every } t > 0.$$

In fact, by continuity this equality holds for all $t \geq 0$, because the integrals converge uniformly with respect to $t \geq 0$. Consequently,

$$\int_0^\infty e^{-ys}\mu_1(dy) = \int_0^\infty e^{-ys}\mu_2(dy) \qquad \text{for all } s \geq 0.$$

Since the measures μ_j are uniquely determined by their Laplace transforms [102, Theorem 15.6], it follows that $\mu_1 = \mu_2$.

(iii) Since $\boldsymbol{W}_{\alpha,\Delta_\lambda}(\cdot)$ is continuous and bounded, the pointwise convergence $\widehat{\mu_n}(\lambda) \to \widehat{\mu}(\lambda)$ follows from the definition of weak convergence. For the restricted measures, we clearly have $\mu_n|_{\mathbb{R}^+} \xrightarrow{w} \mu|_{\mathbb{R}^+}$. Using the well-known Prokhorov's theorem which states that a family of measures on a complete separable metric space is relatively compact in the weak topology if and only if it is tight [102, Theorem 13.29], we see that $\{\mu_n|_{\mathbb{R}^+}\}$ is tight and therefore (by part (i)) $\{\widehat{\mu_n}\}$ is uniformly equicontinuous. Invoking a general result which asserts that if $\{g_n\}$ is a uniformly equicontinuous sequence of functions then $g_n \to g$ pointwise implies that $g_n \to g$ uniformly on compact sets [102, Lemma 15.22], we conclude that the convergence $\widehat{\mu_n} \to \widehat{\mu}$ is uniform on compact sets.

(iv) We only need to show that the sequence $\{\mu_n\}$ is tight. Indeed, if $\{\mu_n\}$ is tight, then Prokhorov's theorem yields that for any subsequence $\{\mu_{n_k}\}$ there exists a further subsequence $\{\mu_{n_{k_j}}\}$ and a measure $\mu \in \mathcal{M}_+(\mathbb{R}_0^+)$ such that $\mu_{n_{k_j}} \xrightarrow{w} \mu$. Then, due to part (iii) and to (3.68), we have $\widehat{\mu}(\lambda) = f(\lambda)$ for all $\lambda \geq 0$, which implies (by part (ii)) that all such subsequences have the same weak limit; consequently, the sequence μ_n itself converges weakly to μ.

To prove the tightness, take $\varepsilon > 0$. Since f is continuous at a neighborhood of zero, we have $\frac{1}{\delta}\int_0^{2\delta}\big(f(0) - f(\lambda)\big)d\lambda \longrightarrow 0$ as $\delta \downarrow 0$; therefore, we can choose $\delta > 0$ such that

$$\frac{1}{\delta}\int_0^{2\delta}\big(f(0) - f(\lambda)\big)d\lambda < \varepsilon.$$

Next we observe that, as a consequence of (3.12) and the dominated convergence theorem, we have $\int_0^{2\delta}\big(1 - \mathbf{W}_{\alpha,\Delta_\lambda}(y)\big)d\lambda \longrightarrow 2\delta$ as $y \to \infty$, meaning that we can pick $M > 0$ such that

$$\int_0^{2\delta}\big(1 - \mathbf{W}_{\alpha,\Delta_\lambda}(y)\big)d\lambda \geq \delta \qquad \text{for all } y > M.$$

By our choice of M and Fubini's theorem,

$$\begin{aligned}
\mu_n\big([M,\infty)\big) &= \frac{1}{\delta}\int_M^\infty \delta\, \mu_n(dy) \\
&\leq \frac{1}{\delta}\int_M^\infty \int_0^{2\delta} \big(1 - \mathbf{W}_{\alpha,\Delta_\lambda}(y)\big)d\lambda\, \mu_n(dy) \\
&\leq \frac{1}{\delta}\int_0^\infty \int_0^{2\delta} \big(1 - \mathbf{W}_{\alpha,\Delta_\lambda}(y)\big)d\lambda\, \mu_n(dy) \\
&= \frac{1}{\delta}\int_0^{2\delta} \big(\widehat{\mu_n}(0) - \widehat{\mu_n}(\lambda)\big)d\lambda.
\end{aligned}$$

Hence, using the dominated convergence theorem,

$$\begin{aligned}\limsup_{n\to\infty} \mu_n\big([M,\infty)\big) &\leq \frac{1}{\delta} \limsup_{n\to\infty} \int_0^{2\delta} \big(\widehat{\mu_n}(0) - \widehat{\mu_n}(\lambda)\big) d\lambda \\ &= \frac{1}{\delta} \int_0^{2\delta} \lim_{n\to\infty} \big(\widehat{\mu_n}(0) - \widehat{\mu_n}(\lambda)\big) d\lambda \\ &= \frac{1}{\delta} \int_0^{2\delta} \big(f(0) - f(\lambda)\big) d\lambda < \varepsilon\end{aligned}$$

due to the choice of δ. Since ε is arbitrary, we conclude that $\{\mu_n\}$ is tight, as desired.

Remark 3.4 Parts (iii) and (iv) of the proposition above show that the following analogue of the Lévy continuity theorem holds for the index Whittaker transform: *the index Whittaker transform is a topological homeomorphism between $\mathcal{P}(\mathbb{R}_0^+)$ with the weak topology and the set $\widehat{\mathcal{P}}$ of index Whittaker transforms of probability measures with the topology of uniform convergence in compact sets.*

3.5 Whittaker Convolution of Measures

The parameter $\alpha < \frac{1}{2}$ will be fixed throughout the rest of this chapter.

Motivated by the connection between the Bessel product formula and the Kingman convolution, we now define the Whittaker convolution in order that the convolution of Dirac measures is the kernel of the product formula (3.36) for the normalized Whittaker W function:

Definition 3.4 Let $\mu, \nu \in \mathcal{M}_{\mathbb{C}}(\mathbb{R}_0^+)$. The measure $\mu \underset{\alpha}{\diamond} \nu$ defined by

$$\int_{\mathbb{R}_0^+} f(x)\, (\mu \underset{\alpha}{\diamond} \nu)(dx) = \int_{\mathbb{R}_0^+} \int_{\mathbb{R}_0^+} (\mathcal{T}_\alpha^y f)(x)\, \mu(dx) \nu(dy), \qquad f \in \mathrm{C_b}(\mathbb{R}_0^+)$$

is called the *Whittaker convolution (of order α)* of the measures μ and ν.

It clearly follows from this definition (and Definition 3.2) that for $x, y > 0$ the Whittaker convolution of Dirac measures $\delta_x \underset{\alpha}{\diamond} \delta_y$ is the absolutely continuous measure defined by

$$(\delta_x \underset{\alpha}{\diamond} \delta_y)(d\xi) = k_\alpha(x, y, \xi)\, r_\alpha(\xi) d\xi.$$

Due to (3.41) and (3.38), the Whittaker convolution of two probability measures $\mu, \nu \in \mathcal{P}(\mathbb{R}_0^+)$ is also a probability measure. Furthermore, Proposition 3.6 below shows that the Whittaker convolution is commutative, associative and such that $(c_1\mu_1 + c_2\mu_2) \underset{\alpha}{\diamond} \nu = c_1(\mu_1 \underset{\alpha}{\diamond} \nu) + c_2(\mu_2 \underset{\alpha}{\diamond} \nu)$ for $\mu_1, \mu_2, \nu \in \mathcal{P}(\mathbb{R}_0^+)$ and $c_1, c_2 \in \mathbb{C}$. Consequently:

Proposition 3.5 *The vector space $\mathcal{M}_{\mathbb{C}}(\mathbb{R}_0^+)$ (with usual addition and scalar multiplication), endowed with the convolution multiplication $\underset{\alpha}{\diamond}$, is a commutative algebra over $\mathbb{C}$ whose multiplicative identity is the Dirac measure δ_0.*

Since $\int_{\mathbb{R}_0^+} f(\xi)(\delta_x \underset{\alpha}{\diamond} \delta_y)(d\xi) = (\mathcal{T}_\alpha^y f)(x)$, the fact that $k_\alpha(x, y, \xi)$ is strictly positive for $x, y, \xi > 0$ yields that $\operatorname{supp}(\delta_x \underset{\alpha}{\diamond} \delta_y) = \mathbb{R}_0^+$ for all $x, y > 0$, in sharp contrast with the compactness axiom H6 which is part of the definition of a hypergroup (Definition 2.4). It is worth mentioning that positive product formulas which lead to convolution operators not satisfying the hypergroup requirements on $\operatorname{supp}(\delta_x \underset{\alpha}{\diamond} \delta_y)$ have also been found for certain families of orthogonal polynomials [39].

We now state the fundamental connection between the Whittaker convolution and the index Whittaker transform (3.62). (The analogous trivialization property for the Kingman convolution was stated in Proposition 2.20.)

Proposition 3.6 *Let $\mu, \mu_1, \mu_2 \in \mathcal{M}_{\mathbb{C}}(\mathbb{R}_0^+)$. We have $\mu = \mu_1 \underset{\alpha}{\diamond} \mu_2$ if and only if*

$$\widehat{\mu}(\lambda) = \widehat{\mu_1}(\lambda)\,\widehat{\mu_2}(\lambda) \qquad \text{for all } \lambda \geq 0.$$

Proof In view of (3.36), we have $\big(\mathcal{T}_\alpha^y \mathbf{W}_{\alpha,\Delta_\lambda}\big)(x) = \mathbf{W}_{\alpha,\Delta_\lambda}(x)\,\mathbf{W}_{\alpha,\Delta_\lambda}(y)$, hence

$$\begin{aligned}
\widehat{\mu_1 \underset{\alpha}{\diamond} \mu_2}(\lambda) &= \int_{\mathbb{R}_0^+} \mathbf{W}_{\alpha,\Delta_\lambda}(x)\,(\mu_1 \underset{\alpha}{\diamond} \mu_2)(dx) \\
&= \int_{\mathbb{R}_0^+}\int_{\mathbb{R}_0^+} \big(\mathcal{T}_\alpha^y \mathbf{W}_{\alpha,\Delta_\lambda}\big)(x)\,\mu_1(dx)\mu_2(dy) \\
&= \int_{\mathbb{R}_0^+}\int_{\mathbb{R}_0^+} \mathbf{W}_{\alpha,\Delta_\lambda}(x)\,\mathbf{W}_{\alpha,\Delta_\lambda}(y)\,\mu_1(dx)\mu_2(dy) = \widehat{\mu_1}(\lambda)\widehat{\mu_2}(\lambda), \qquad \lambda \geq 0.
\end{aligned}$$

This proves the "only if" part, and the converse follows from the uniqueness property in Proposition 3.4(ii).

Remark 3.5 It was noted above that the Whittaker convolution cannot be interpreted as a particular case of the axiomatic framework of hypergroups. One can show that, in addition, the Whittaker convolution also does not constitute an example of an Urbanik convolution algebra (cf. Definition 2.3), because the homogeneity axiom U5 fails to hold for the Whittaker convolution.

Indeed, let $a \in \mathbb{R}^+ \setminus \{1\}$ and assume that the identity $\Theta_a(\delta_x \underset{\alpha}{\diamond} \delta_y) = \Theta_a(\delta_x) \underset{\alpha}{\diamond} \Theta_a(\delta_y)$ holds for all $x, y > 0$. Then

$$\begin{aligned}
\mathbf{W}_{\alpha,\Delta_\lambda}(ax)\,\mathbf{W}_{\alpha,\Delta_\lambda}(ay) &= \widehat{\Theta_a(\delta_x)}(\lambda) \cdot \widehat{\Theta_a(\delta_y)}(\lambda) \\
&= \big[\Theta_a(\delta_x) \underset{\alpha}{\diamond} \Theta_a(\delta_y)\big]^{\widehat{}}\,(\lambda)
\end{aligned}$$

$$
\begin{aligned}
&= [\Theta_a(\delta_x \underset{\alpha}{\diamond} \delta_y)]\widehat{}(\lambda) \\
&= \int \boldsymbol{W}_{\alpha,\Delta_\lambda}(a\xi)\,(\delta_x \underset{\alpha}{\diamond} \delta_y)(d\xi) \\
&= \int \boldsymbol{W}_{\alpha,\Delta_\lambda}(a\xi)\,k_\alpha(x,y,\xi)\,r_\alpha(\xi)d\xi.
\end{aligned}
$$

Therefore, $x \mapsto \boldsymbol{W}_{\alpha,\Delta_\lambda}(ax)$ is, for each $\lambda \geq 0$, a solution of the functional equation (3.56). However, it follows from Lemma 3.5 that any bounded solution of this functional equation is of the form $\boldsymbol{W}_{\alpha,\sigma}(x)$ for some $\sigma \in \mathbb{C}$ with $|\mathrm{Re}\,\sigma| \leq \frac{1}{2} - \alpha$. One can check (using e.g. the asymptotic expansion (3.13), see also (3.74) below) that the identity $\boldsymbol{W}_{\alpha,\Delta_\lambda}(ax) \equiv \boldsymbol{W}_{\alpha,\sigma}(x)$ does not hold for any pair $(\lambda, \sigma) \in \mathbb{R}_0^+ \times \mathbb{C}$, so we obtain a contradiction.

As advertised in the introduction to this chapter, the Whittaker convolution $\underset{\alpha}{\diamond}$ can be extended to a more general family of convolutions associated with the differential operators $\gamma x^2 \frac{d^2}{dx^2} + \gamma(c + 2(1-\alpha)x)\frac{d}{dx}$ $(\gamma, c > 0)$. We will see in Remark 3.11 that this is achieved via the rescaled convolutions

$$
\mu \underset{\alpha,c}{\diamond} \nu := \Theta_c\big((\Theta_{1/c}\mu) \underset{\alpha}{\diamond} (\Theta_{1/c}\nu)\big) \qquad \big(\neq \mu \underset{\alpha}{\diamond} \nu \quad \text{if } c \neq 1\big).
$$

3.5.1 *Infinitely Divisible Distributions*

The set of $\underset{\alpha}{\diamond}$-*infinitely divisible distributions* is defined as

$$
\mathcal{P}_{\alpha,\mathrm{id}} = \big\{\mu \in \mathcal{P}(\mathbb{R}_0^+) \;\big|\; \text{for all } n \in \mathbb{N} \text{ there exists } \nu_n \in \mathcal{P}(\mathbb{R}_0^+) \text{ such that } \mu = (\nu_n)^{\diamond_\alpha n}\big\}, \tag{3.72}
$$

where $(\nu_n)^{\diamond_\alpha n}$ denotes the n-fold Whittaker convolution of ν_n with itself.

Lemma 3.6 *Let $\mu \in \mathcal{P}_{\alpha,\mathrm{id}}$. Then $0 < \widehat{\mu}(\lambda) \leq 1$ for all $\lambda \geq 0$. Moreover, μ has no nontrivial idempotent divisors, i.e., if $\mu = \vartheta \underset{\alpha}{\diamond} \nu$ (with $\vartheta, \nu \in \mathcal{P}(\mathbb{R}_0^+)$) where ϑ is idempotent with respect to the Whittaker convolution (that is, it satisfies $\vartheta = \vartheta \underset{\alpha}{\diamond} \vartheta$), then $\vartheta = \delta_0$.*

Proof The inequality $\widehat{\mu}(\lambda) \leq 1$ is obvious from (3.67). The positivity can be proved as follows (the argument is similar to that of [166, Lemma 7.5]): for every $n \in \mathbb{N}$ there exists ν_n such that $\widehat{\mu}(\lambda) = \widehat{\nu_n}(\lambda)^n$. We can define

$$
\phi(\lambda) := \lim_{n\to\infty} \widehat{\nu_n}(\lambda) = \begin{cases} 1, & \text{if } \widehat{\mu}(\lambda) \neq 0, \\ 0, & \text{if } \widehat{\mu}(\lambda) = 0 \end{cases}
$$

and then (by the continuity of $\widehat{\mu}$, Proposition 3.4(i)) we have $\phi(\lambda) = 1$ in a neighbourhood of 0. Therefore (Proposition 3.4(iv)) ϕ is the index Whittaker transform of a probability measure; in particular, it is continuous on $\mathbb{R}_0^+$, so we conclude that $\phi \equiv 1$. Thus $\widehat{\mu}$ has no zeros, and by continuity it follows that $\widehat{\mu}(\lambda) > 0$ for all λ.

Assume that $\mu = \vartheta \underset{\alpha}{\diamond} \nu$ with ϑ idempotent. Then $\bigl(\widehat{\vartheta}(\lambda)\bigr)^2 = \widehat{\vartheta}(\lambda)$ for all λ, and consequently $\widehat{\vartheta}(\lambda)$ only takes the values 0 and 1. However, $\widehat{\mu}(\lambda) = \widehat{\vartheta}(\lambda)\,\widehat{\nu}(\lambda) \neq 0$; hence $\widehat{\vartheta}(\lambda) = 1$ for all λ, i.e., $\vartheta = \delta_0$.

The first part of the lemma shows that the index Whittaker transform of any measure $\mu \in \mathcal{P}_{\alpha,\mathrm{id}}$ is of the form

$$\widehat{\mu}(\lambda) = e^{-\psi_\mu(\lambda)},$$

where $\psi_\mu(\lambda)$ $(\lambda \geq 0)$ is a positive continuous function such that $\psi_\mu(0) = 0$, which we shall call the *log-Whittaker transform* of μ. The next result shows that the log-Whittaker transform of an infinitely divisible distribution grows at most linearly:

Proposition 3.7 *Let $\mu \in \mathcal{P}_{\alpha,\mathrm{id}}$. Then*

$$\psi_\mu(\lambda) \leq C_\mu(1+\lambda) \qquad \textit{for all } \lambda \geq 0$$

for some constant $C_\mu > 0$ which is independent of λ.

The proof relies on the following lemma:

Lemma 3.7 *The normalized Whittaker W function satisfies the inequality*

$$1 - \mathbf{W}_{\alpha,\nu}(y) \leq \bigl((\tfrac{1}{2}-\alpha)^2 - \nu^2\bigr)y \qquad \textit{for each } y \geq 0 \textit{ and } \nu \in [0, \tfrac{1}{2}-\alpha] \cup i\mathbb{R}.$$

Proof By Proposition 3.1, $\mathbf{W}_{\alpha,\nu}(\cdot)$ solves the equation $-\mathcal{A}_\alpha u = \bigl((\frac{1}{2}-\alpha)^2 - \nu^2\bigr)u$, and thus we have

$$-\frac{d}{d\xi}\biggl[\xi^{2-2\alpha} e^{-1/\xi} \frac{d}{d\xi}\mathbf{W}_{\alpha,\nu}(\xi)\biggr] = \bigl((\tfrac{1}{2}-\alpha)^2 - \nu^2\bigr)\,\xi^{-2\alpha} e^{-1/\xi}\,\mathbf{W}_{\alpha,\nu}(\xi).$$

Recalling that $\mathbf{W}_{\alpha,\nu}(x)$ satisfies the boundary conditions given in (3.8), after integrating both sides between 0 and y and then between 0 and x we obtain

$$1 - \mathbf{W}_{\alpha,\nu}(x) = \bigl((\tfrac{1}{2}-\alpha)^2 - \nu^2\bigr)\int_0^x y^{2\alpha-2} e^{1/y} \int_0^y \xi^{-2\alpha} e^{-1/\xi}\,\mathbf{W}_{\alpha,\nu}(\xi)\,d\xi\,dy. \tag{3.73}$$

Using (3.67) and the inequality $(\frac{\xi}{y})^{2-2\alpha} \leq 1$ (which holds for $0 < \xi \leq y$ due to the assumption $\alpha < \frac{1}{2}$), we thus find that

$$\begin{aligned}1 - \boldsymbol{W}_{\alpha,\nu}(x) &\leq \big((\tfrac{1}{2} - \alpha)^2 - \nu^2\big) \int_0^x y^{2\alpha-2} e^{1/y} \int_0^y \xi^{2\alpha} e^{-1/\xi}\, d\xi\, dy \\ &\leq \big((\tfrac{1}{2} - \alpha)^2 - \nu^2\big) \int_0^x e^{1/y} \int_0^y \xi^{-2} e^{-1/\xi}\, d\xi\, dy \\ &= \big((\tfrac{1}{2} - \alpha)^2 - \nu^2\big) x\end{aligned}$$

as required.

Proof of Proposition 3.7 Let $\nu_n \in \mathcal{P}(\mathbb{R}_0^+)$ be defined as in (3.72), so that $\widehat{\nu_n}(\lambda) \equiv \exp(-\frac{1}{n}\psi_\mu(\lambda))$. Due to the inequality $1 - e^{-\tau} \leq \tau$ $(\tau \geq 0)$ and the fact that $\lim_n n(1 - e^{-k/n}) = k$ for each $k \in \mathbb{R}$, we have

$$n\big(1 - \widehat{\nu_n}(\lambda)\big) \leq \psi_\mu(\lambda) \text{ for all } n \in \mathbb{N}, \qquad \lim_{n\to\infty} n\big(1 - \widehat{\nu_n}(\lambda)\big) = \psi_\mu(\lambda).$$

Pick $\lambda_1 > 0$. It follows from the asymptotic expansion (3.13) (which can be differentiated term by term, cf. [145, §2.1(iii)]) that

$$\frac{d^n}{dy^n}\boldsymbol{W}_{\alpha,\nu}(y) \xrightarrow[y\to 0]{} (-1)^n \big(\tfrac{1}{2}-\alpha+\nu\big)_n \big(\tfrac{1}{2}-\alpha-\nu\big)_n \qquad (n = 0, 1, 2, \ldots). \tag{3.74}$$

In particular, $\lim_{y\to 0} \frac{d}{dy}\boldsymbol{W}_{\alpha,\Delta_\lambda}(y) = -\lambda$, hence there exists $\varepsilon > 0$ such that $\frac{d}{dy}\boldsymbol{W}_{\alpha,\Delta_{\lambda_1}}(y) \leq -\frac{\lambda_1}{2}$ for all $0 < y \leq \varepsilon$, and then we have

$$\frac{1}{\lambda_1}\big(1 - \boldsymbol{W}_{\alpha,\Delta_{\lambda_1}}(y)\big) = -\frac{1}{\lambda_1}\int_0^y \frac{d}{dx}\boldsymbol{W}_{\alpha,\Delta_{\lambda_1}}(x)\, dy \geq \frac{y}{2} \qquad \text{for all } 0 \leq y \leq \varepsilon. \tag{3.75}$$

Using also Lemma 3.7, we get

$$\begin{aligned}n\int_{[0,\varepsilon)} \big(1 - \boldsymbol{W}_{\alpha,\Delta_\lambda}(y)\big)\nu_n(dy) &\leq \lambda n \int_{[0,\varepsilon)} y\, \nu_n(dy) \\ &\leq \frac{2\lambda n}{\lambda_1}\int_{[0,\varepsilon)} \big(1 - \boldsymbol{W}_{\alpha,\Delta_{\lambda_1}}(y)\big)\nu_n(dy) \qquad (3.76) \\ &\leq \frac{2\lambda n}{\lambda_1}\big(1 - \widehat{\nu_n}(\lambda_1)\big) \leq \frac{2\lambda}{\lambda_1}\psi_\mu(\lambda_1).\end{aligned}$$

Next, from the asymptotic expansion given in (3.18) we easily see that there exists $\lambda_2 > 0$ such that

$$\left|\boldsymbol{W}_{\alpha,\Delta_{\lambda_2}}(y)\right| \le \frac{1}{2} \qquad \text{for all } y \ge \varepsilon$$

and using (3.67), we obtain

$$\begin{aligned} n\int_{[\varepsilon,\infty)} \bigl(1-\boldsymbol{W}_{\alpha,\Delta_\lambda}(y)\bigr)\nu_n(dy) &\le 2n\int_{[\varepsilon,\infty)} \nu_n(dy) \\ &\le 4n\int_{[\varepsilon,\infty)} \bigl(1-\boldsymbol{W}_{\alpha,\Delta_{\lambda_2}}(y)\bigr)\nu_n(dy) \\ &\le 4n\bigl(1-\widehat{\nu_n}(\lambda_2)\bigr) \le 4\psi_\mu(\lambda_2). \end{aligned} \tag{3.77}$$

Combining (3.76) and (3.77), one sees that

$$n(1-\widehat{\nu_n}(\lambda)) = n\int_{\mathbb{R}_0^+} \bigl(1-\boldsymbol{W}_{\alpha,\Delta_{\lambda_2}}(y)\bigr)\nu_n(dy) \le \frac{2\lambda}{\lambda_1}\psi_\mu(\lambda_1)+4\psi_\mu(\lambda_2) \le C_\mu(1+\lambda), \tag{3.78}$$

where $C_\mu = \max\bigl\{\frac{2}{\lambda_1}\psi_\mu(\lambda_1), 4\psi_\mu(\lambda_2)\bigr\}$. Taking the limit $n \to \infty$ in the inequality (3.78) yields

$$\psi_\mu(\lambda) \le C_\mu(1+\lambda)$$

which completes the proof.

3.5.2 *Lévy–Khintchine Type Representation*

One can prove that an analogue of the classical Lévy–Khintchine formula holds for the log-Whittaker transforms of $\underset{\alpha}{\diamond}$-infinitely divisible distributions. To establish this result, one needs to adapt the notions of compound Poisson and Gaussian measures to the context of the Whittaker convolution algebra:

Definition 3.5 Let $\mu \in \mathcal{P}(\mathbb{R}_0^+)$ and $a > 0$. The measure $\mathbf{e}_\alpha(a\mu)$ defined by

$$\mathbf{e}_\alpha(a\mu) = e^{-a}\sum_{n=0}^{\infty} \frac{a^n}{n!}\mu^{\diamond_\alpha n}$$

(the infinite sum converging in the weak topology) is said to be the $\underset{\alpha}{\diamond}$*-compound Poisson measure* associated with $a\mu$.

This definition is completely analogous to that of the classical compound Poisson measure. From the definition it immediately follows that $\mathbf{e}_\alpha(a\mu) \in \mathcal{P}(\mathbb{R}_0^+)$. Moreover, its index Whittaker transform can be easily deduced using Proposition 3.6:

$$\widehat{\mathbf{e}_\alpha(a\mu)}(\lambda) = e^{-a} \sum_{n=0}^{\infty} \frac{a^n}{n!} \widehat{\mu^{\diamond_\alpha n}}(\lambda) = e^{-a} \sum_{n=0}^{\infty} \frac{a^n}{n!} \big(\widehat{\mu}(\lambda)\big)^n = \exp\big(a(\widehat{\mu}(\lambda) - 1)\big).$$

Since $\mathbf{e}_\alpha((a+b)\mu) = \mathbf{e}_\alpha(a\mu) \underset{\alpha}{\diamond} \mathbf{e}_\alpha(b\mu)$, every $\underset{\alpha}{\diamond}$-compound Poisson measure belongs to $\mathcal{P}_{\alpha,\text{id}}$.

Definition 3.6 A measure $\mu \in \mathcal{P}(\mathbb{R}_0^+)$ is called a $\underset{\alpha}{\diamond}$-*Gaussian measure* if $\mu \in \mathcal{P}_{\alpha,\text{id}}$ and

$$\mu = \mathbf{e}_\alpha(a\nu) \underset{\alpha}{\diamond} \vartheta \quad \big(a > 0,\ \nu \in \mathcal{P}(\mathbb{R}_0^+),\ \vartheta \in \mathcal{P}_{\alpha,\text{id}}\big) \qquad \Longrightarrow \qquad \nu = \delta_0.$$

Remark 3.6 This definition is similar to the definition of Gaussian measures on locally compact abelian groups as in [146, Chapter IV]. It is analogous with the classical notion of a Gaussian measure on $\mathbb{R}$ by the following result:

Let $\mathfrak{e}(a\nu) := e^{-a} \sum_{k=0}^{\infty} \frac{a^k}{k!} \mu^{*k}$, *where* $*$ *is the ordinary convolution. For a measure* $\mu \in \mathcal{P}(\mathbb{R})$, *the following conditions are equivalent:*

(i) $\mu(dx) = \frac{1}{\sqrt{2\pi}\,\sigma} \exp\big(-\frac{(x-c)^2}{2\sigma^2}\big) dx$ *for some* $c \in \mathbb{R}$ *and* $\sigma > 0$;

(ii) μ *is infinitely divisible, and if* $\mu = \mathfrak{e}(a\nu) * \vartheta$ *(with* $a > 0$, $\nu \in \mathcal{P}(\mathbb{R})$ *and* $\vartheta \in \mathcal{P}(\mathbb{R})$ *infinitely divisible), then* $\nu = \delta_0$.

(The implication (i) $\Longrightarrow$ (ii) is a consequence of the Lévy-Cramer theorem [125, §III.1] which asserts that if $\mu \in \mathcal{P}(\mathbb{R})$ is a Gaussian measure (in the sense of (i)) and $\mu = \mu_1 * \mu_2$ with $\mu_1, \mu_2 \in \mathcal{P}(\mathbb{R})$, then μ_1 and μ_2 are also Gaussian measures. The converse implication follows from the fact that if an infinitely divisible $\mu \in \mathcal{P}(\mathbb{R})$ is such that $\mu = \mathfrak{e}(a\nu) * \vartheta$ implies $\nu = \delta_0$, then the Lévy measure in the classical Lévy–Khintchine formula must be the zero measure, which means that μ is a Gaussian measure; see the discussion after Equation (16.8) in [102].)

The Lévy–Khintchine type representation for measures $\mu \in \mathcal{P}_{\alpha,\text{id}}$ reads as follows:

Theorem 3.3 *The log-Whittaker transform of a measure* $\mu \in \mathcal{P}_{\alpha,\text{id}}$ *can be represented in the form*

$$\psi_\mu(\lambda) = \psi_\gamma(\lambda) + \int_{\mathbb{R}^+} \big(1 - \mathbf{W}_{\alpha,\Delta_\lambda}(x)\big)\nu(dx), \tag{3.79}$$

where ν is a σ-finite measure on $\mathbb{R}^+$ which is finite on (ε, ∞) for all $\varepsilon > 0$ and such that

$$\int_{\mathbb{R}^+} \big(1 - \mathbf{W}_{\alpha,\Delta_\lambda}(x)\big)\nu(dx) < \infty$$

and γ is a $\underset{\alpha}{\diamond}$-Gaussian measure with log-Whittaker transform $\psi_\gamma(\lambda)$. Conversely, each function of the form (3.79) *is a log-Whittaker transform of some $\mu \in \mathcal{P}_{\alpha,\text{id}}$.*

This theorem is a particular case of the Lévy–Khintchine type theorem for stochastic convolutions stated in Proposition 2.18. To keep this chapter more self-contained, a sketch of the proof is presented below; it relies on an algebraic-topological technique which has been earlier used to establish a canonical representation for infinitely divisible distributions on locally compact abelian groups [146, 147].

Proof Let $\mu \in \mathcal{P}_{\alpha,\text{id}}$, let $\infty > a_1 > a_2 > \ldots$ with $\lim a_n = 0$, and let $I_n = [0, a_n)$, $J_n = [a_n, \infty)$. Consider the set Q_μ of all proper divisors of μ of the form $\mathbf{e}_\alpha(\pi)$ such that $\pi(I_1) = 0$. (A measure $\nu \in \mathcal{P}(\mathbb{R}_0^+)$ is said to be a *proper divisor* of $\mu \in \mathcal{P}_{\alpha,\text{id}}$ if $\nu \in \mathcal{P}_{\alpha,\text{id}}$ and $\mu = \nu \underset{\alpha}{\diamond} \theta$ for some $\theta \in \mathcal{P}_{\alpha,\text{id}}$.) Using Proposition 3.4 and the properties of the normalized Whittaker W function, one can prove (see [198, Corollary 1]) that the set $\mathrm{D}(\mathfrak{P})$ of all divisors (with respect to the Whittaker convolution) of measures $\nu \in \mathfrak{P}$ is relatively compact whenever $\mathfrak{P} \subset \mathcal{P}(\mathbb{R}_0^+)$ is relatively compact. This, in turn, implies that $\sup_{\mathbf{e}_\alpha(\pi) \in Q_\mu} \big[\int_{\mathbb{R}_0^+} (1 - \mathbf{W}_{\alpha,\Delta_\lambda}(x))\pi(dx)\big] < \infty$ and therefore, by compactness of the set of proper divisors of μ, there exists a divisor $\mu_1 = \mathbf{e}_\alpha(\pi_1) \in Q_\mu$ such that $\pi_1(J_1)$ is maximal among all elements of Q_μ. Write

$$\mu = \mu_1 \underset{\alpha}{\diamond} \beta_1 \qquad (\beta_1 \in \mathcal{P}_{\alpha,\text{id}}).$$

Applying the same reasoning to β_1 with I_1 replaced by I_2, we get $\beta_1 = \mu_2 \underset{\alpha}{\diamond} \beta_2 = \mathbf{e}_\alpha(\pi_2) \underset{\alpha}{\diamond} \beta_2$. If we perform this successively, we get

$$\mu = \beta_n \underset{\alpha}{\diamond} \theta_n, \qquad \text{where } \theta_n = \mu_1 \underset{\alpha}{\diamond} \mu_2 \underset{\alpha}{\diamond} \ldots \mu_n, \qquad \mu_k = \mathbf{e}_\alpha(\pi_k)$$

with $\pi_k(I_k) = 0$ and $\pi_k(J_k)$ having the specified maximality property. The sequences $\{\beta_n\}$ and $\{\theta_n\}$ are relatively compact; letting β and θ be limit points, we have

$$\mu = \beta \underset{\alpha}{\diamond} \theta \qquad (\beta, \theta \in \mathcal{P}_{\alpha,\text{id}}).$$

Suppose, by contradiction, that β is not $\underset{\alpha}{\diamond}$-Gaussian, and let $\mathbf{e}_\alpha(\eta)$, with $\eta \neq \delta_0$, be a divisor of β. Clearly $\eta(J_k) > 0$ for some k; given that each β_n divides β_{n-1}, we have $\beta_k = \mathbf{e}_\alpha(\eta) \underset{\alpha}{\diamond} \nu$ ($\nu \in \mathcal{P}_{\alpha,\mathrm{id}}$). If we let $\widetilde{\eta}$ be the restriction of η to the interval J_k, then

$$\beta_{k-1} = \mathbf{e}_\alpha(\pi_k + \widetilde{\eta}) \underset{\alpha}{\diamond} \mathbf{e}_\alpha(\eta - \widetilde{\eta}) \underset{\alpha}{\diamond} \nu,$$

which is absurd (because $(\pi_k + \widetilde{\eta})(J_k) > \pi_k(J_k)$, contradicting the maximality property which defines π_k). To determine the log-Whittaker transform of θ, note that $\theta_n = \mathbf{e}_\alpha(\Pi_n)$ is the $\underset{\alpha}{\diamond}$-compound Poisson measure associated with $\Pi_n := \sum_{k=1}^n \pi_k$, thus $\psi_{\theta_n}(\lambda) = \int_{\mathbb{R}_0^+} \bigl(1 - \boldsymbol{W}_{\alpha,\Delta_\lambda}(x)\bigr)\Pi_n(dx)$. Since $\{\Pi_n\}$ is an increasing sequence of measures and each $\mathbf{e}_\alpha(\Pi_n)$ dividing μ, there exists a σ-finite measure ν such that

$$\begin{aligned}\psi_\theta(\lambda) &= \lim_n \int_{\mathbb{R}_0^+} \bigl(1 - \boldsymbol{W}_{\alpha,\Delta_\lambda}(x)\bigr)\Pi_n(dx)\\ &= \int_{\mathbb{R}_0^+} \bigl(1 - \boldsymbol{W}_{\alpha,\Delta_\lambda}(x)\bigr)\nu(dx) < \infty\end{aligned}$$

($\mu \in \mathcal{P}_{\alpha,\mathrm{id}}$ ensures the finiteness of the integral); from the relative compactness of $\mathrm{D}(\{\mu\})$ it is possible to conclude that $\nu(J_k) < \infty$ for all k.

For the converse, let ν_n be the restriction of ν to the interval J_n defined as above. It is verified without difficulty that the right-hand side of (3.79) is continuous at zero, hence by Proposition 3.4(d) $\beta \underset{\alpha}{\diamond} \mathbf{e}_\alpha(\nu_n) \xrightarrow{w} \mu \in \mathcal{P}(\mathbb{R}_0^+)$, and $\mu \in \mathcal{P}_{\alpha,\mathrm{id}}$ because $\mathcal{P}_{\alpha,\mathrm{id}}$ is closed under weak convergence.

Remark 3.7 Proposition 3.11 below ensures that the function $\psi_c(\lambda) = -c\lambda$ is, for each $c > 0$, the log-Whittaker transform of a $\diamond$-Gaussian measure. However, the above Lévy–Khintchine representation provides no information on whether the log-Whittaker transform of the $\diamond$-Gaussian measure γ must be of the form $\psi_\gamma(\lambda) = -c\lambda$ for some $c > 0$. Such a characterization of Gaussian measures has been established for Urbanik convolution algebras [190, 191] and for Sturm–Liouville hypergroups on $\mathbb{R}_0^+$ [30, 163]; however, the proofs of these results depend on assumptions which are not satisfied by the Whittaker convolution. (The proof for Urbanik convolutions relies on the homogeneity axiom U5 of Definition 2.3, while the proof for Sturm–Liouville hypergroups depends on a regularity property of the associated Sturm–Liouville type integral transform which cannot be easily extended to the index Whittaker transform.) A related characterization of Gaussian measures has also been established on spaces endowed with a generalized characteristic function $\int \omega_\lambda(x)\mu(dx)$ having properties similar to those of Proposition 3.4, and in which there exists not only a convolution with respect to the variable x but also a positivity-preserving convolution with respect to the dual variable λ (see [198]). We leave open the problem of extending these characterizations to the Whittaker convolution.

3.6 Lévy Processes with Respect to the Whittaker Convolution

3.6.1 Convolution Semigroups

Having in mind the study of Lévy-like processes on the Whittaker convolution algebra, we first introduce the notion of a Whittaker convolution semigroup.

Definition 3.7 A family $\{\mu_t\}_{t\geq 0} \subset \mathcal{P}(\mathbb{R}_0^+)$ is called a $\underset{\alpha}{\diamond}$-*convolution semigroup* if it satisfies the conditions

- $\mu_s \underset{\alpha}{\diamond} \mu_t = \mu_{s+t}$ for all $s, t \geq 0$;
- $\mu_0 = \delta_0$;
- $\mu_t \overset{w}{\longrightarrow} \delta_0$ as $t \downarrow 0$.

Remark 3.8 Similarly to the classical case (cf. [166, Section 7]), the $\underset{\alpha}{\diamond}$-infinitely divisible distributions are in one-to-one correspondence with the $\underset{\alpha}{\diamond}$-convolution semigroups:

(i) *If* $\{\mu_t\}$ *is a* $\underset{\alpha}{\diamond}$*-convolution semigroup, then* μ_t *is (for each* $t \geq 0$*) a* $\underset{\alpha}{\diamond}$*-infinitely divisible distribution.*
(Indeed, for each $n \in \mathbb{N}$ the measure $\mu_{t/n} \in \mathcal{P}(\mathbb{R}_0^+)$ is such that $(\mu_{t/n})^{\diamond_\alpha n} = \mu_t$.)

(ii) *If* μ *is a* $\underset{\alpha}{\diamond}$*-infinitely divisible distribution with log-Whittaker transform* $\psi_\mu(\lambda)$*, then the semigroup* $\{\mu_t\}$ *defined by* $\widehat{\mu_t}(\lambda) = \exp(-t\,\psi_\mu(\lambda))$ *is the unique* $\underset{\alpha}{\diamond}$*-convolution semigroup such that* $\mu_1 = \mu$.
(To prove this, it suffices to justify that $\exp(-t\,\psi_\mu(\lambda))$ is, for each $t > 0$, the index Whittaker transform of a probability measure. If $t = \frac{p}{q} \in \mathbb{Q}$, this is true because $\widehat{(\nu_q)^{\diamond_\alpha p}}(\lambda) = \exp\bigl(-\frac{p}{q}\,\psi_\mu(\lambda)\bigr)$, where $\nu_q \in \mathcal{P}(\mathbb{R}_0^+)$ is defined as in (3.72). If $t > 0$ is irrational, let $\bigl\{\frac{p_n}{q_n}\bigr\} \subset \mathbb{Q}$ be a sequence converging to t and define $\mu_t \in \mathcal{P}(\mathbb{R}_0^+)$ as the weak limit of the measures $(\nu_{q_n})^{\diamond_\alpha p_n}$; the existence of the weak limit follows from Proposition 3.4(iv), and it is clear that $\widehat{\mu_t}(\lambda) = \exp(-t\,\psi_\mu(\lambda))$.)

From this it follows, in particular, that $\underset{\alpha}{\diamond}$-convolution semigroups admit a Lévy–Khintchine type representation (Theorem 3.3).

Unsurprisingly, each $\underset{\alpha}{\diamond}$-convolution semigroup is associated with a conservative Feller semigroup of operators which commute with the Whittaker translation:

Proposition 3.8 *Let* $\{\mu_t\}_{t\geq 0}$ *be a* $\underset{\alpha}{\diamond}$*-convolution semigroup. Then the family* $\{T_t\}_{t\geq 0}$ *of convolution operators defined by*

$$T_t : \mathrm{C_b}(\mathbb{R}_0^+) \longrightarrow \mathrm{C_b}(\mathbb{R}_0^+), \qquad T_t f := \mathcal{T}_\alpha^{\mu_t} f, \tag{3.80}$$

where $\mathcal{T}_\alpha^\nu$ ($\nu \in \mathcal{M}_\mathbb{C}(\mathbb{R}_0^+)$) is the operator defined by

$$(\mathcal{T}_\alpha^\nu f)(x) := \int_{\mathbb{R}_0^+} (\mathcal{T}_\alpha^y f)(x)\, \nu(dy),$$

is a conservative Feller semigroup. Furthermore, we have $T_t \mathcal{T}_\alpha^\nu f = \mathcal{T}_\alpha^\nu T_t f$ for all $t \geq 0$ and $\nu \in \mathcal{M}_\mathbb{C}(\mathbb{R}_0^+)$ (in particular, $T_t \mathcal{T}_\alpha^x f = \mathcal{T}_\alpha^x T_t f$ for $x \geq 0$).

One should note that this result is similar to the Feller property for Kingman convolution semigroups, stated in Proposition 2.21.

Proof For $\mu, \nu \in \mathcal{P}(\mathbb{R}_0^+)$ we have $\int (\mathcal{T}_\alpha^\mu f)(x)\nu(dx) = \iint (\mathcal{T}_\alpha^y f)(x)\mu(dx)\nu(dy) = \int f(x)\big(\mu \underset{\alpha}{\diamond} \nu\big)(dx)$. Therefore, by associativity and commutativity of the Whittaker convolution,

$$\begin{aligned}
(\mathcal{T}_\alpha^\mu(\mathcal{T}_\alpha^\nu f))(x) &= \int_{\mathbb{R}_0^+} \mathcal{T}_\alpha^\mu(\mathcal{T}_\alpha^\nu f)\, d\delta_x \\
&= \int_{\mathbb{R}_0^+} \mathcal{T}_\alpha^\nu f\, d\big(\mu \underset{\alpha}{\diamond} \delta_x\big) = \int_{\mathbb{R}_0^+} f\, d\big(\nu \underset{\alpha}{\diamond} (\mu \underset{\alpha}{\diamond} \delta_x)\big) \\
&= \int_{\mathbb{R}_0^+} f\, d\big((\mu \underset{\alpha}{\diamond} \nu) \underset{\alpha}{\diamond} \delta_x\big) = \big(\mathcal{T}_\alpha^{\mu \underset{\alpha}{\diamond} \nu} f\big)(x) \qquad (\mu, \nu \in \mathcal{P}(\mathbb{R}_0^+)),
\end{aligned} \tag{3.81}$$

and the convolution semigroup property yields that $T_0 = \mathrm{Id}$ and $T_t T_s = T_{t+s}$ for $t, s \geq 0$. The property $T_t\big(\mathrm{C}_0(\mathbb{R}_0^+)\big) \subset \mathrm{C}_0(\mathbb{R}_0^+)$ follows at once from (3.51) and the dominated convergence theorem. The positivity and conservativeness of T_t follows from the corresponding property of the Whittaker translation (Proposition 3.2(a)). To prove the strong continuity of the semigroup, let $f \in \mathrm{C}_0(\mathbb{R}_0^+)$ and $x \geq 0$. From the definition of weak convergence and the fact that $(\mathcal{T}_\alpha^0 f)(x) = f(x)$ we deduce that

$$\begin{aligned}
\big|(T_t f)(x) - f(x)\big| &= \left| \int_{\mathbb{R}_0^+} \big((\mathcal{T}_\alpha^y f)(x) - f(x)\big)\mu_t(dy) \right| \\
&\underset{t\downarrow 0}{\longrightarrow} \left| \int_{\mathbb{R}_0^+} \big((\mathcal{T}_\alpha^y f)(x) - f(x)\big)\delta_0(dy) \right| = 0.
\end{aligned}$$

and therefore $\|T_t f - f\|_\infty \longrightarrow 0$ as $t \downarrow 0$ (cf. Proposition 2.2). The concluding statement is a consequence of (3.81).

Proposition 3.9 *Let $\{T_t\}$ be a Feller semigroup determined by the $\underset{\alpha}{\diamond}$-convolution semigroup $\{\mu_t\}_{t\geq 0}$. Then, for each $1 \leq p < \infty$, $\bigl\{T_t|_{\mathrm{C_c}(\mathbb{R}_0^+)}\bigr\}$ has an extension $\{T_t^{(p)}\}$ which is a strongly continuous contraction semigroup on $L^p(r_\alpha)$. Moreover, the operators $T_t^{(p)}$ are given by*

$$\bigl(T_t^{(p)} f\bigr)(x) = (\mathcal{T}_\alpha^{\mu_t} f)(x) := \int_{\mathbb{R}_0^+} (\mathcal{T}_\alpha^y f)(x)\, \mu_t(dy) \qquad \bigl(f \in L^p(r_\alpha)\bigr). \tag{3.82}$$

Proof By Proposition 3.2(b) and Minkowski's integral inequality, we have

$$\begin{aligned} \|T_t^{(p)} f\|_{p,\alpha} &\leq \biggl[\int_0^\infty \biggl(\int_{\mathbb{R}_0^+} |(\mathcal{T}_\alpha^y f)(x)|\, \mu_t(dy)\biggr)^p r_\alpha(x)dx\biggr]^{\frac{1}{p}} \\ &\leq \int_{\mathbb{R}_0^+} \|\mathcal{T}_\alpha^y f\|_{p,\alpha}\, \mu_t(dy) \\ &\leq \|f\|_{p,\alpha}, \end{aligned} \tag{3.83}$$

showing that the operators $T_t^{(p)}$ defined by (3.82) are contractions on $L^p(r_\alpha)$. To prove the strong continuity, let $f \in L^p(r_\alpha)$, $\varepsilon > 0$ and choose $g \in \mathrm{C}_c^\infty(\mathbb{R}^+)$ such that $\|f - g\|_{p,\alpha} \leq \varepsilon$. Then it follows from (3.83) and the strong continuity of the Feller semigroup $\{T_t\}$ that

$$\begin{aligned} &\limsup_{t\downarrow 0} \bigl\|T_t^{(p)} f - f\bigr\|_{p,\alpha} \\ &\leq \limsup_{t\downarrow 0} \Bigl(\bigl\|T_t^{(p)} f - T_t^{(p)} g\bigr\|_{p,\alpha} + \|f - g\|_{p,\alpha} + \|T_t g - g\|_{p,\alpha}\Bigr) \\ &\leq 2\varepsilon + C \cdot \limsup_{t\downarrow 0} \|T_t g - g\|_\infty \;=\; 2\varepsilon, \end{aligned}$$

where $C = [\int_{\mathrm{supp}(g)} r_\alpha(x)dx]^{1/p}$ ($C < \infty$ because the support $\mathrm{supp}(g) \subset \mathbb{R}^+$ is compact). Since ε is arbitrary, we find that $\lim_{t\downarrow 0} \bigl\|T_t^{(p)} f - f\bigr\|_{p,\alpha} = 0$ for each $f \in L^p(r_\alpha)$.

It is worth pointing out that, taking advantage of the correspondence between functions $f \in L^2(r_\alpha)$ and their index Whittaker transforms (Proposition 3.3), the action of the L^2-Markov semigroup $\{T_t^{(2)}\}$ can be explicitly written as

$$\mathcal{W}_\alpha(T_t^{(2)} f)(\tau) = e^{-t\,\psi(\tau^2 + (\frac{1}{2}-\alpha)^2)} \cdot (\mathcal{W}_\alpha f)(\tau), \qquad f \in L^2(r_\alpha), \tag{3.84}$$

where ψ is the log-Whittaker transform of the $\underset{\alpha}{\diamond}$-convolution semigroup $\{\mu_t\}$ (i.e., of the measure μ_1). Indeed, for $f \in \mathrm{C_c}(\mathbb{R}_0^+)$ and $\mu \in \mathcal{M}_\mathbb{C}(\mathbb{R}_0^+)$ we have

$$\begin{aligned}\big(\mathcal{W}_\alpha(\mathcal{T}_\alpha^\mu f)\big)(\tau) &= \int_0^\infty \boldsymbol{W}_{\alpha,i\tau}(x) \int_{\mathbb{R}_0^+} (\mathcal{T}_\alpha^y f)(x)\mu(dy)\, r_\alpha(x)dx \\ &= \int_{\mathbb{R}_0^+} \big(\mathcal{W}_\alpha(\mathcal{T}_\alpha^y f)\big)(\tau)\, \mu(dy) \\ &= \int_{\mathbb{R}_0^+}\int_{\mathbb{R}_0^+}\int_{\mathbb{R}_0^+} \boldsymbol{W}_{\alpha,i\tau}(x) k_\alpha(x,y,\xi)\, r_\alpha(x)dx\, f(\xi) r_\alpha(\xi)d\xi\, \mu(dy) \\ &= \int_{\mathbb{R}_0^+}\int_{\mathbb{R}_0^+} \boldsymbol{W}_{\alpha,i\tau}(y) \boldsymbol{W}_{\alpha,i\tau}(\xi) f(\xi) r_\alpha(\xi)d\xi\, \mu(dy) \\ &= \widehat{\mu}(\tau^2 + (\tfrac{1}{2}-\alpha)^2)\cdot(\mathcal{W}_\alpha f)(\tau)\end{aligned}$$

(the second, third and fourth equalities being obtained by changing the order of integration and using (3.36)). The identity $\big(\mathcal{W}_\alpha(\mathcal{T}_\alpha^\mu f)\big)(\tau) = \widehat{\mu}(\tau^2 + (\frac{1}{2}-\alpha)^2)\cdot(\mathcal{W}_\alpha f)(\tau)$ extends, by continuity, to all $f \in L^2(r_\alpha)$, and then Remark 3.8 yields (3.84).

The index Whittaker transform also allows us to give the following characterization of the generator of the semigroup $\{T_t^{(2)}\}$:

Proposition 3.10 *Let $\{\mu_t\}$ be a $\underset{\alpha}{\diamond}$-convolution semigroup with log-Whittaker transform ψ and let $\{T_t^{(2)}\}$ be the associated Markovian semigroup on $L^2(r_\alpha)$. Then the infinitesimal generator $(\mathcal{G}^{(2)}, \mathcal{D}(\mathcal{G}^{(2)}))$ of the semigroup $\{T_t^{(2)}\}$ is the self-adjoint operator given by*

$$\big(\mathcal{W}_\alpha(\mathcal{G}^{(2)} f)\big)(\tau) = -\psi(\tau^2 + (\tfrac{1}{2}-\alpha)^2)\cdot(\mathcal{W}_\alpha f)(\tau), \qquad f \in \mathcal{D}(\mathcal{G}^{(2)}),$$

where

$$\mathcal{D}(\mathcal{G}^{(2)}) = \biggl\{ f \in L^2(r_\alpha) \biggm| \int_0^\infty \big|\psi(\tau^2 + (\tfrac{1}{2}-\alpha)^2)\big|^2 \big|(\mathcal{W}_\alpha f)(\tau)\big|^2 \rho_\alpha(\tau)d\tau < \infty \biggr\}.$$

Proof The proof is very similar to that of the corresponding result for the Fourier transform and the generator of an ordinary convolution semigroup (see [16, Theorem 12.16]), so we only give a sketch.

It follows from (3.52) that $\langle T_t^{(2)} f, g\rangle = \langle f, T_t^{(2)} g\rangle$ for $f, g \in L^2(r_\alpha)$, i.e. $\{T_t^{(2)}\}$ is a semigroup of symmetric operators in $L^2(r_\alpha)$. It is well-known from the theory of semigroups on Hilbert spaces that the generators of self-adjoint contraction semigroups are the operators $-\mathfrak{A}$ where $\mathfrak{A}$ is a positive self-adjoint operator [42, Theorem 4.6]. Hence, in particular, the generator $(\mathcal{G}^{(2)}, \mathcal{D}(\mathcal{G}^{(2)}))$ is self-adjoint.

Letting $f \in \mathcal{D}(\mathcal{G}^{(2)})$, so that $L^2\text{-}\lim_{t\downarrow 0} \frac{1}{t}(T_t^{(2)} f - f) = \mathcal{G}^{(2)} f \in L^2(r_\alpha)$, from (3.84) we get

$$L^2\text{-}\lim_{t\downarrow 0} \frac{1}{t}\bigl(e^{-t\widetilde{\psi}} - 1\bigr) \cdot \mathcal{W}_\alpha f = \mathcal{W}_\alpha(\mathcal{G}^{(2)} f)$$

(here we write $\widetilde{\psi}(\tau) := \psi(\tau^2 + (\frac{1}{2} - \alpha)^2))$. The convergence holds almost everywhere along a sequence $\{t_n\}_{n\in\mathbb{N}}$ such that $t_n \to 0$, so we conclude that $\mathcal{W}_\alpha(\mathcal{G}^{(2)} f) = -\widetilde{\psi} \cdot \mathcal{W}_\alpha f \in L^2\bigl(\mathbb{R}^+; \rho_\alpha(\tau)d\tau\bigr)$.

Conversely, if we let $f \in L^2(r_\alpha)$ with $-\widetilde{\psi} \cdot \mathcal{W}_\alpha f \in L^2\bigl(\mathbb{R}^+; \rho_\alpha(\tau)d\tau\bigr)$, then we have

$$L^2\text{-}\lim_{t\downarrow 0} \frac{1}{t}\bigl(\mathcal{W}_\alpha(T_t^{(2)} f) - \mathcal{W}_\alpha f\bigr) = -\widetilde{\psi} \cdot \mathcal{W}_\alpha f \in L^2\bigl(\mathbb{R}^+; \rho_\alpha(\tau)d\tau\bigr)$$

and the isometry gives that $L^2\text{-}\lim_{t\downarrow 0} \frac{1}{t}\bigl(T_t^{(2)} f - f\bigr) \in L^2(r_\alpha)$, meaning that $f \in \mathcal{D}(\mathcal{G}^{(2)})$.

3.6.2 *Lévy and Gaussian Processes*

Definition 3.8 Let $\{\mu_t\}_{t\geq 0}$ be a $\underset{\alpha}{\diamond}$-convolution semigroup. An $\mathbb{R}_0^+$-valued Markov process $X = \{X_t\}_{t\geq 0}$ is said to be a $\underset{\alpha}{\diamond}$-*Lévy process* associated with $\{\mu_t\}_{t\geq 0}$ if its transition probabilities are given by

$$P\bigl[X_t \in B | X_s = x\bigr] = (\mu_{t-s} \underset{\alpha}{\diamond} \delta_x)(B), \qquad 0 \leq s \leq t,\ x \geq 0,\ B \text{ a Borel subset of } \mathbb{R}_0^+.$$

In other words, a $\underset{\alpha}{\diamond}$-Lévy process is a Feller process associated with the Feller semigroup defined in (3.80). Consequently, the general connection between Feller semigroups and Feller processes (Sect. 2.1) ensures that for each (initial) distribution $\nu \in \mathcal{P}(\mathbb{R}_0^+)$ and $\underset{\alpha}{\diamond}$-convolution semigroup $\{\mu_t\}_{t\geq 0}$ there exists a $\underset{\alpha}{\diamond}$-Lévy process X associated with $\{\mu_t\}_{t\geq 0}$ and such that $P[X_0 \in \cdot] = \nu$. Being a Feller process, any $\underset{\alpha}{\diamond}$-Lévy process is stochastically continuous and has a càdlàg modification (Proposition 2.3).

As expected (cf. Corollary 2.3 for the Kingman convolution), the $\underset{\alpha}{\diamond}$-Lévy processes are a subclass of Feller processes which includes the Shiryaev process generated by the Sturm–Liouville operator (3.2):

Proposition 3.11 *The Shiryaev process $\{Y_t\}_{t\geq 0}$ is a $\underset{\alpha}{\diamond}$-Lévy process.*

Proof For $t, x \geq 0$ let us write $p_{t,x}(dy) \equiv P_x[Y_t \in dy]$. According to Corollary 2.2, we have

$$p_{t,x}(dy) = \int_{\mathbb{R}_0^+} e^{-t(\tau^2+(\frac{1}{2}-\alpha)^2)} \mathbf{W}_{\alpha,i\tau}(x)\, \mathbf{W}_{\alpha,i\tau}(y)\, \rho_\alpha(\tau) d\tau\, r_\alpha(y) dy, \qquad t, x > 0, \tag{3.85}$$

where the integral converges absolutely. Consequently, by Proposition 3.3,

$$\widehat{p_{t,x}}(\lambda) = e^{-t\lambda} \mathbf{W}_{\alpha,\Delta_\lambda}(x), \qquad t, x \geq 0 \tag{3.86}$$

(the weak continuity of $p_{t,x}$ justifies that the equality also holds for $t = 0$ and for $x = 0$). This shows that $p_{t,x} = p_{t,0} \underset{\alpha}{\diamond} \delta_x$ where $\widehat{p_{t,0}}(\lambda) = e^{-t\lambda}$. It is clear from the properties of the index Whittaker transform of measures that $\{p_{t,0}\}_{t\geq 0}$ is a $\underset{\alpha}{\diamond}$-convolution semigroup; therefore, Y is a $\underset{\alpha}{\diamond}$-Lévy process.

Remark 3.9 The proposition above ensures that there exists a $\underset{\alpha}{\diamond}$-convolution semigroup $\{\mu_t\}$ such that $\widehat{\mu_t}(\lambda) = e^{-t\lambda}$. An interesting problem, which we do not address in this work, is to prove or disprove the existence of convolution semigroups $\{\mu_t^\beta\}_{t\geq 0}$ such that $\widehat{\mu_t^\beta}(\lambda) = e^{-t\lambda^\beta}$, where $0 < \beta < 1$. A positive answer has been given for Urbanik convolution algebras (see [190]), but the proof depends on the homogeneity axiom U5 which is not satisfied by the Whittaker convolution.

In the context of Urbanik convolution algebras, if ν is a measure with generalized characteristic function $e^{-t\lambda^\beta}$, then ν satisfies the property

$$\text{given } a, b > 0, \text{ there exists } c > 0 \text{ such that } \quad \Theta_a \nu \diamond \Theta_b \nu = \Theta_c \nu.$$

Such measures are therefore called stable measures with respect to the generalized convolution. (This is analogous to the classical notion of a strictly stable probability distribution on $\mathbb{R}$ as a measure $\nu \in \mathcal{P}(\mathbb{R})$ such that given $a, b \in \mathbb{R}$ there exists $c \in \mathbb{R}$ for which we have $aX_1 + bX_2 \overset{d}{=} cX$, where X, X_1, X_2 are mutually independent random variables with common distribution ν and $\overset{d}{=}$ denotes equality in distribution; see e.g. [60, Chapter VI] for the basics on stable distributions on $\mathbb{R}$.) For a discussion of the notion of stable measures in the context of generalized (hypergroup) convolutions not satisfying the homogeneity axiom, we refer to [211].

Since $\underset{\alpha}{\diamond}$-Lévy processes are Feller processes, they can be characterized as the solution of the corresponding martingale problem, as stated in Theorem 2.1. The next proposition provides some additional equivalent martingale characterizations of $\underset{\alpha}{\diamond}$-Lévy processes. For an $\mathbb{R}_0^+$-valued càdlàg process $X = \{X_t\}_{t\geq 0}$, a linear operator $A : \mathcal{D} \longrightarrow \mathrm{C}(\mathbb{R}_0^+)$ with domain $\mathcal{D} \subset \mathrm{C}(\mathbb{R}_0^+)$ and a function $f \in \mathcal{D}$, we introduce the notation $Z_X^{A,f} = \{Z_{X,t}^{A,f}\}_{t\geq 0}$, where

$$Z_{X,t}^{A,f} := f(X_t) - f(X_0) - \int_0^t (Af)(X_s)\, ds. \tag{3.87}$$

Proposition 3.12 *Let $\{\mu_t\}_{t\geq 0}$ be a $\underset{\alpha}{\diamond}$-convolution semigroup with log-Whittaker transform ψ and let $(A, \mathcal{D}_A)$ be the infinitesimal generator of the Feller semigroup determined by $\{\mu_t\}$ (cf. Proposition 3.8). Let X be an $\mathbb{R}_0^+$-valued càdlàg Markov process. The following assertions are equivalent:*

(i) *X is a $\underset{\alpha}{\diamond}$-Lévy process associated with $\{\mu_t\}$;*
(ii) *$\{e^{t\psi(\lambda)}\mathbf{W}_{\alpha,\Delta_\lambda}(X_t)\}_{t\geq 0}$ is a martingale for each $\lambda \geq 0$;*
(iii) *$\bigl\{\mathbf{W}_{\alpha,\Delta_\lambda}(X_t) - \mathbf{W}_{\alpha,\Delta_\lambda}(X_0) + \psi(\lambda)\int_0^t \mathbf{W}_{\alpha,\Delta_\lambda}(X_s)\, ds\bigr\}_{t\geq 0}$ is a martingale for each $\lambda \geq 0$;*
(iv) *$Z_X^{A,\mathbf{W}_{\alpha,\Delta_\lambda}(\cdot)}$ is a martingale for each $\lambda \geq 0$;*
(v) *$Z_X^{A,f}$ is a martingale for each $f \in \mathcal{D}_A$.*

Proof The proof is identical to that of the corresponding martingale characterization for Lévy processes on commutative hypergroups [163, Theorem 3.4].

A $\underset{\alpha}{\diamond}$-convolution semigroup $\{\mu_t\}_{t\geq 0}$ such that μ_1 is a $\underset{\alpha}{\diamond}$-Gaussian measure will be called a *$\underset{\alpha}{\diamond}$-Gaussian convolution semigroup*, and a $\underset{\alpha}{\diamond}$-Lévy process associated with a $\underset{\alpha}{\diamond}$-Gaussian convolution semigroup is said to be a *$\underset{\alpha}{\diamond}$-Gaussian process*. An alternative characterization of $\underset{\alpha}{\diamond}$-Gaussian convolution semigroups (which in particular implies that any $\underset{\alpha}{\diamond}$-Gaussian convolution semigroup is fully composed of $\underset{\alpha}{\diamond}$-Gaussian measures) is given in the next lemma.

Lemma 3.8 *Let $\mu \in \mathcal{P}_{\alpha,\mathrm{id}}$ and let $\{\mu_t\}$ be the $\underset{\alpha}{\diamond}$-convolution semigroup $\{\mu_t\}$ such that $\mu_1 = \mu$. Then, the following conditions are equivalent:*

(i) *μ is a $\underset{\alpha}{\diamond}$-Gaussian measure;*
(ii) *$\lim_{t\downarrow 0} \frac{1}{t}\mu_t[\varepsilon, \infty) = 0$ for every $\varepsilon > 0$;*
(iii) *$\lim_{t\downarrow 0} \frac{1}{t}(\mu_t \underset{\alpha}{\diamond} \delta_x)\bigl(\mathbb{R}_0^+ \setminus (x-\varepsilon, x+\varepsilon)\bigr) = 0$ for every $x \geq 0$ and $\varepsilon > 0$.*

Proof ***(i)$\Longrightarrow$(ii):*** Let $\{t_n\}_{n\in\mathbb{N}}$ be a sequence such that $t_n \to 0$ as $n \to \infty$, and let $\nu_n = \mathbf{e}_\alpha\bigl(\frac{1}{t_n}\mu_{t_n}\bigr)$. We have

$$\lim_{n\to\infty} \widehat{\nu_n}(\lambda) = \lim_{n\to\infty} \exp\left[\frac{1}{t_n}\bigl(\widehat{\mu_1}(\lambda)^{t_n} - 1\bigr)\right] = \widehat{\mu_1}(\lambda), \qquad \lambda > 0 \tag{3.88}$$

and therefore, by Proposition 3.4(iv), $\nu_n \overset{w}{\longrightarrow} \mu_1$ as $n \to \infty$. From this it follows, cf. [196], that if π_n denotes the restriction of $\frac{1}{t_n}\mu_{t_n}$ to $[a, b) \setminus \mathcal{V}_a$, then $\{\pi_n\}$ is relatively compact; if π is a limit point, then $\mathbf{e}_\alpha(\pi)$ is a divisor of μ_1. Since μ_1 is Gaussian, $\mathbf{e}_\alpha(\pi) = \delta_a$, hence π must be the zero measure, showing that (ii) holds.

(ii)$\Longrightarrow$(i): As in (3.88), for $\lambda > 0$ we have

$$\begin{aligned}\widehat{\mu_1}(\lambda) &= \lim_{n\to\infty} \exp\biggl[\frac{1}{t_n}\int_{[a,b)} \bigl(\boldsymbol{W}_{\alpha,\Delta_\lambda}(x) - 1\bigr)\mu_{t_n}(dx)\biggr] \\ &= \lim_{n\to\infty} \exp\biggl[\frac{1}{t_n}\int_{\mathcal{V}_a} \bigl(\boldsymbol{W}_{\alpha,\Delta_\lambda}(x) - 1\bigr)\mu_{t_n}(dx)\biggr],\end{aligned}$$

where the second equality is due to (ii), noting that $\frac{1}{t_n}\int_{[a,b)\setminus\mathcal{V}_a}(\boldsymbol{W}_{\alpha,\Delta_\lambda}(x) - 1)\mu_{t_n}(dx) \leq \frac{2}{t}\mu_{t_n}\bigl([a,b)\setminus\mathcal{V}_a\bigr)$. Given that $\nu_n = \mathbf{e}_\alpha\bigl(\frac{1}{t_n}\mu_{t_n}\bigr) \xrightarrow{w} \mu_1$, we have (again, see [196])

$$\widehat{\mu}_1(\lambda) = \exp\biggl[\int_{(a,b)} \bigl(\boldsymbol{W}_{\alpha,\Delta_\lambda}(x) - 1\bigr)\,\eta(dx)\biggr], \qquad \lambda > 0$$

for some σ-finite measure η on (a,b) which, by the above, vanishes on the complement of any neighbourhood of the point a. Therefore, μ_1 is Gaussian.

(ii) $\Longleftrightarrow$ (iii): To prove the nontrivial direction, assume that (ii) holds, and fix $x, \varepsilon > 0$ with $0 < x < \varepsilon$. Write $E_\varepsilon = \mathbb{R}_0^+ \setminus (x-\varepsilon, x+\varepsilon)$, and let $\mathbb{1}_\varepsilon$ denote its indicator function. We start the proof by establishing an upper bound for the function $(\mathcal{T}_\alpha^x \mathbb{1}_\varepsilon)(y)$, with $y > 0$ small. Using the estimate (3.39), together with the inequalities

$$\begin{aligned}&\Bigl(1 + \tfrac{x}{\xi} + \tfrac{y}{\xi}\Bigr)^{2\alpha} \leq (1+\xi^{-1})(1+x+\delta), \\ &\qquad\qquad\qquad\qquad\qquad\qquad\qquad\qquad (x,\xi > 0,\ y < \delta), \\ &\tfrac{|x-\xi|^5}{(8xy\xi)^{\frac{5}{2}}} \exp\Bigl(-\tfrac{|x-\xi|^2}{8xy\xi}\Bigr) \leq 1\end{aligned}$$

it is easily seen that

$$\begin{aligned} q(x,y,\xi) r_\alpha(\xi) &\leq C_1 y^2 |x-\xi|^{-5} \xi(1+\xi) \exp\biggl(-\frac{1}{2\xi} - \frac{y}{4x\xi} - \frac{(x-\xi)^2}{8xy\xi}\biggr) \\ &\leq C_2 y^2 \xi(1+\xi)\exp\biggl(-\frac{1}{2\xi} - \frac{(x-\xi)^2}{8\delta x\xi}\biggr) \qquad (y \leq \delta,\ \xi \in E_\varepsilon),\end{aligned}$$

where the constants $C_1, C_2 > 0$ depends only on x, δ and ε. Consequently, for $y < \delta$ we have

$$\begin{aligned}(\mathcal{T}_\alpha^x \mathbb{1}_\varepsilon)(y) &= \int_{E_\varepsilon} k_\alpha(x, y, \xi) r_\alpha(\xi) d\xi \\ &\leq C_2\, y^2 \int_0^\infty \xi(1+\xi) \exp\Bigl(-\frac{1}{2\xi} - \frac{(x-\xi)^2}{8\delta x \xi}\Bigr) d\xi \\ &\leq C_3\, y^2, \end{aligned} \tag{3.89}$$

the convergence of the integral justifying that the last inequality holds for a possibly larger constant C_2.

Let $\lambda > 0$ be arbitrary. If $\delta > 0$ is sufficiently small, then from (3.75) and (3.89) it follows that

$$\frac{(\mathcal{T}_\alpha^x \mathbb{1}_\varepsilon)(y)}{1 - \boldsymbol{W}_{\alpha,\Delta_\lambda}(y)} \leq 2C_3 \lambda^{-1} y, \qquad y \leq \delta,$$

and therefore, there exists $\delta' > 0$ (which depends on λ) such that $(\mathcal{T}_\alpha^x \mathbb{1}_\varepsilon)(y) \leq 1 - \boldsymbol{W}_{\alpha,\Delta_\lambda}(y)$ for all $y \in [0, \delta')$. We then estimate

$$\begin{aligned}\frac{1}{t}(\mu_t \underset{\alpha}{\diamond} \delta_x)(E_\varepsilon) &= \frac{1}{t} \int_{\mathbb{R}_0^+} (\mathcal{T}_\alpha^x \mathbb{1}_\varepsilon)(y) \mu_t(dy) \\ &\leq \frac{1}{t} \int_{[0,\delta')} \bigl(1 - \boldsymbol{W}_{\alpha,\Delta_\lambda}(y)\bigr) \mu_t(dy) + \frac{1}{t} \mu_t[\delta', \infty) \\ &\leq \frac{1}{t} \int_{\mathbb{R}_0^+} \bigl(1 - \boldsymbol{W}_{\alpha,\Delta_\lambda}(y)\bigr) \mu_t(dy) + \frac{1}{t} \mu_t[\delta', \infty) \\ &= \frac{1}{t}\bigl(1 - \widehat{\mu_t}(\lambda)\bigr) + \frac{1}{t} \mu_t[\delta', \infty). \end{aligned}$$

Since we are assuming that (ii) holds and we know that $\lim_{t \downarrow 0} \frac{1}{t}\bigl(1 - \widehat{\mu_t}(\lambda)\bigr) = -\log \widehat{\mu}(\lambda)$ (cf. proof of Proposition 3.7), the above inequality gives

$$\limsup_{t \downarrow 0} \frac{1}{t} (\mu_t \underset{\alpha}{\diamond} \delta_x)(E_\varepsilon) \leq -\log \widehat{\mu}(\lambda).$$

By the properties of the index Whittaker transform, the right-hand side is continuous and vanishes for $\lambda = 0$, so from the arbitrariness of λ we see that $\lim_{t \downarrow 0} \frac{1}{t}(\mu_t \underset{\alpha}{\diamond} \delta_x)(E_\varepsilon) = 0$, as desired.

Denoting, as in (3.85), the law of the Shiryaev process started at x by $p_{t,x}$, it follows from Propositions 2.4 and 2.7 that $\lim_{t \downarrow 0} \frac{1}{t} p_{t,x}\bigl(\mathbb{R}_0^+ \setminus (x - \varepsilon, x + \varepsilon)\bigr) = 0$ for any $x \geq 0$ and $\varepsilon > 0$, meaning that the Shiryaev process is a $\underset{\alpha}{\diamond}$-Gaussian process.

It turns out that, as a consequence of the previous lemma, any other $\underset{\alpha}{\diamond}$-Gaussian process is also a one-dimensional diffusion:

Corollary 3.2 *Let $X = \{X_t\}_{t\geq 0}$ be a $\underset{\alpha}{\diamond}$-Gaussian process associated with the $\underset{\alpha}{\diamond}$-Gaussian convolution semigroup $\{\mu_t\}_{t\geq 0}$. Then:*

(i) *X has a modification whose paths are a.s. continuous;*
(ii) *Let $(\mathcal{G}, \mathcal{D}(\mathcal{G}))$ be the infinitesimal generator of the Feller semigroup determined by $\{\mu_t\}$. Then $\mathcal{G}$ is a local operator, i.e., $(\mathcal{G}f)(x) = (\mathcal{G}g)(x)$ whenever $f, g \in \mathcal{D}(\mathcal{G})$ and $f = g$ on some neighborhood of $x \geq 0$.*

Proof We know from Lemma 3.8 that the associated $\underset{\alpha}{\diamond}$-Gaussian convolution semigroup is such that $\lim_{t\downarrow 0} \frac{1}{t}(\mu_t \underset{\alpha}{\diamond} \delta_x)\big(\mathbb{R}_0^+ \setminus (x-\varepsilon, x+\varepsilon)\big) = 0$ for every $x \geq 0$ and $\varepsilon > 0$. Using Proposition 2.4, we conclude that the càdlàg modification of X has a.s. continuous sample paths. The locality of the generator is then proved by applying Proposition 2.5 to the $\mathbb{R}$-valued process $\widetilde{X} = \{\widetilde{X}_t\}_{t\geq 0}$ which is the extension of X obtained by setting $\widetilde{X}_t(\omega) = x$ whenever the initial distribution is $\nu = \delta_x$, $x < 0$.

3.6.3 Some Auxiliary Results on the Whittaker Translation

In this subsection we return to the Whittaker translation operator (3.40), which we will now interpret as an operator on the space

$$\mathfrak{X} := \left\{ f \in L_0(\mathbb{R}^+) \;\middle|\; \begin{array}{c} |f(x)| \leq b_1 \exp\!\Big(\frac{1}{x} + b_2(x^{-\beta} + x^{\beta})\Big) \\ \text{for some } b_1, b_2 \geq 0 \text{ and } 0 \leq \beta < 1 \end{array} \right\}, \tag{3.90}$$

being $L_0(\mathbb{R}^+)$ the space of Lebesgue measurable functions $f : \mathbb{R}^+ \longrightarrow \mathbb{C}$. The goal of this digression is to determine some properties which will be useful for introducing (in the next subsection) the notion of moment functions with respect to the Whittaker convolution.

We first note that the condition $f \in \mathfrak{X}$ ensures that for each $x, y > 0$ the Whittaker translation $(\mathcal{T}_\alpha^y f)(x) = \int_0^\infty f(\xi) k_\alpha(x, y, \xi)\, r_\alpha(\xi) d\xi$ exists as an absolutely convergent integral, as can be verified using (3.39).

Lemma 3.9 *Fix $y, M > 0$. Let $f \in \mathfrak{X}$ and $p \in \mathbb{N}_0$. Then, for each $\varepsilon > 0$ there exists $\delta, M_0 > 0$ such that*

$$\int_{E_M} |f(\xi)| \left|\frac{\partial^p}{\partial x^p} k_\alpha(x, y, \xi)\right| r_\alpha(\xi)\, d\xi < \varepsilon \qquad \textit{for all } x \in (0, \delta] \textit{ and } M \geq M_0,$$

where $E_M = (0, \frac{1}{M}] \cup [M, \infty)$.

Proof Fix $k \geq -\frac{1}{2} + \max\{\alpha, 0\}$. Note that if $\sigma \in \mathbb{C}$ then (after a new choice of b_2 and β) the function $\xi \mapsto \xi^\sigma f(\xi)$ also belongs to $\mathfrak{X}$. Let $\delta < \frac{y}{4}$. If $|\xi - y| \geq 2\delta$ and $x \leq \delta$, using (3.39), the boundedness of the function $|t|^{k+\frac{1}{2}} e^{-|t|}$ and the inequalities

$$\frac{(x+y+\xi)^{2\alpha}}{|x+\xi-y|^{2k+1}} \leq \left(1+\frac{2y}{\delta}\right)^{2\alpha}\left(\frac{\delta}{4}\right)^{2\alpha-2k-1}, \qquad \alpha \geq 0,$$

$$\frac{(x+y+\xi)^{2\alpha}}{|x+\xi-y|^{2k+1}} \leq \frac{y^{2\alpha}}{\delta^{2k+1}}, \qquad \alpha \leq 0,$$

we find that

$$\begin{aligned} x^{-k}&|f(\xi)|\, k_\alpha(x,y,\xi)\, r_\alpha(\xi) \\ &\leq \frac{C}{(xy\xi)^{k+\frac{1}{2}}}(x+y+\xi)^{2\alpha} \exp\left(b_2(\xi^{-\beta}+\xi^\beta) - \frac{(x+\xi-y)^2}{4xy\xi}\right) \\ &\leq C\exp\left(b_2(\xi^{-\beta}+\xi^\beta) - \frac{(x+\xi-y)^2}{8xy\xi}\right), \qquad |\xi-y| \geq 2\delta,\ x \leq \delta, \end{aligned} \tag{3.91}$$

where C depends only on y and δ. Since $0 \leq \beta < 2$, the integral $\int_0^\infty \exp\Big\{b_2(\xi^{-\beta} + \xi^\beta) - \frac{(x+\xi-y)^2}{8xy\xi}\Big\} d\xi$ converges uniformly in $x \in [0, \delta]$. Combining this with the inequality (3.91), we conclude that $M_0 > 0$ can be chosen so large that

$$\int_{E_M} x^{-k}|f(\xi)|\, k_\alpha(x,y,\xi)\, r_\alpha(\xi)\, d\xi < \delta \qquad \text{for all } 0 < x < \left(\tfrac{\delta}{2}\right)^{1/2} \text{ and } M \geq M_0. \tag{3.92}$$

Using the identity (3.61), one can deduce (by induction) that the function $f(\xi)\,\frac{\partial^p}{\partial x^p} k_\alpha(x,y,\xi)$ can be rewritten as a finite sum of the form $\sum_j C_j x^{-k_j} g_j(\xi)\, k_{\alpha_j}(x,y,\xi)$, where $g_j \in \mathfrak{X}$ and $k_j \geq -1 + \max\{2\alpha_j, 0\}$ for all j. Therefore, the conclusion of the lemma follows from (3.92).

Lemma 3.10 *Let* $f \in \mathfrak{X} \cap \mathrm{C}^2(\mathbb{R}^+)$ *and* $y > 0$. *Then:*

(i) $\lim_{x\to 0} (\mathcal{T}_\alpha^y f)(x) = f(y)$;
(ii) $\lim_{x\to 0} p_\alpha(x) \frac{\partial}{\partial x}(\mathcal{T}_\alpha^y f)(x) = 0.$

Proof

(i) We will first show that it is enough to prove the result for $f \in \mathrm{C}^2_{\mathrm{c}}(\mathbb{R}^+)$. Suppose that part *(i)* of the lemma holds for $f \in \mathrm{C}^2_{\mathrm{c}}(\mathbb{R}^+)$. Let $g \in \mathfrak{X} \cap \mathrm{C}^2(\mathbb{R}^+)$ and $\varepsilon, M > 0$; then, choose $\delta > 0$ and $g_{\mathrm{c}} \in \mathrm{C}^2_{\mathrm{c}}(\mathbb{R}^+)$ such that $g(\xi) = g_{\mathrm{c}}(\xi)$ for all $\xi \in [\frac{1}{M}, M]$ and

$$|(\mathcal{T}^y_\alpha g)(x) - (\mathcal{T}^y_\alpha g_{\mathrm{c}})(x)| < \varepsilon \qquad \text{for all } x \in (0, \delta]$$

(to see that this is possible, apply the case $p = 0$ of Lemma 3.9). If $y \in [\frac{1}{M}, M]$, we obtain

$$\limsup_{x\to 0} |(\mathcal{T}^y_\alpha g)(x) - g(y)| \leq \varepsilon + \lim_{x\to 0} |(\mathcal{T}^y_\alpha g_{\mathrm{c}})(x) - g_{\mathrm{c}}(y)| = \varepsilon.$$

As M and ε are arbitrary, we conclude that $\lim_{x\to 0}(\mathcal{T}^y_\alpha g)(x) = g(y)$ for all $y > 0$ and $g \in \mathfrak{X} \cap \mathrm{C}^2(\mathbb{R}^+)$.

Let us now prove that $\lim_{x\to 0}(\mathcal{T}^y_\alpha f)(x) = f(y)$ holds for $f \in \mathrm{C}^2_{\mathrm{c}}(\mathbb{R}^+)$. Using the integral representation for $k_\alpha(x, y, \xi)$ given in (3.55), we write

$$\begin{aligned}(\mathcal{T}^y_\alpha f)(x) &= \int_0^\infty f(\xi) \int_0^\infty \boldsymbol{W}_{\alpha,i\tau}(x)\, \boldsymbol{W}_{\alpha,i\tau}(y)\, \boldsymbol{W}_{\alpha,i\tau}(\xi)\, \rho_\alpha(\tau) d\tau\, r_\alpha(\xi) d\xi \\ &= \int_0^\infty (\mathcal{W}_\alpha f)(\tau)\, \boldsymbol{W}_{\alpha,i\tau}(x)\, \boldsymbol{W}_{\alpha,i\tau}(y)\, \rho_\alpha(\tau) d\tau,\end{aligned} \tag{3.93}$$

where the second equality is obtained by changing the order of integration, which is valid because f has compact support. It was noted above that the index Whittaker transform $\mathcal{W}_\alpha$ is a particular case of the Sturm–Liouville integral transform (2.21)–(2.22), thus it follows from Lemma 2.4(b) that $(\mathcal{W}_\alpha f)(\tau)\, \boldsymbol{W}_{\alpha,i\tau}(y) \in L^1\big(\mathbb{R}^+; \rho_\alpha(\tau) d\tau\big)$. Recalling also that $|\boldsymbol{W}_{\alpha,i\tau}(x)| \leq 1$ $(x, \tau \geq 0)$ and $|\boldsymbol{W}_{\alpha,i\tau}(0)| = 1$, cf. (3.67) and (3.74), by dominated convergence we obtain that

$$\lim_{x\to 0} (\mathcal{T}^y_\alpha f)(x) = \int_0^\infty (\mathcal{W}_\alpha f)(\tau) \Big(\lim_{x\to 0} \boldsymbol{W}_{\alpha,i\tau}(x)\Big)\, \boldsymbol{W}_{\alpha,i\tau}(y)\, \rho_\alpha(\tau) d\tau = f(y),$$

concluding the proof.

(ii) Identical reasoning as in part (i) shows that it is enough to prove the result for $f \in \mathrm{C}^2_{\mathrm{c}}(\mathbb{R}^+)$. Taking $f \in \mathrm{C}^2_{\mathrm{c}}(\mathbb{R}^+)$, differentiation of (3.93) under the integral sign gives

$$p_\alpha(x) \frac{\partial}{\partial x} (\mathcal{T}^y_\alpha f)(x) = \int_0^\infty (\mathcal{W}_\alpha f)(\tau) \big(p_\alpha \boldsymbol{W}'_{\alpha,i\tau}\big)(x)\, \boldsymbol{W}_{\alpha,i\tau}(y)\, \rho_\alpha(\tau) d\tau.$$

If we now apply (3.74), by dominated convergence we conclude that $\lim_{x\to 0} p_\alpha(x) \frac{\partial}{\partial x}(\mathcal{T}_\alpha^y f)(x) = 0$.

3.6.4 Moment Functions

Moment functions for generalized convolutions are functions having the same additivity property which is satisfied by the monomials under the classical convolution. Such functions have been applied to the study of limit theorems for hypergroup convolution structures (see the discussion in [19, pp. 530–531]). Let us introduce, in a similar way, the notion of moment functions with respect to the Whittaker convolution:

Definition 3.9 The sequence of functions $\{\varphi_k\}_{k=1,\dots,n}$ is said to be a $\underset{\alpha}{\diamond}$*-moment sequence (of length n)* if $\varphi_k \in \mathfrak{X}$ for $k = 1, \dots, n$ (cf. (3.90)) and

$$(\mathcal{T}_\alpha^y \varphi_k)(x) = \sum_{j=0}^{k} \binom{k}{j} \varphi_j(x)\varphi_{k-j}(y) \qquad (k = 1, \dots, n;\ x, y \geq 0), \qquad (3.94)$$

where $\varphi_0(x) := 1$ $(x \geq 0)$.

It is worth recalling that for $x, y > 0$ the left-hand side of (3.94) is given by the integral $\int_0^\infty \varphi_k(\xi) k_\alpha(x, y, \xi)\, r_\alpha(\xi) d\xi$, which converges absolutely. This actually implies that $\underset{\alpha}{\diamond}$-moment functions are necessarily smooth:

Lemma 3.11 *If $\{\varphi_k\}_{k=1,\dots,n}$ is a $\underset{\alpha}{\diamond}$-moment sequence, then $\varphi_k \in \mathrm{C}^\infty(\mathbb{R}^+)$ for all k.*

Proof Let $M > 0$ and $1 \leq k \leq n$. Let $f \in \mathrm{C}_\mathrm{c}^\infty(2M, 3M)$ be such that $\int_{2M}^{3M} f(x)\, r_\alpha(x) dx = 1$, and set $f(x) = 0$ for $x \notin (2M, 3M)$. Then

$$\begin{aligned}\sum_{j=0}^{k} \binom{k}{j} \varphi_j(y) \int_0^\infty \varphi_{k-j}(x) f(x)\, r_\alpha(x) dx &= \int_0^\infty (\mathcal{T}_\alpha^y \varphi_k)(x)\, f(x)\, r_\alpha(x) dx \\ &= \int_0^\infty \varphi_k(x)\, (\mathcal{T}_\alpha^y f)(x)\, r_\alpha(x) dx,\end{aligned}$$

where the second equality follows from the identity (3.52), which is easily seen to hold also for $f \in \mathrm{C}_\mathrm{c}^\infty(\mathbb{R}^+)$ and $g \in \mathfrak{X}$. Hence if we prove that the right-hand side is an infinitely differentiable function of $0 < y < M$, then by induction it follows that each $\varphi_k \in \mathrm{C}^\infty(0, M)$ and, by arbitrariness of M, $\varphi_k \in \mathrm{C}^\infty(\mathbb{R}^+)$.

By (3.39), we have

$$
\begin{aligned}
&(\mathcal{T}_\alpha^y f)(x) \\
&\leq C_1 \|f\|_\infty (xy)^{-\frac{1}{2}} \int_{2M}^{3M} \xi^{-\frac{1}{2}-2\alpha} (x+y+\xi)^{2\alpha} \exp\left(\frac{1}{2y} - \frac{1}{2\xi} - \frac{x}{4y\xi} - \frac{(y-\xi)^2}{4xy\xi}\right) d\xi \\
&\leq C_2 \|f\|_\infty (xy)^{-\frac{1}{2}} (1+x^{2\alpha}) \exp\left(\frac{1}{2y} - \frac{x}{12M^2} - \frac{1}{12x}\right),
\end{aligned}
$$

where C_1 and C_2 are constants depending only on M. Since $\varphi_k \in \mathfrak{X}$, we find that

$$
\varphi_k(x)\, (\mathcal{T}_\alpha^y f)(x)\, r_\alpha(x) \leq C\, (xy)^{-\frac{1}{2}}\, (1+x^{-2\alpha}) \exp\left(\frac{1}{2y} + b_2(x^\beta + x^{-\beta}) - \frac{x}{12M^2} - \frac{1}{12x}\right), \tag{3.95}
$$

where $C > 0$, $b_2 \geq 0$ and $0 \leq \beta < 1$ do not depend on y. Denoting the right-hand side of (3.95) by $J(x, y)$, it is easily seen that the integral $\int_0^\infty J(x, y) dx$ converges locally uniformly and, therefore, $\int_0^\infty \varphi_k(x)\, (\mathcal{T}_\alpha^y f)(x)\, \mathrm{m}(x) dx$ is a continuous function of $0 < y < M$. Using the identity (3.61) (with x and y interchanged) and similar arguments, one can derive an upper bound for the derivatives $\frac{\partial^n}{\partial y^n} (\mathcal{T}_\alpha^y f)(x)$ ($n = 1, 2, \ldots$) and then deduce that $\int_0^\infty \varphi_k(x)\, (\mathcal{T}_\alpha^y f)(x)\, r_\alpha(x) dx$ is n times continuously differentiable.

Proposition 3.13 *$\{\varphi_k\}_{k=1,\ldots,n}$ is a $\underset{\alpha}{\diamond}$-moment sequence if and only if there exist $\lambda_1, \ldots, \lambda_n \in \mathbb{R}$ such that*

$$
\mathcal{A}_\alpha \varphi_k(x) = \sum_{j=1}^{k} \binom{k}{j} \lambda_j \varphi_{k-j}(x), \qquad \varphi_k(0) = 0, \qquad (p_\alpha \varphi_k')(0) = 0, \tag{3.96}
$$

for $k = 1, \ldots, n$, where $\varphi_0 \equiv 1$ and $\mathcal{A}_\alpha$ is the differential operator (3.2).

Proof Let $\{\varphi_k\}_{k=1,\ldots,n}$ be a $\underset{\alpha}{\diamond}$-moment sequence. First we will show that $\varphi_k \in \mathrm{C}^\infty(\mathbb{R}^+) \cap \mathrm{C}^1(\mathbb{R}_0^+)$ with $\varphi_k(0) = (p_\alpha \varphi_k')(0) = 0$. By Lemma 3.11, $\varphi_k \in \mathfrak{X} \cap \mathrm{C}^\infty(\mathbb{R}^+)$. It thus follows from Lemma 3.10 that for fixed $y > 0$ we have $\lim_{x \to 0} (\mathcal{T}_\alpha^y \varphi_k)(x) = \varphi_k(y)$ and $\lim_{x \to 0} p_\alpha(x) \frac{\partial}{\partial x} (\mathcal{T}_\alpha^y \varphi_k)(x) = 0$. If we rewrite (3.94) as

$$
\varphi_k(x) = (\mathcal{T}_\alpha^y \varphi_k)(x) - \sum_{j=0}^{k-1} \binom{k}{j} \varphi_j(x) \varphi_{k-j}(y) \tag{3.97}
$$

and let $x \to 0$ on the right-hand side, we deduce (by induction on k) that $\lim_{x \to 0} \varphi_k(x) = 0$ for all k. After differentiating both sides of (3.97), we similarly find that $\lim_{x \to 0} (p_\alpha \varphi_k')(x) = 0$ for each k.

We now prove that φ_k satisfies $\mathcal{A}_\alpha \varphi_k = \sum_{j=1}^{k} \binom{k}{j} \lambda_j \varphi_{k-j}$, omitting the details which are similar to the proof of [184, Theorem 4.5]. We know from (3.58) that $\mathcal{A}_{\alpha,x} k_\alpha(x, y, \xi) = \mathcal{A}_{\alpha,y} k_\alpha(x, y, \xi)$. Moreover, from the identity (3.61) it follows that the integral defining $(\mathcal{T}_\alpha^y \varphi_k)(x)$ can be differentiated under the integral sign. Therefore, the right-hand side of (3.94) is, for each k, a solution of $\mathcal{A}_{\alpha,x} u = \mathcal{A}_{\alpha,y} u$, i.e.

$$\sum_{j=0}^{k} \binom{k}{j} (\mathcal{A}_\alpha \varphi_j)(x)\, \varphi_{k-j}(y) = \sum_{j=0}^{k} \binom{k}{j} \varphi_j(x)\, (\mathcal{A}_\alpha \varphi_{k-j})(y).$$

Assume by induction that $\mathcal{A}_\alpha \varphi_\ell(x) = \sum_{j=1}^{\ell} \binom{\ell}{j} \lambda_j \varphi_{\ell-j}(x)$ for $\ell = 1, \ldots, k-1$. Using the induction hypothesis and rearranging the terms in a suitable way, we find that

$$(\mathcal{A}_\alpha \varphi_k)(x) - \sum_{j=1}^{k-1} \lambda_j \varphi_{k-j}(x) = (\mathcal{A}_\alpha \varphi_k)(y) - \sum_{j=1}^{k-1} \lambda_j \varphi_{k-j}(y) \qquad \text{for all } x, y > 0$$

and, consequently,

$$(\mathcal{A}_\alpha \varphi_k)(x) - \sum_{j=1}^{k-1} \lambda_j \varphi_{k-j}(x) = \lambda_k$$

for some $\lambda_k \in \mathbb{R}$.

For the converse, suppose that $\{\varphi_k\}_{k=1,\ldots,n}$ are solutions of (3.96). Integrating, we obtain $\varphi_k(x) = -\int_0^x \frac{1}{p_\alpha(y)} \int_0^y r_\alpha(\xi) \bigl[\sum_{j=1}^{k} \binom{k}{j} \lambda_j \varphi_{k-j}(\xi)\bigr] d\xi\, dy$. Straightforward bounds on this integral yield that $\varphi_k \in \mathfrak{X}$ (see the proof of Proposition 3.15). We can assume by induction that

$$(\mathcal{T}_\alpha^y \varphi_r)(x) = \sum_{\ell=0}^{r} \binom{r}{\ell} \varphi_\ell(x) \varphi_{r-\ell}(y) \qquad \text{for } r = 1, \ldots, k-1$$

and the goal is to prove that $\Phi_{k,y}(x) := (\mathcal{T}_\alpha^y \varphi_k)(x) - \sum_{j=0}^{k} \binom{k}{j} \varphi_j(x) \varphi_{k-j}(y)$ vanishes identically. We compute

$$\mathcal{A}_{\alpha,x}\bigl[\Phi_{k,y}(x)\bigr] = \mathcal{T}_\alpha^y (\mathcal{A}_\alpha \varphi_k)(x) - \sum_{j=0}^{k} \binom{k}{j} (\mathcal{A}_\alpha \varphi_j)(x) \varphi_{k-j}(y)$$

$$= \sum_{j=1}^{k} \binom{k}{j} \lambda_j (\mathcal{T}_\alpha^y \varphi_{k-j})(x) - \sum_{j=0}^{k} \binom{k}{j} \varphi_{k-j}(y) \sum_{\ell=1}^{k} \binom{j}{\ell} \lambda_\ell \varphi_{j-\ell}(x)$$

$$= \sum_{j=1}^{k} \binom{k}{j} \lambda_j \sum_{\ell=0}^{k-j} \binom{k-j}{\ell} \varphi_\ell(x)\varphi_{k-j-\ell}(y) - \sum_{j=0}^{k} \binom{k}{j} \varphi_{k-j}(y) \sum_{\ell=1}^{k} \binom{j}{\ell} \lambda_\ell \varphi_{j-\ell}(x)$$

$$= 0.$$

Here, the first equality follows from the identity $\mathcal{T}_\alpha^y(\mathcal{A}_\alpha\varphi)(x) \equiv \mathcal{A}_{\alpha,x}(\mathcal{T}_\alpha^y\varphi)(x)$, which can be verified using (3.58) and integration by parts; the second equality applies (3.96); the induction hypothesis gives the third equality; and the last step is obtained by rearranging the sums. Furthermore, $\lim_{x\to 0} \Phi_{k,y}(x) = \lim_{x\to 0} p_\alpha(x) \frac{\partial}{\partial x} \Phi_{k,y}(x) = 0$ (due to Lemma 3.10); by uniqueness of solution, $\Phi_{k,y}(x) \equiv 0$, showing that (3.94) holds.

The functions $\widetilde{\varphi}_{\alpha,k}$ $(k \in \mathbb{N})$ defined as the unique solution of

$$\begin{aligned} \mathcal{A}_\alpha \widetilde{\varphi}_{\alpha,k}(x) &= -k(1-2\alpha)\widetilde{\varphi}_{\alpha,k-1}(x) - k(k-1)\widetilde{\varphi}_{\alpha,k-2}(x), \\ \widetilde{\varphi}_{\alpha,k}(0) &= 0, \\ (p_\alpha \widetilde{\varphi}'_{\alpha,k})(0) &= 0 \end{aligned} \tag{3.98}$$

(where $\widetilde{\varphi}_{\alpha,-1}(x) := 0$ and $\widetilde{\varphi}_{\alpha,0}(x) := 1$) are said to be the *canonical* $\underset{\alpha}{\diamond}$*-moment functions*. By integration of the differential equation, we find the explicit recursive expression

$$\widetilde{\varphi}_{\alpha,k}(x) = k \int_0^x \frac{1}{p_\alpha(y)} \int_0^y r_\alpha(\xi) \bigl[(1-2\alpha)\widetilde{\varphi}_{\alpha,k-1}(\xi) + (k-1)\widetilde{\varphi}_{\alpha,k-2}(\xi)\bigr]\, d\xi\, dy. \tag{3.99}$$

Moreover, as a consequence of the uniqueness of solution for (3.98) and the Laplace representation (3.63), the canonical moment functions can also be represented as

$$\begin{aligned} \widetilde{\varphi}_{\alpha,k}(x) &= \frac{\partial^k}{\partial \sigma^k}\Big|_{\sigma=\frac{1}{2}-\alpha} \boldsymbol{W}_{\alpha,\sigma}(x) \\ &= \int_{-\infty}^{\infty} \biggl[\frac{\partial^k(e^{\sigma s})}{\partial \sigma^k}\Big|_{\sigma=\frac{1}{2}-\alpha}\biggr] \eta_{\alpha,x}(s) ds \\ &= \int_{-\infty}^{\infty} s^k e^{(\frac{1}{2}-\alpha)s} \eta_{\alpha,x}(s) ds. \end{aligned}$$

The first (canonical) moment function can be written in closed form:

Proposition 3.14 *We have*

$$\widetilde{\varphi}_{\alpha,1}(x) = \frac{1}{\Gamma(1-2\alpha)} G_{23}^{31}\biggl(\frac{1}{x} \biggm| \begin{matrix} 0,1 \\ 0,0,1-2\alpha \end{matrix}\biggr) \tag{3.100}$$

where $G^{mn}_{pq}\big(z \,|\, {}^{a_1,\dots,a_p}_{b_1,\dots,b_q}\big)$ *denotes the Meijer-G function [157, Section 8.2]. In the particular case* $\alpha = 0$*, we have* $\widetilde{\varphi}_{\alpha,1}(x) = e^{\frac{1}{x}}\Gamma(0, \frac{1}{x})$*, where* $\Gamma(a, z)$ *is the incomplete Gamma function [56, Chapter IX].*

Proof We know from (3.99) that $\widetilde{\varphi}_{\alpha,1}(x) = (1-2\alpha)\int_0^x \frac{1}{p_\alpha(y)}\int_0^y r_\alpha(\xi)d\xi\, dy$. Consequently,

$$\begin{aligned}\widetilde{\varphi}_{\alpha,1}(x) &= (1-2\alpha)\int_{\frac{1}{x}}^{\infty} v^{-2\alpha}e^{v}\int_v^{\infty} w^{2\alpha-2}e^{-w}dw\, dv\\ &= (1-2\alpha)\int_{\frac{1}{x}}^{\infty} v^{-2\alpha}e^{v}\Gamma(-1+2\alpha, v)dv\\ &= \frac{1}{\Gamma(1-2\alpha)}\int_{\frac{1}{x}}^{\infty} G^{21}_{12}\biggl(v \,\bigg|\, {-1 \atop -2\alpha, -1}\biggr)dv\\ &= \frac{1}{\Gamma(1-2\alpha)}G^{31}_{23}\biggl(\frac{1}{x} \,\bigg|\, {0, 1 \atop 0, 0, 1-2\alpha}\biggr),\end{aligned}$$

where the first equality is obtained via a change of variables, the second equality follows from the definition of the incomplete Gamma function, the third step is due to [157, Relations 8.2.2.15 and 8.4.16.13] and the final step applies [128, Equation 5.6.4(6)]. The result for $\alpha = 0$ follows from the identity $G^{31}_{23}\big(\frac{1}{x} \,|\, {}^{0,1}_{0,0,1}\big) = e^{\frac{1}{x}}\Gamma(0, \frac{1}{x})$, cf. [157, Relations 8.2.2.9 and 8.4.16.13].

Actually, the right-hand side of (3.100) can be written (for $\alpha < \frac{1}{2}$) as a sum of simpler special functions. Such representation can be obtained by applying [7, Equation (A13)].

Returning to moment functions of general order, it is clear from the explicit representation (3.99) that $\widetilde{\varphi}_{\alpha,k}(x) > 0$ for all $x > 0$ and $k \in \mathbb{N}$. We note that $\widetilde{\varphi}_{\alpha,2} \ge \widetilde{\varphi}^2_{\alpha,1}$ (by Jensen's inequality applied to $\widetilde{\varphi}_{\alpha,k}(x) = \int_{-\infty}^{\infty} s^k e^{(\frac{1}{2}-\alpha)s}\eta_{\alpha,x}(s)ds$) and that the Taylor expansions of the first two moment functions as $x \to 0$ are

$$\begin{aligned}\widetilde{\varphi}_{\alpha,1}(x) &= (1-2\alpha)x - (1-2\alpha)(1-\alpha)x^2 + o(x^2),\\ \widetilde{\varphi}_{\alpha,2}(x) &= 2x - (1+2\alpha-4\alpha^2)x^2 + o(x^2)\end{aligned} \tag{3.101}$$

(these relations can be deduced from the asymptotic expansion (3.13), taking into account that $\widetilde{\varphi}_{\alpha,k}(x) = \frac{\partial^k}{\partial\sigma^k}\big|_{\sigma=\frac{1}{2}-\alpha} \mathbf{W}_{\alpha,\sigma}(x)$). Concerning the growth of the moment functions as $x \to \infty$, we have:

Proposition 3.15 *Let* $\varepsilon > 0$*. For each* $k \in \mathbb{N}$*,* $\widetilde{\varphi}_{\alpha,k}(x) = O(x^\varepsilon)$ *as* $x \to \infty$*.*

Proof Due to (3.12), it suffices to prove that $\widetilde{\varphi}_{\alpha,k}(x) = O\big(\mathbf{W}_{\alpha,\frac{1}{2}-\alpha+\varepsilon}(x)\big)$ as $x \to \infty$ for each $k \in \mathbb{N}$. This is trivial for $k = 0$ since $\widetilde{\varphi}_{\alpha,0} \equiv 1 = O(x^\varepsilon) = O\big(\mathbf{W}_{\alpha,\frac{1}{2}-\alpha+\varepsilon}(x)\big)$. By induction, suppose that $\widetilde{\varphi}_{\alpha,j}(x) = O\big(\mathbf{W}_{\alpha,\frac{1}{2}-\alpha+\varepsilon}(x)\big)$ for

$j = 0, \ldots, k-1$. This implies that $\widetilde{\varphi}_{\alpha,j}(x) \leq C \cdot \boldsymbol{W}_{\alpha,\frac{1}{2}-\alpha+\varepsilon}(x)$ for all $x \geq 0$ and $j = 0, \ldots, k-1$ (where $C > 0$ does not depend on x). Recalling (3.73) and (3.99), we find

$$\begin{aligned}\widetilde{\varphi}_{\alpha,k}(x) &\leq C \int_0^x \frac{1}{p_\alpha(y)} \int_0^y r_\alpha(\xi)\, \boldsymbol{W}_{\alpha,\frac{1}{2}-\alpha+\varepsilon}(\xi) d\xi\, dy \\ &= C \cdot \frac{1}{\varepsilon(1-2\alpha+\varepsilon)} \big(\boldsymbol{W}_{\alpha,\frac{1}{2}-\alpha+\varepsilon}(x) - 1\big)\end{aligned}$$

and therefore $\widetilde{\varphi}_{\alpha,k}(x) = O\big(\boldsymbol{W}_{\alpha,\frac{1}{2}-\alpha+\varepsilon}(x)\big)$, proving the proposition.

The previous proposition shows that the *modified moments* $\mathbb{E}[\widetilde{\varphi}_{\alpha,k}(X)]$ will only diverge if the tails of the random variable X are very heavy. The next result shows that the modified moments can be computed via the index Whittaker transform:

Proposition 3.16 *Let* $\mu \in \mathcal{P}(\mathbb{R}_0^+)$ *and* $k \in \mathbb{N}$. *The following assertions are equivalent:*

(i) $\int_{\mathbb{R}_0^+} \widetilde{\varphi}_{\alpha,k}(x)\mu(dx) < \infty$;

(ii) $\sigma \mapsto \int_{\mathbb{R}_0^+} \boldsymbol{W}_{\alpha,\sigma}(x)\mu(dx)$ *is k times differentiable on* $[0, \frac{1}{2}-\alpha]$.

If (i) and (ii) hold, then $\int_{\mathbb{R}_0^+} \frac{\partial^k \boldsymbol{W}_{\alpha,\sigma}(x)}{\partial\sigma^k} \mu(dx) = \frac{\partial^k}{\partial\sigma^k}\big[\int_{\mathbb{R}_0^+} \boldsymbol{W}_{\alpha,\sigma}(x)\mu(dx)\big]$ *for all* $\sigma \in [0, \frac{1}{2}-\alpha]$ *and, in particular,* $\int_{\mathbb{R}_0^+} \widetilde{\varphi}_{\alpha,k}(x)\mu(dx) = \frac{\partial^k}{\partial\sigma^k}\big|_{\sigma=\frac{1}{2}-\alpha}\big[\int_{\mathbb{R}_0^+} \boldsymbol{W}_{\alpha,\sigma}(x)\mu(dx)\big]$.

Proof The following proof is similar to that of the corresponding result for Sturm–Liouville hypergroups (see [208, Theorem 4.11] for further details). Write $\boldsymbol{W}_{\alpha,\sigma}^{\{k\}}(x) := \frac{\partial^k \boldsymbol{W}_{\alpha,\sigma}(x)}{\partial\sigma^k}$. By the Laplace representation (3.63),

$$\boldsymbol{W}_{\alpha,\sigma}^{\{k\}}(x) = \int_0^\infty s^k (e^{\sigma s} + (-1)^k e^{-\sigma s})\, \eta_{\alpha,x}(s) ds.$$

Since sinh and cosh are both increasing and convex functions on $\mathbb{R}^+$, we have

$$0 \leq \boldsymbol{W}_{\alpha,\sigma_1}^{\{k\}}(x) \leq \boldsymbol{W}_{\alpha,\sigma_2}^{\{k\}}(x) \leq \widetilde{\varphi}_{\alpha,k}(x) \qquad \text{for } 0 \leq \sigma_1 \leq \sigma_2 \leq \tfrac{1}{2} - \alpha, \tag{3.102}$$

$$\frac{\widetilde{\varphi}_{\alpha,k}(x) - \boldsymbol{W}_{\alpha,\sigma_1}^{\{k\}}(x)}{\frac{1}{2} - \alpha - \sigma_1} \leq \frac{\widetilde{\varphi}_{\alpha,k}(x) - \boldsymbol{W}_{\alpha,\sigma_2}^{\{k\}}(x)}{\frac{1}{2} - \alpha - \sigma_2} \qquad \text{for } 0 \leq \sigma_1 \leq \sigma_2 < \tfrac{1}{2} - \alpha. \tag{3.103}$$

Moreover, from the inequalities $\sinh y \leq y \cosh y$ and $\cosh y \leq y \sinh y + \mathbb{1}_{[0,2]}(y)$ we can deduce that $y^k(e^y + (-1)^k e^{-y}) \leq y^{k+1}(e^y + (-1)^{k+1}e^{-y}) + 2^k$ for all $k \in \mathbb{N}$ and, therefore,

$$\widetilde{\varphi}_{\alpha,k}(x) \leq (\tfrac{1}{2} - \alpha)\widetilde{\varphi}_{\alpha,k+1}(x) + \left(\frac{2}{\frac{1}{2} - \alpha}\right)^k \qquad \text{for } x \geq 0,\ k \in \mathbb{N}. \tag{3.104}$$

Suppose that (i) holds. By (3.104), $\int_{\mathbb{R}_0^+} \widetilde{\varphi}_{\alpha,j}(x)\mu(dx) < \infty$ for $j = 1, \ldots, k$, so we can assume by induction that $\sigma \mapsto \int_{\mathbb{R}_0^+} \mathbf{W}_{\alpha,\sigma}(x)\mu(dx)$ is $k-1$ times differentiable on $[0, \frac{1}{2} - \alpha]$ and $\int_{\mathbb{R}_0^+} \frac{\partial^{k-1}\mathbf{W}_{\alpha,\sigma}(x)}{\partial\sigma^{k-1}}\mu(dx) = \frac{\partial^{k-1}}{\partial\sigma^{k-1}}\bigl[\int_{\mathbb{R}_0^+} \mathbf{W}_{\alpha,\sigma}(x)\mu(dx)\bigr]$. Combining (3.103) with the dominated convergence theorem (and the well-known corollary on the differentiation of Lebesgue integrals under the integral sign), we can then conclude that (ii) holds and $\int_{\mathbb{R}_0^+} \frac{\partial^k \mathbf{W}_{\alpha,\sigma}(x)}{\partial\sigma^k}\mu(dx) = \frac{\partial^k}{\partial\sigma^k}\bigl[\int_{\mathbb{R}_0^+} \mathbf{W}_{\alpha,\sigma}(x)\mu(dx)\bigr]$.

Conversely, suppose that (ii) holds and assume by induction that $\int_{\mathbb{R}_0^+} \widetilde{\varphi}_{\alpha,j}(x)\,\mu(dx) < \infty$ for $j = 1, \ldots, k-1$. Observe that from (3.103) it follows that $\frac{\widetilde{\varphi}_{\alpha,k-1}(x) - \mathbf{W}_{\alpha,\sigma}^{\{k-1\}}(x)}{\frac{1}{2}-\alpha-\sigma}$ is an increasing function of σ which tends to $\widetilde{\varphi}_{\alpha,k}(x)$ as $\sigma \uparrow \frac{1}{2} - \alpha$. Using the monotone convergence theorem, we thus find that

$$\begin{aligned}\int_{\mathbb{R}_0^+} \widetilde{\varphi}_{\alpha,k}(x)\mu(dx) &= \lim_{\sigma \uparrow \frac{1}{2}-\alpha} \int_{\mathbb{R}_0^+} \frac{\widetilde{\varphi}_{\alpha,k-1}(x) - \mathbf{W}_{\alpha,\sigma}^{\{k-1\}}(x)}{\frac{1}{2} - \alpha - \sigma}\mu(dx) \\ &= \frac{\partial^k}{\partial\sigma^k}\bigg|_{\sigma=\frac{1}{2}-\alpha}\left[\int_{\mathbb{R}_0^+} \mathbf{W}_{\alpha,\sigma}(x)\mu(dx)\right],\end{aligned}$$

so that (i) holds.

The martingale property of $\underset{\alpha}{\diamond}$-moment functions applied to $\underset{\alpha}{\diamond}$-Lévy processes is given below. (See [210, Proposition 6.11] for similar results on hypergroup convolution structures.)

Proposition 3.17 *Let $\{\varphi_k\}_{k=1,2}$ be a pair of $\underset{\alpha}{\diamond}$-moment functions. Let $X = \{X_t\}_{t\geq 0}$ be a $\underset{\alpha}{\diamond}$-Lévy process. Then:*

(a) *If $\mathbb{E}[\varphi_1(X_t)]$ exists for all $t > 0$, then the process $\bigl\{\varphi_1(X_t) - \mathbb{E}[\varphi_1(X_t)]\bigr\}_{t\geq 0}$ is a martingale;*
(b) *If, in addition, $\mathbb{E}[\varphi_2(X_t)]$ exists for all $t > 0$, then the process*

$$\bigl\{\varphi_2(X_t) - 2\varphi_1(X_t)\mathbb{E}[\varphi_1(X_t)] - \mathbb{E}[\varphi_2(X_t)] + 2\mathbb{E}[\varphi_1(X_t)]^2\bigr\}_{t\geq 0}$$

is a martingale.

In particular, if we let Y be the Shiryaev process started at $Y_0 = 0$ and let λ_1, λ_2 be as in Proposition 3.13, then the processes $\{\varphi_1(Y_t) + \lambda_1 t\}_{t\geq 0}$ and $\{\varphi_2(Y_t) + 2\lambda_1 t\varphi_1(Y_t) + \lambda_2 t + \lambda_1^2 t^2\}_{t\geq 0}$ are martingales.

Proof To prove (a), we let $0 \leq s < t$ and compute

$$\mathbb{E}[\varphi_1(X_t) \mid X_s] = \int_{\mathbb{R}_0^+} \varphi_1 \, d\big(\mu_{t-s} \underset{\alpha}{\diamond} \delta_{X_s}\big) = (\mathcal{T}_\alpha^{\mu_{t-s}} \varphi_1)(X_s) = \int_{\mathbb{R}_0^+} \varphi_1 \, d\mu_{t-s} + \varphi_1(X_s).$$

Taking the expectation of both sides yields $\int_{\mathbb{R}_0^+} \varphi_1 \, d\mu_{t-s} = \mathbb{E}[\varphi_1(X_t)] - \mathbb{E}[\varphi_1(X_s)]$; consequently,

$$\begin{aligned}\mathbb{E}\Big[\varphi_1(X_t) - \mathbb{E}[\varphi_1(X_t)] \,\Big|\, \varphi_1(X_s) - \mathbb{E}[\varphi_1(X_s)]\Big] &= \mathbb{E}\big[\varphi_1(X_t) - \mathbb{E}[\varphi_1(X_t)] \mid X_s\big] \\ &= \varphi_1(X_s) - \mathbb{E}[\varphi_1(X_s)],\end{aligned}$$

which shows that $\varphi_1(X_t) - \mathbb{E}[\varphi_1(X_t)]$ is a martingale. Part (b) can be proved by similar arguments.

Let Y be the Shiryaev process started at zero and $\{p_{t,0}\}_{t\geq 0}$ be the associated $\underset{\alpha}{\diamond}$-convolution semigroup. Then by (3.86) we have $\int_{\mathbb{R}_0^+} \boldsymbol{W}_{\alpha,\sigma}(x)\, p_{t,0}(dx) = e^{t(\sigma^2 - (\frac{1}{2}-\alpha)^2)}$ for $\sigma \in [0, \frac{1}{2} - \alpha]$, and it therefore follows from Proposition 3.16 that $\mathbb{E}[\widetilde{\varphi}_{\alpha,1}(X_t)] = \int \widetilde{\varphi}_{\alpha,1}\, dp_{t,0} = (1-2\alpha)t$ and $\mathbb{E}[\widetilde{\varphi}_{\alpha,2}(X_t)] = 2t + (1-2\alpha)^2 t^2$. It follows from Proposition 3.13 that $\varphi_1 = -\frac{\lambda_1}{1-2\alpha}\widetilde{\varphi}_{\alpha,1}$ and $\varphi_2 = \frac{\lambda_1^2}{(1-2\alpha)^2}\widetilde{\varphi}_{\alpha,2} - \big(\frac{2\lambda_1^2}{(1-2\alpha)^3} + \frac{\lambda_2}{1-2\alpha}\big)\widetilde{\varphi}_{\alpha,1}$. Consequently, $\mathbb{E}[\varphi_1(Y_t)] = -\lambda_1 t$ and $\mathbb{E}[\varphi_2(Y_t)] = \lambda_1^2 t^2 - \lambda_2 t$, so that the final statement holds.

3.6.5 Lévy-Type Characterization of the Shiryaev Process

In this subsection we will show that the martingale property given in the last statement of the previous proposition is in fact a (Lévy-type) characterization of the Shiryaev process. For this purpose, it is convenient to focus on the moment functions ϕ_1 and ϕ_2 that correspond to the choice $\lambda_1 = -1$ and $\lambda_2 = 0$, i.e.

$$\phi_1(x) \equiv \phi_{\alpha,1}(x) = \int_0^x \frac{1}{p_\alpha(y)} \int_0^y r_\alpha(\xi)\, d\xi\, dy = \frac{1}{1-2\alpha}\widetilde{\varphi}_{\alpha,1}(x),$$

$$\begin{aligned}\phi_2(x) \equiv \phi_{\alpha,2}(x) &= 2\int_0^x \frac{1}{p_\alpha(y)} \int_0^y r_\alpha(\xi)\, \phi_1(\xi)\, d\xi\, dy \\ &= \frac{1}{(1-2\alpha)^2}\Big[\widetilde{\varphi}_{\alpha,2}(x) - \frac{2}{1-2\alpha}\widetilde{\varphi}_{\alpha,1}(x)\Big].\end{aligned}$$

In the following results, we write

$$\mathrm{C}^{k,\ell}(\mathbb{R}_0^+) := \bigl\{f \in \mathrm{C}^k(\mathbb{R}_0^+) : f\big|_{[0,\varepsilon)} \in \mathrm{C}^\ell[0,\varepsilon) \text{ for some } \varepsilon > 0\bigr\}.$$

Lemma 3.12

(a) If $f \in \mathrm{C}^{2,4}(\mathbb{R}_0^+)$ with $f'(0) = f'''(0) = 0$, then there exists $h \in \mathrm{C}^2(\mathbb{R}_0^+)$ with $f(x) = h(\phi_1(x^2))$ for $x \geq 0$.

(b) There exists a unique function $h_0 \in \mathrm{C}^2(\mathbb{R}_0^+)$ such that $h_0(\phi_1(x)) = \phi_2(x)$ for $x \geq 0$, and it satisfies $h_0''(x) > 0$ for all $x \geq 0$.

Proof From (3.101) we find that the Taylor expansions of the functions $\phi_1(x^2)$ and $\phi_2(x^2)$ as $x \to 0$ are of the form $\phi_1(x^2) = c_1x^2 + c_2x^4 + o(x^4)$ and $\phi_2(x^2) = c_3x^4 + o(x^4)$, with $c_1, c_3 > 0$. Consequently, part (a) can be proved using the same arguments as in [163, Lemma 5.7]. Letting $f(x) = \phi_2(x^2)$, we deduce that in particular there exists $h_0 \in \mathrm{C}^2(\mathbb{R}_0^+)$ such that $h_0(\phi_1(x)) = \phi_2(x)$ for all $x \geq 0$. A straightforward adaptation of the proof of [163, Lemma 5.8] yields that $h_0''(x) > 0$ for $x \geq 0$.

Lemma 3.13 *Let $X = \{X_t\}_{t\geq 0}$ be an $\mathbb{R}_0^+$ valued process with a.s. continuous paths and such that the processes $Z_X^{\mathcal{A}_\alpha,\phi_j}$ defined by (3.87) are local martingales for $j = 1, 2$. Then*

$$\bigl[Z_X^{\mathcal{A}_\alpha,\phi_1}\bigr]_t = 2\int_0^t X_s^2(\phi_1'(X_s))^2\,ds \qquad \textit{almost surely.}$$

Moreover, $Z_X^{\mathcal{A}_\alpha,g}$ is a local martingale whenever $g \in \mathrm{C}^{2,4}(\mathbb{R}_0^+)$.

Proof This proof is analogous to that of [163, Lemma 6.2], to which we refer for further details.

Let $h \in \mathrm{C}^2(\mathbb{R}_0^+)$. Given that $\mathcal{A}_\alpha\phi_1 = 1$, an application of the chain rule shows that $\mathcal{A}_\alpha(h(\phi_1))(x) = x^2h''(\phi_1(x))\,(\phi_1'(x))^2 + h'(\phi_1(x))$. Since $Z_X^{\mathcal{A}_\alpha,\phi_1}$ is a local martingale, we can apply Itô's formula for continuous semimartingales [138, Theorem 6.2] to the process $h(\phi_1(X_t))$ and deduce that $d(h(\phi_1(X_t))) = h'(\phi_1(X_t))\,d\phi_1(X_t) + \frac{1}{2}h''(\phi_1(X_t))\,d[\phi_1(X)]_t$. Consequently,

$$\begin{aligned} &d(h(\phi_1(X_t))) - \mathcal{A}_\alpha(h(\phi_1(X_t)))\,dt \\ &\quad = h''(\phi_1(X_t))\Bigl(\tfrac{1}{2}d[\phi_1(X)]_t - X_t^2(\phi_1'(X_t))^2dt\Bigr) + dV_t^h, \end{aligned} \tag{3.105}$$

where $\bigl\{V_t^h := \int_0^t h'(\phi_1(X_s))(d\phi_1(X_s) - ds)\bigr\}$ is a local martingale (cf. [138, Proposition 2.63]; note that $d\phi_1(X_t) - dt = dZ_{X,t}^{\mathcal{A}_\alpha,\phi_1}$ is the differential of a local martingale). If, in particular, h is the function h_0 from Lemma 3.12(b), then

$\int_0^t d(h_0(\phi_1(X_s))) - \mathcal{A}_\alpha(h_0(\phi_1(X_s)))\,ds = Z_{X,t}^{\mathcal{A}_\alpha,\phi_2}$ is also a local martingale, and from (3.105) we find that

$$\int_0^t h_0''(\phi_1(X_s))\Big(\tfrac{1}{2}d[\phi_1(X)]_s - X_s^2(\phi_1'(X_s))^2 ds\Big) \qquad \text{is a local martingale.}$$

But $\int_0^t h_0''(\phi_1(X_s))\big(\frac{1}{2}d[\phi_1(X)]_s - X_s^2(\phi_1'(X_s))^2 ds\big)$ is also a process of locally finite variation (cf. [138, Proposition 2.73]; note that $\frac{1}{2}d[\phi_1(X)]_s - X_s^2(\phi_1'(X_s))^2 ds$ is the differential of a process of locally finite variation), hence it is a.s. equal to zero (see [138, Theorem 2.11]). Consequently, taking into account that $h_0'' > 0$ (Lemma 3.12(b)), we have $d[Z_X^{\mathcal{A}_\alpha,\phi_1}]_t - 2X_t^2(\phi_1'(X_t))^2 dt = d[\phi_1(X)]_t - 2X_t^2(\phi_1'(X_t))^2 dt = 0$ a.s., proving the first assertion.

The result just proved, combined with (3.105), implies that $\big\{h(\phi_1(X_t)) - \int_0^t \mathcal{A}_\alpha(h(\phi_1))(X_s)ds\big\}_{t\geq 0}$ is, for each $h \in \mathrm{C}^2(\mathbb{R}_0^+)$, a local martingale. Applying Lemma 3.12(a) with $f(x) := g(x^2) \in \mathrm{C}^{2,4}(\mathbb{R}_0^+)$, we find that $g(x) \equiv h(\phi_1(x))$ for some $h \in \mathrm{C}^2(\mathbb{R}_0^+)$, and this proves the second assertion.

We are finally ready to establish the martingale characterization of the Shiryaev process. (We call it a *Lévy-type characterization* because it resembles the Lévy characterization of Brownian motion stated in Theorem 2.2. A parallel result for hypergroup structures is given in [163, Theorem 6.3].)

Theorem 3.4 (Lévy-Type Characterization for the Shiryaev Process) *Let $Y = \{Y_t\}_{t\geq 0}$ be an $\mathbb{R}_0^+$-valued Markov process with a.s. continuous paths. The following assertions are equivalent:*

(i) *Y is the Shiryaev process;*
(ii) *$\{\phi_1(Y_t) - t\}_{t\geq 0}$ and $\{\phi_2(Y_t) - 2t\phi_1(Y_t) + t^2\}_{t\geq 0}$ are martingales (or local martingales);*
(iii) *$Z_Y^{\mathcal{A}_\alpha,\phi_1}$ is a local martingale with $[Z_Y^{\mathcal{A}_\alpha,\phi_1}]_t = 2\int_0^t Y_s^2(\phi_1'(Y_s))^2\,ds$.*

Proof **(i) $\Longrightarrow$ (ii):** This follows from Proposition 3.17.

(ii) $\Longrightarrow$ (iii): Assume that (ii) is true. Since $dZ_{Y,t}^{\mathcal{A}_\alpha,\phi_1} = d\phi_1(Y_t) - dt$, the process $Z_Y^{\mathcal{A}_\alpha,\phi_1}$ is a local martingale. Furthermore,

$$dZ_{Y,t}^{\mathcal{A}_\alpha,\phi_2} = d\phi_2(Y_t) - 2\phi_1(Y_t)dt = d\big(\phi_2(Y_t) - 2t\phi_1(Y_t) + t^2\big) + 2t\big(d\phi_1(Y_t) - dt\big)$$

(where integration by parts [138, Proposition 2.28] gives the second equality) and therefore the process $Z_Y^{\mathcal{A}_\alpha,\phi_2}$ is also a local martingale. By Lemma 3.13, $[Z_Y^{\mathcal{A}_\alpha,\phi_1}]_t = 2\int_0^t Y_s^2(\phi_1'(Y_s))^2\,ds$.

(iii) $\Longrightarrow$ (i): Assuming that (iii) holds, Eq. (3.105) and the proof of Lemma 3.13 show that, for each $\lambda \geq 0$, $Z_Y^{\mathcal{A}_\alpha, W_{\alpha,\Delta_\lambda}(\cdot)}$ is a local martingale and (by boundedness on compact time intervals, cf. [138, Corollary 1.145]) a true martingale. Proposition 3.12 now yields that Y is the Shiryaev process.

Remark 3.10 In this section we have focused on continuous-time stochastic processes which are additive with respect to the Whittaker convolution. In a similar way, one can introduce the discrete-time counterparts of the processes studied above.

An $\mathbb{R}_0^+$-valued Markov chain $\{S_n\}_{n\in\mathbb{N}_0}$ with $S_0 = 0$ is said to be $\underset{\alpha}{\diamond}$-*additive* if there exist measures $\mu_n \in \mathcal{P}(\mathbb{R}_0^+)$ such that

$$P[S_n \in B | S_{n-1} = x] = (\mu_n \underset{\alpha}{\diamond} \delta_x)(B), \qquad n \in \mathbb{N}, x \geq 0, B \text{ a Borel subset of } \mathbb{R}_0^+.$$

If $\mu_n = \mu$ for all n, then $\{S_n\}$ is said to be a $\underset{\alpha}{\diamond}$-*random walk*. One can give an explicit construction for $\underset{\alpha}{\diamond}$-additive Markov chains, cf. [19, Section 7.1] (see also Sect. 4.5.3 below).

In the context of hypergroups, moment functions have been successfully applied to the study of the limiting behaviour of additive Markov chains (cf. [19, Chapter 7]). Parallel results hold for the Whittaker convolution. For instance, letting $\{S_n\}$ be a $\underset{\alpha}{\diamond}$-additive Markov chain constructed as above, the following strong laws of large numbers are established as in [19, Theorems 7.3.21 and 7.3.24]:

(a) *If* $\{r_n\}_{n\in\mathbb{N}}$ *is a sequence of positive numbers such that* $\lim_n r_n = \infty$ *and* $\sum_{n=1}^{\infty} \frac{1}{r_n}\bigl(\mathbb{E}[\widetilde{\varphi}_{\alpha,2}(X_n)] - \mathbb{E}[\widetilde{\varphi}_{\alpha,1}(X_n)]^2\bigr) < \infty$, *then*

$$\lim_n \frac{1}{\sqrt{r_n}}\bigl(\widetilde{\varphi}_{\alpha,1}(S_n) - \mathbb{E}[\widetilde{\varphi}_{\alpha,1}(S_n)]\bigr) = 0 \qquad P\text{-a.s.}$$

(b) *If* $\{S_n\}$ *is a* $\underset{\alpha}{\diamond}$*-random walk such that* $\mathbb{E}[\widetilde{\varphi}_{\alpha,2}(X_1)^{\theta/2}] < \infty$ *for some* $1 \leq \theta < 2$, *then* $\mathbb{E}[\widetilde{\varphi}_{\alpha,1}(X_1)] < \infty$ *and*

$$\lim_n \frac{1}{n^{1/\theta}}\bigl(\widetilde{\varphi}_{\alpha,1}(S_n) - n\mathbb{E}[\widetilde{\varphi}_{\alpha,1}(X_1)]\bigr) = 0 \qquad P\text{-a.s.}$$

3.7 Whittaker Convolution of Functions

After having studied the probabilistic properties of the Whittaker convolution, we return to the study of the basic properties of this convolution, which we shall now regard as a binary operator on weighted L^p spaces.

Definition 3.10 Let $f, g : \mathbb{R}^+ \to \mathbb{C}$ be complex-valued functions. If the double integral

$$(f \underset{\alpha}{\diamond} g)(x) := \int_0^\infty (\mathcal{T}_\alpha^x f)(\xi)\, g(\xi)\, r_\alpha(\xi) d\xi$$

$$= \int_0^\infty \int_0^\infty k_\alpha(x, y, \xi)\, f(y)\, g(\xi)\, r_\alpha(y) dy\, r_\alpha(\xi) d\xi$$

exists for almost every $0 < x < \infty$, then we call it the *Whittaker convolution (of order α)* of the functions f and g.

Note that this definition is obtained from Definition 3.4 by letting μ and ν be the absolutely continuous measures defined by $\mu(dx) = f(x)\, r_\alpha(x) dx$ and $\nu(dx) = g(x)\, r_\alpha(x) dx$. The Whittaker convolution of functions is positivity-preserving (i.e., $f \underset{\alpha}{\diamond} g \geq 0$ whenever $f, g \geq 0$) and commutative (i.e., $f \underset{\alpha}{\diamond} g = g \underset{\alpha}{\diamond} f$). Moreover, it is a generalization of the Kontorovich–Lebedev convolution: indeed, for $\alpha = 0$ we have

$$(f \underset{0}{\diamond} g)(x) = (f \underset{KL}{\diamond} g)(x) = \frac{1}{4} \pi^{-\frac{1}{2}} x^{-\frac{3}{2}} e^{\frac{1}{2x}} (\mathfrak{f} \underset{KL}{*} \mathfrak{g})(\tfrac{1}{2x}),$$

where $\underset{KL}{\diamond}$ is the normalized Kontorovich–Lebedev convolution (3.7), $\underset{KL}{*}$ is the classical Kontorovich–Lebedev convolution (3.4), $\mathfrak{f}(x) = x^{-\frac{3}{2}} e^{-x} f(\frac{1}{2x})$ and $\mathfrak{g}(x) = x^{-\frac{3}{2}} e^{-x} g(\frac{1}{2x})$.

3.7.1 Mapping Properties in the Spaces $L^p(r_\alpha)$

The well-known Young convolution inequality has a natural analogue for the Whittaker convolution of functions belonging to the family of L^p spaces defined in (3.42). (The following result should also be compared with the Young inequality for the Hankel convolution stated in Proposition 2.22.)

Proposition 3.18 (Young Inequality for the Whittaker Convolution) *Let $p_1, p_2 \in [1, \infty]$ such that $\frac{1}{p_1} + \frac{1}{p_2} \geq 1$. For $f \in L^{p_1}(r_\alpha)$ and $g \in L^{p_2}(r_\alpha)$, the $\mathcal{L}$-convolution $f \underset{\alpha}{\diamond} g$ is well-defined and, for $s \in [1, \infty]$ defined by $\frac{1}{s} = \frac{1}{p_1} + \frac{1}{p_2} - 1$, it satisfies*

$$\|f \underset{\alpha}{\diamond} g\|_{s,\alpha} \leq \|f\|_{p_1,\alpha} \|g\|_{p_2,\alpha}$$

(in particular, $f \underset{\alpha}{\diamond} g \in L^s(r_\alpha)$). Consequently, the Whittaker convolution is a continuous bilinear operator from $L^{p_1}(r_\alpha) \times L^{p_2}(r_\alpha)$ into $L^s(r_\alpha)$.

Proof The proof is analogous to that of the Young inequality for the ordinary convolution. Define $\frac{1}{t_1} = \frac{1}{p_1} - \frac{1}{s}$ and $\frac{1}{t_2} = \frac{1}{p_2} - \frac{1}{s}$. Observe that

$$|(\mathcal{T}_\alpha^x f)(y)|\, |g(y)| \leq |(\mathcal{T}_\alpha^x f)(y)|^{p_1/t_1}\, |g(y)|^{p_2/t_2} \big[|(\mathcal{T}_\alpha^x f)(y)|^{p_1}\, |g(y)|^{p_2} \big]^{1/s}.$$

Since $\frac{1}{s}+\frac{1}{t_1}+\frac{1}{t_2}=1$, we have by Hölder's inequality and Proposition 3.2(b)

$$\begin{aligned}
&\int_0^\infty |(\mathcal{T}_\alpha^x f)(y)|\,|g(y)|\,r_\alpha(y)dy \\
&\leq \left(\int_0^\infty |(\mathcal{T}_\alpha^x f)(y)|^{p_1} r_\alpha(y)dy\right)^{\frac{1}{t_1}} \left(\int_0^\infty |g(y)|^{p_2} r_\alpha(y)dy\right)^{\frac{1}{t_2}} \\
&\quad\times \left(\int_0^\infty |(\mathcal{T}_\alpha^x f)(y)|^{p_1} |g(y)|^{p_2} r_\alpha(y)dy\right)^{\frac{1}{s}} \\
&\leq \|f\|_{p_1}^{p_1/t_1} \|g\|_{p_2}^{p_2/t_2} \left(\int_0^\infty |(\mathcal{T}_\alpha^x f)(y)|^{p_1} |g(y)|^{p_2} r_\alpha(y)dy\right)^{1/s}.
\end{aligned}$$

Using again Proposition 3.2(b) we conclude that

$$\|f \underset{\alpha}{\diamond} g\|_{s,\alpha} \leq \|f\|_{p_1,\alpha}^{p_1/t_1} \|g\|_{p_2,\alpha}^{p_2/t_2} \|f\|_{p_1,\alpha}^{p_1/s} \|g\|_{p_2,\alpha}^{p_2/s} = \|f\|_{p_1,\alpha} \|g\|_{p_2,\alpha}.$$

Another analogue of a well-known property of the ordinary convolution is the fact that the Whittaker convolution of functions belonging to L^p spaces with conjugate exponents defines a continuous function:

Proposition 3.19 *Let $p, q \in [1,\infty]$ with $\frac{1}{p}+\frac{1}{q}=1$. If $f \in L^p(r_\alpha)$ and $g \in L^q(r_\alpha)$, then $f \underset{\alpha}{\diamond} g \in C_b(\mathbb{R}^+)$.*

Proof The previous proposition ensures the boundedness of $f \underset{\alpha}{\diamond} g$. For the continuity, let $x_0 > 0$; then for $1 < p < \infty$ we have

$$\begin{aligned}
\left|(f \underset{\alpha}{\diamond} g)(x) - (f \underset{\alpha}{\diamond} g)(x_0)\right| &= \left|\int_0^\infty \big((\mathcal{T}_\alpha^x f)(\xi) - (\mathcal{T}_\alpha^{x_0} f)(\xi)\big) g(\xi)\, r_\alpha(\xi) d\xi\right| \\
&\leq \|\mathcal{T}_\alpha^x f - \mathcal{T}_\alpha^{x_0} f\|_{p,\alpha} \|g\|_{q,\alpha} \to 0 \qquad \text{as } x \to x_0
\end{aligned}$$

by Hölder's inequality and Proposition 3.2(c). In the case $p=\infty$ (and by symmetry $p=1$), the continuity of $f \underset{\alpha}{\diamond} g$ follows by dominated convergence, using parts (a) and (c) of Proposition 3.2.

Some fundamental connections between the Whittaker convolution (and translation), the index Whittaker transform and the differential operator $\mathcal{A}_\alpha$ are given in the following proposition.

Proposition 3.20 *Let $y > 0$ and $\tau \geq 0$. Then:*

(a) *If $f \in L^2(r_\alpha)$, then $\big(\mathcal{W}_\alpha(\mathcal{T}_\alpha^y f)\big)(\tau) = \mathbf{W}_{\alpha,i\tau}(y)\,(\mathcal{W}_\alpha f)(\tau)$;*
(b) *If $f \in L^2(r_\alpha)$ and $g \in L^1(r_\alpha)$, then $\big(\mathcal{W}_\alpha(f \underset{\alpha}{\diamond} g)\big)(\tau) = (\mathcal{W}_\alpha f)(\tau)\,(\mathcal{W}_\alpha g)(\tau)$;*

(c) If $f \in L^2(r_\alpha)$ and $g \in L^1(r_\alpha)$, then $\mathcal{T}_\alpha^y(f \underset{\alpha}{\diamond} g) = (\mathcal{T}_\alpha^y f) \underset{\alpha}{\diamond} g$;

(d) If $f \in \mathcal{D}_\alpha^{(2)}$, then $\mathcal{T}_\alpha^y f \in \mathcal{D}_\alpha^{(2)}$ and $\mathcal{A}_\alpha(\mathcal{T}_\alpha^y f) = \mathcal{T}_\alpha^y(\mathcal{A}_\alpha f)$;

(e) If $f \in \mathcal{D}_\alpha^{(2)}$ and $g \in L^1(r_\alpha)$, then $f \underset{\alpha}{\diamond} g \in \mathcal{D}_\alpha^{(2)}$ and $\mathcal{A}_\alpha(f \underset{\alpha}{\diamond} g) = (\mathcal{A}_\alpha f) \underset{\alpha}{\diamond} g$.

Proof

(a) Let $f \in L^1(r_\alpha) \cap L^2(r_\alpha)$. Using Fubini's theorem and the product formula (3.36), we compute

$$\begin{aligned}
\big(\mathcal{W}_\alpha(\mathcal{T}_\alpha^y f)\big)(\tau) &= \int_0^\infty \int_0^\infty f(\xi) k_\alpha(x,y,\xi)\, r_\alpha(\xi) d\xi\, \mathbf{W}_{\alpha,i\tau}(x)\, r_\alpha(x) dx \\
&= \mathbf{W}_{\alpha,i\tau}(y) \int_0^\infty f(\xi)\, \mathbf{W}_{\alpha,i\tau}(\xi)\, r_\alpha(\xi) d\xi \\
&= \mathbf{W}_{\alpha,i\tau}(y)\, (\mathcal{W}_\alpha f)(\tau).
\end{aligned}$$

By denseness and continuity, the equality extends to all $f \in L^2(r_\alpha)$, as required.

(b) For $f \in L^1(r_\alpha) \cap L^2(r_\alpha)$ and $g \in L^1(r_\alpha)$ we have

$$\begin{aligned}
(\mathcal{W}_\alpha(f \underset{\alpha}{\diamond} g))(\tau) &= \int_0^\infty \int_0^\infty (\mathcal{T}_\alpha^x f)(\xi)\, g(\xi)\, r_\alpha(\xi) d\xi\, \mathbf{W}_{\alpha,i\tau}(x)\, r_\alpha(x) dx \\
&= \int_0^\infty g(\xi) \big(\mathcal{W}_\alpha(\mathcal{T}_\alpha^\xi f)\big)(\tau)\, r_\alpha(\xi) d\xi \\
&= (\mathcal{W}_\alpha f)(\tau) \int_0^\infty g(\xi)\, \mathbf{W}_{\alpha,i\tau}(\xi)\, r_\alpha(\xi) d\xi = (\mathcal{W}_\alpha f)(\tau)\, (\mathcal{W}_\alpha g)(\tau),
\end{aligned}$$

where we have used Fubini's theorem and part (a). Again, denseness yields the result.

(c) By the previous properties,

$$\mathcal{W}_\alpha\big[\mathcal{T}_\alpha^y(f \underset{\alpha}{\diamond} g)\big](\tau) = \mathcal{W}_\alpha\big[(\mathcal{T}_\alpha^y f) \underset{\alpha}{\diamond} g\big](\tau) = \mathbf{W}_{\alpha,i\tau}(y)\, (\mathcal{W}_\alpha f)(\tau)\, (\mathcal{W}_\alpha g)(\tau).$$

Since both $\mathcal{T}_\alpha^y(f \underset{\alpha}{\diamond} g)$ and $(\mathcal{T}_\alpha^y f) \underset{\alpha}{\diamond} g$ are elements of the space $L^2(r_\alpha)$ (see Proposition 3.21 below), this implies that $\mathcal{T}_\alpha^y(f \underset{\alpha}{\diamond} g) = (\mathcal{T}_\alpha^y f) \underset{\alpha}{\diamond} g$.

(d) Recalling the inequality (3.67), it is evident that $\mathcal{T}_\alpha^y\big(\mathcal{D}_\alpha^{(2)}\big) \in \mathcal{D}_\alpha^{(2)}$. Since the index Whittaker transforms of $\mathcal{A}_\alpha(\mathcal{T}_\alpha^y f)$ and $\mathcal{T}_\alpha^y(\mathcal{A}_\alpha f)$ are both equal to $\big(\tau^2 + (\frac{1}{2} - \alpha)^2\big)\mathbf{W}_{\alpha,i\tau}(y)(\mathcal{W}_\alpha f)(\tau)$, the result follows.

(e) The proof is similar to that of (d).

We have seen in Proposition 3.18 that if $f \in L^2(r_\alpha)$ and $g \in L^p(r_\alpha)$ $(1 \leq p < 2)$ then the Whittaker convolution $f \underset{\alpha}{\diamond} g$ exists and belongs to $L^{\frac{2p}{2-p}}(r_\alpha)$. Using the index Whittaker transform, this result can be strengthened as follows:

Proposition 3.21 *Let $f \in L^2(r_\alpha)$ and $g \in L^p(r_\alpha)$ $(1 \leq p < 2)$. Then $f \underset{\alpha}{\diamond} g \in L^2(r_\alpha)$, and we have*

$$\|f \underset{\alpha}{\diamond} g\|_{2,\alpha} \leq C_p \|f\|_{2,\alpha} \|g\|_{p,\alpha},$$

where $C_p = \big\| W_{\alpha,0} \big\|_{q,\alpha} < \infty$ (being $\frac{1}{p} + \frac{1}{q} = 1$).

Proof The fact that $\| W_{\alpha,0}\|_{q,\alpha}$ is finite for each $2 < q \leq \infty$ is easily verified using the limiting forms (3.10), (3.12). Now, for $f, g \in \mathrm{C_c}(\mathbb{R}^+)$ we have

$$\begin{aligned}
\|f \underset{\alpha}{\diamond} g\|_{2,\alpha} &= \big\|(\mathcal{W}_\alpha f) \cdot (\mathcal{W}_\alpha g)\big\|_{L^2(\rho_\alpha)} \\
&\leq \sup_{\tau \geq 0} \big|(\mathcal{W}_\alpha g)(\tau)\big| \cdot \big\|\mathcal{W}_\alpha f\big\|_{L^2(\rho_\alpha)} \\
&\leq \big\| W_{\alpha,0} \big\|_{q,\alpha} \|g\|_{p,\alpha} \|f\|_{2,\alpha},
\end{aligned}$$

where we denoted $L^2(\rho_\alpha) = L^2\big(\mathbb{R}^+; \rho_\alpha(\tau)d\tau\big)$; we have used the isometric property of the index Whittaker transform, and the final step relies on the inequality $|W_{\alpha,i\tau}(x)| \leq W_{\alpha,0}(x)$ (proved in (3.66)) and on Hölder's inequality. As usual, the result for $f \in L^2(r_\alpha)$ and $g \in L^p(r_\alpha)$ follows from the denseness of $\mathrm{C_c}(\mathbb{R}^+)$ in these L^p spaces.

Corollary 3.3

(a) *If $f, g \in L^2(r_\alpha)$, then $f \underset{\alpha}{\diamond} g \in L^q(r_\alpha)$ for all $2 < q \leq \infty$, with*

$$\|f \underset{\alpha}{\diamond} g\|_{q,\alpha} \leq C_q \|f\|_{2,\alpha} \|g\|_{2,\alpha}$$

being $C_q = \big\| W_{\alpha,0} \big\|_{q,\alpha}$.

(b) *Let $1 \leq p_1 < 2$ and $1 \leq p_2 \leq 2$ such that $\frac{1}{p_1} + \frac{1}{p_2} \leq \frac{3}{2}$. Let t be defined by $\frac{1}{t} = \frac{1}{p_1} + \frac{1}{p_2} - 1$. If $f \in L^{p_1}(r_\alpha)$ and $g \in L^{p_2}(r_\alpha)$, then $f \underset{\alpha}{\diamond} g \in L^s(r_\alpha)$ for all $s \in [2, t]$.*

Proof The following proof is adapted from [61, Section 5].

(a) Let $h \in L^p(r_\alpha)$ $(\frac{1}{p} + \frac{1}{q} = 1)$ and $f, g \in \mathrm{C_c}(\mathbb{R}^+)$. Using Proposition 3.21 and Fubini's theorem, we obtain

$$\begin{aligned}\left|\int_0^\infty (f \underset{\alpha}{\diamond} g)(x)h(x)r_\alpha(x)dx\right| &\leq \int_0^\infty (|f| \underset{\alpha}{\diamond} |h|)(x)|g(x)|r_\alpha(x)dx \\ &\leq \|g\|_{2,\alpha} \big\||f| \underset{\alpha}{\diamond} |h|\big\|_{2,\alpha} \leq C_q \|g\|_{2,\alpha}\|f\|_{2,\alpha}\|h\|_{p,\alpha}.\end{aligned}$$

Therefore

$$\|f \underset{\alpha}{\diamond} g\|_{q,\alpha} = \sup_{\substack{h\in L^p(r_\alpha)\\ \|h\|_{p,\alpha}\leq 1}} \left|\int_0^\infty (f \underset{\alpha}{\diamond} g)(x)h(x)r_\alpha(x)dx\right| \leq C_q\|f\|_{2,\alpha}\|g\|_{2,\alpha}$$

and the usual continuity argument yields the result.

(b) By the Young inequality (Proposition 3.18) $f \underset{\alpha}{\diamond} g \in L^t(r_\alpha)$, and we know that $L^2(r_\alpha) \cap L^t(r_\alpha) \subset L^s(r_\alpha)$, thus we just need to show that $f \underset{\alpha}{\diamond} g \in L^2(r_\alpha)$. Observe that $g_1 := g \cdot \mathbb{1}_{\{|g|\geq 1\}} \in L^1(r_\alpha) \cap L^{p_2}(r_\alpha)$ and $g_2 := g \cdot \mathbb{1}_{\{|g|<1\}} \in L^{p_2}(r_\alpha) \cap L^\infty(r_\alpha)$. It follows from Propositions 3.18 and 3.21 that, respectively, $f \underset{\alpha}{\diamond} g_1 \in L^{p_1}(r_\alpha) \cap L^t(r_\alpha) \subset L^2(r_\alpha)$ and $f \underset{\alpha}{\diamond} g \in L^2(r_\alpha)$; consequently, $f \underset{\alpha}{\diamond} g = f \underset{\alpha}{\diamond} g_1 + f \underset{\alpha}{\diamond} g_2 \in L^2(r_\alpha)$.

3.7.2 The Convolution Banach Algebra $L_{\alpha,\nu}$

In this subsection we focus on the properties of the Whittaker convolution in the family of spaces $\{L_{\alpha,\nu}\}_{\nu\geq 0}$, where

$$L_{\alpha,\nu} := L^1\big(\mathbb{R}^+,\, W_{\alpha,\nu}(x)\, r_\alpha(x)dx\big) \qquad (\alpha < \tfrac{1}{2},\ \nu \geq 0).$$

We observe that, by the limiting forms of the Whittaker W function,

$$f \in L_{\alpha,\nu} \text{ if and only if } f \in L^1\big((0,1],\, x^{-2\alpha}e^{-\frac{1}{x}}dx\big) \cap L^1\big([1,\infty),\, x^{-\frac{1}{2}-\alpha+\nu}dx\big) \quad \text{for } \nu > 0,$$

$$f \in L_{\alpha,0} \text{ if and only if } f \in L^1\big((0,1],\, x^{-2\alpha}e^{-\frac{1}{x}}dx\big) \cap L^1\big([1,\infty),\, x^{-\frac{1}{2}-\alpha}\log x\, dx\big), \tag{3.106}$$

and therefore the spaces $L_{\alpha,\nu}$ are ordered:

$$L_{\alpha,\nu_1} \subset L_{\alpha,\nu_2} \quad \text{whenever } \nu_1 > \nu_2.$$

It is also interesting to note that the family $\{L_{\alpha,\nu}\}_{\nu\geq 0}$ contains the space $L^1(r_\alpha) \equiv L_{\alpha,\frac{1}{2}-\alpha}$. (Recall that by (3.17) we have $\mathbf{W}_{\alpha,\frac{1}{2}-\alpha}(y) = 1$.)

The following lemma collects some properties of the index Whittaker transform in the spaces $L_{\alpha,\nu}$ ($\alpha < \frac{1}{2}$, $\nu \geq 0$):

Lemma 3.14 *If $f \in L_{\alpha,\nu}$, then its index Whittaker transform $(\mathcal{W}_\alpha f)(\tau) = \int_0^\infty f(y)\, \mathbf{W}_{\alpha,i\tau}(y)\, r_\alpha(y)dy$ is, for every τ belonging to the complex strip $|\mathrm{Im}\,\tau| \leq \nu$, well-defined as an absolutely convergent integral, and it satisfies*

$$(\mathcal{W}_\alpha f)(\tau) \xrightarrow[\tau\to\infty]{} 0 \qquad \textit{uniformly in the strip } |\mathrm{Im}\,\tau| \leq \nu. \tag{3.107}$$

Moreover, if $(\mathcal{W}_\alpha f)(\tau) = 0$ for all $\tau \geq 0$, then $f(x) = 0$ for almost every $x > 0$.

Proof The absolute convergence of the integral defining $\mathcal{W}_\alpha f$ is clear from the inequality (3.66). It follows from (3.63) and the Riemann-Lebesgue lemma that for each $y > 0$ we have $\mathbf{W}_{\alpha,i\tau}(y) \longrightarrow 0$ as $\tau \to \infty$ uniformly in the strip $|\mathrm{Im}\,\tau| \leq \nu$, hence dominated convergence gives (3.107). Letting μ be the (possibly unbounded) measure $\mu(dx) = f(x)\, r_\alpha(x)dx$, the same proof of Proposition 3.4(ii) shows that if $\widehat{\mu}(\tau^2 + (\frac{1}{2} - \alpha)^2) \equiv (\mathcal{W}_\alpha f)(\tau) = 0$ for all $\tau \geq 0$, then μ is the zero measure, so that $f(x) = 0$ a.e.

Proposition 3.22 *For $f, g \in L_{\alpha,\nu}$, the Whittaker convolution $f \underset{\alpha}{\diamond} g$ is well-defined and satisfies*

$$\|f \underset{\alpha}{\diamond} g\|_{L_{\alpha,\nu}} \leq \|f\|_{L_{\alpha,\nu}} \|g\|_{L_{\alpha,\nu}}$$

(in particular, $f \underset{\alpha}{\diamond} g \in L_{\alpha,\nu}$). Moreover, properties (a) and (b) in Proposition 3.20 are valid when f and g belong to $L_{\alpha,\nu}$ and τ is a complex number such that $|\mathrm{Im}\,\tau| \leq \nu$.

Proof We compute

$$\begin{aligned}
&\|f \underset{\alpha}{\diamond} g\|_{L_{\alpha,\nu}} \\
&\leq \int_0^\infty \int_0^\infty \int_0^\infty |f(y)| k_\alpha(x, y, \xi) r_\alpha(y)dy\, |g(\xi)| r_\alpha(\xi)d\xi\, \mathbf{W}_{\alpha,\nu}(x) r_\alpha(x)dx \\
&= \int_0^\infty \int_0^\infty \int_0^\infty \mathbf{W}_{\alpha,\nu}(x) k_\alpha(x, y, \xi) r_\alpha(x)dx\, |f(y)| r_\alpha(y)dy\, |g(\xi)| r_\alpha(\xi)d\xi \\
&= \int_0^\infty |f(y)|\, \mathbf{W}_{\alpha,\nu}(y) r_\alpha(y)dy \int_0^\infty |g(\xi)|\, \mathbf{W}_{\alpha,\nu}(\xi) r_\alpha(\xi)d\xi \\
&= \|f\|_{L_{\alpha,\nu}} \|g\|_{L_{\alpha,\nu}},
\end{aligned}$$

where the positivity of the integrand justifies the change of order of integration, and the second equality follows from the product formula (3.36). The final statement is proved using the same calculations as before.

Corollary 3.4 *The Banach space $L_{\alpha,\nu}$, equipped with the convolution multiplication $f \cdot g \equiv f \underset{\alpha}{\diamond} g$, is a commutative Banach algebra without identity element.*

Proof Proposition 3.22 shows that the Whittaker convolution defines a binary operation on $L_{\alpha,\nu}$ for which the norm is submultiplicative. The commutativity and associativity of the Whittaker convolution in the space $L_{\alpha,\nu}$ follows from the property $\big(\mathcal{W}_\alpha(f \underset{\alpha}{\diamond} g)\big)(\tau) = (\mathcal{W}_\alpha f)(\tau)\, (\mathcal{W}_\alpha g)(\tau)$ and the injectivity property of Lemma 3.14.

Suppose now that there exists e $\in L_{\alpha,\nu}$ such that $f \underset{\alpha}{\diamond} \mathrm{e} = f$ for all $f \in L_{\alpha,\nu}$. This means that

$$(\mathcal{W}_\alpha f)(\tau)\, (\mathcal{W}_\alpha \mathrm{e})(\tau) = (\mathcal{W}_\alpha f)(\tau) \qquad \text{for all } f \in L_{\alpha,\nu} \text{ and } \tau \geq 0.$$

Clearly, this implies that $(\mathcal{W}_\alpha \mathrm{e})(\tau) = 1$ for all $\tau \geq 0$, which contradicts Lemma 3.14. This shows that there exists no identity element for the Whittaker convolution on the space $L_{\alpha,\nu}$.

We will see that the index Whittaker transform on the Banach algebra $L_{\alpha,\nu}$ admits a Wiener-Lévy type theorem which resembles the classical Wiener-Lévy theorem on integral equations with difference kernel (cf. [71, §17, Corollary 1], [111, p. 164]). For this, we will need the following lemma:

Lemma 3.15 *Let $J : L_{\alpha,\nu} \longrightarrow \mathbb{C}$ be a linear functional satisfying*

$$J(f \underset{\alpha}{\diamond} g) = J(f) \cdot J(g) \qquad \textit{for all } f, g \in L_{\alpha,\nu}. \tag{3.108}$$

Then $J(f) = \int_0^\infty f(\xi)\, \mathbf{W}_{\alpha,i\tau}(\xi)\, r_\alpha(\xi) d\xi$ for some τ belonging to the complex strip $|\mathrm{Im}\, \tau| \leq \nu$, including infinity.

Notice that $\tau = \infty$ corresponds, by (3.107), to the zero functional on $L_{\alpha,\nu}$.

Proof By the standard theorem on duality of L^p spaces,

$$J(f) = \int_0^\infty f(\xi)\, \omega(\xi)\, r_\alpha(\xi) d\xi,$$

where $\frac{\omega}{W_{\alpha,\nu}} \in L^\infty(r_\alpha)$, i.e., (3.57) holds. Since $J(f \underset{\alpha}{\diamond} g) = J(f) \cdot J(g)$, for $f, g \in L_{\alpha,\nu}$ we have

$$\int_0^\infty f(\xi)\, \omega(\xi)\, r_\alpha(\xi) d\xi \cdot \int_0^\infty g(\xi)\, \omega(\xi)\, r_\alpha(\xi) d\xi$$

$$= \int_0^\infty \int_0^\infty (\mathcal{T}_\alpha^\xi f)(y)\, g(y) r_\alpha(y) dy\, \omega(\xi) r_\alpha(\xi) d\xi$$

$$= \int_0^\infty \int_0^\infty (\mathcal{T}_\alpha^y f)(\xi)\, \omega(\xi) r_\alpha(\xi) d\xi\; g(y) r_\alpha(y) dy$$

$$= \int_0^\infty \int_0^\infty (\mathcal{T}_\alpha^y \omega)(\xi)\, f(\xi) r_\alpha(\xi) d\xi\; g(y) r_\alpha(y) dy,$$

where the last equality follows from the commutativity of the Whittaker convolution, cf. Corollary 3.4. (The commutativity easily extends to $f \in L_{\alpha,\nu}$ and $\frac{\omega}{W_{\alpha,\nu}} \in L^\infty(r_\alpha)$ via a continuity argument.) Since f and g are arbitrary,

$$\omega(x)\,\omega(y) = (\mathcal{T}_\alpha^y \omega)(x) \equiv \int_0^\infty \omega(\xi)\, k_\alpha(x, y, \xi)\, r_\alpha(\xi) d\xi \qquad \text{for a.e. } x, y > 0$$

and the conclusion follows from Lemma 3.5.

Theorem 3.5 (Wiener-Lévy Type Theorem) *Let $f \in L_{\alpha,\nu}$ ($\alpha < \frac{1}{2}$, $\nu \geq 0$) and $\varrho \in \mathbb{C}$. The following assertions are equivalent:*

(i) *$\varrho + (\mathcal{W}_\alpha f)(\tau) \neq 0$ for all τ belonging to the complex strip $|\mathrm{Im}\,\tau| \leq \nu$, including infinity;*
(ii) *There exists a unique function $g \in L_{\alpha,\nu}$ such that*

$$\frac{1}{\varrho + (\mathcal{W}_\alpha f)(\tau)} = \varrho^{-1} + (\mathcal{W}_\alpha g)(\tau) \qquad (|\mathrm{Im}\,\tau| \leq \nu). \tag{3.109}$$

Before the proof, we need to introduce some relevant notions from Gelfand's theory of maximal ideals in commutative Banach algebras (cf. e.g. [170, Chapter 6]). Let V be a commutative Banach algebra with identity element. A *proper ideal* on V is a nonempty linear subspace $\mathcal{I} \subsetneqq V$ such that $v \cdot x = x \cdot v \in \mathcal{I}$ whenever $v \in V$ and $x \in \mathcal{I}$. A proper ideal $\mathcal{I}$ is said to be *maximal* in V if $\mathcal{I} = \mathcal{J}$ whenever $\mathcal{J}$ is a proper ideal such that $\mathcal{I} \subset \mathcal{J}$. A linear functional $F : V \longrightarrow \mathbb{C}$ is said to be a *multiplicative linear functional* if $F \not\equiv 0$ and $F(x \cdot y) = F(x)F(y)$ for all $x, y \in V$. (This implies that $F(\mathrm{e}) = 1$, where e is the identity element in V.) We will make use of the following basic results [170, Proposition 6.1.12 and Theorem 6.2.2]:

- If $v \in V$ is not invertible, then v is contained in a maximal ideal of V;
- If F is a multiplicative linear functional on V, then $\mathcal{I} = \mathrm{Ker}(F) \equiv \{v \in V \mid F(v) = 0\}$ is a maximal ideal on V, and conversely if $\mathcal{I}$ is a maximal ideal on V then there exists a unique multiplicative linear functional $F : V \longrightarrow \mathbb{C}$ such that $\mathcal{I} = \mathrm{Ker}(F)$.

Proof of Theorem 3.5 **(i) $\Longrightarrow$ (ii):** Let $V_{\alpha,\nu}$ be the Banach algebra obtained from $L_{\alpha,\nu}$ by formally adjoining an identity element e, that is, $V_{\alpha,\nu} := \{\varrho\mathrm{e} + f(\cdot) \mid \varrho \in \mathbb{C},\ f \in L_{\alpha,\nu}\}$ endowed with the norm $\|\varrho\mathrm{e} + f\| = |\varrho| + \|f\|_{L_{\alpha,\nu}}$. The index Whittaker transform is naturally extended to $V_{\alpha,\nu}$ as $\big(\mathcal{W}_\alpha(\varrho\mathrm{e} + f)\big)(\tau) := \varrho + (\mathcal{W}_\alpha f)(\tau)$ ($|\mathrm{Im}\,\tau| \leq \nu$). It follows from Proposition 3.22 that

$$J_{\alpha,\tau} : V_{\alpha,\nu} \longrightarrow \mathbb{C}, \qquad J_{\alpha,\tau}(\varrho\mathrm{e} + f) := \big(\mathcal{W}_\alpha(\varrho\mathrm{e} + f)\big)(\tau) \tag{3.110}$$

is, for each τ in the strip $|\text{Im}\,\tau| \leq \nu$ (including infinity), a multiplicative linear functional on $V_{\alpha,\nu}$. We claim that there are no multiplicative linear functionals in $V_{\alpha,\nu}$ other than the functionals $J_{\alpha,\tau}$ defined in (3.110). Indeed, if $J : V_{\alpha,\nu} \longrightarrow \mathbb{C}$ is a multiplicative linear functional, then restricting to $L_{\alpha,\nu}$ we obtain a functional $f \mapsto J(f)$ on $L_{\alpha,\nu}$ such that (3.108) holds. By Lemma 3.15, $J(f) = (\mathcal{W}_\alpha f)(\tau)$ for some τ in the strip $|\text{Im}\,\tau| \leq \nu$ (including infinity), and thus by linearity $J(\varrho\mathrm{e} + f) = \varrho + (\mathcal{W}_\alpha f)(\tau)$. Hence $J = J_{\alpha,\tau}$, as we had claimed.

Assume that $\varrho + (\mathcal{W}_\alpha f)(\tau) \neq 0$ for all τ with $|\text{Im}\,\tau| \leq \nu$. By the above, we have $\varrho\mathrm{e} + f \notin \text{Ker}(J)$ for all multiplicative linear functionals $J : V_{\alpha,\nu} \longrightarrow \mathbb{C}$, and using the results stated before the proof we deduce that $\varrho\mathrm{e} + f$ is invertible on $V_{\alpha,\nu}$. Denoting the inverse by $\varrho'\mathrm{e} + g$ ($\varrho' \in \mathbb{C}$, $g \in L_{\alpha,\nu}$), we obtain

$$\big(\varrho + (\mathcal{W}_\alpha f)(\tau)\big) \cdot \big(\varrho' + (\mathcal{W}_\alpha g)(\tau)\big) = 1 \qquad (|\text{Im}\,\tau| \leq \nu).$$

We know that $\lim_{\tau\to\infty} \mathbf{W}_{\alpha,i\tau}(y) = 0$ for $y > 0$, hence as in Lemma 3.14 it follows that the left hand side equals $\varrho\varrho'$ when $\tau = \infty$. We thus have $\varrho' = \varrho^{-1}$, so that (3.109) holds.

(ii) $\Longrightarrow$ (i): This implication is straightforward: given that $g \in L_{\alpha,\nu}$, Lemma 3.14 ensures that $|(\mathcal{W}_\alpha g)(\tau)| \leq (\mathcal{W}_\alpha |g|)(i\nu) < \infty$ for all τ in the strip $|\text{Im}\,\tau| \leq \nu$. Since (3.109) holds, it follows that $\varrho + (\mathcal{W}_\alpha f)(\tau) = \frac{1}{\varrho^{-1} + (\mathcal{W}_\alpha g)(\tau)} \neq 0$ for all τ with $|\text{Im}\,\tau| \leq \nu$.

3.8 Convolution-Type Integral Equations

In this final section of the chapter we demonstrate that the Whittaker convolution, and especially the analogue of the Wiener-Lévy theorem proved above, can be used to study the existence of solution for integral equations of the second kind which can be represented as Whittaker convolution equations, in the sense defined as follows:

Definition 3.11 The integral equation of the second kind

$$f(x) + \int_0^\infty J(x, y) f(y)\, dy = h(x), \tag{3.111}$$

where h is a known function and f is to be determined, is said to be a *Whittaker convolution equation* if there exists $\alpha < \frac{1}{2}$ and $\theta \in L_{\alpha,0}$ such that $J(x, y) = (\mathcal{T}_\alpha^x \theta)(y)\, r_\alpha(y) \equiv (\mathcal{T}_\alpha^x \theta)(y)\, y^{-2\alpha} e^{-1/y}$. In other words, (3.111) is a Whittaker convolution equation if it can be written in the form

$$f(x) + (f \underset{\alpha}{\diamond} \theta)(x) = h(x) \tag{3.112}$$

for some $\alpha < \frac{1}{2}$ and $\theta \in L_{\alpha,0}$.

Suppose that $h, \theta \in L_{\alpha,\nu}$ (being $\alpha < \frac{1}{2}$ and $\nu \geq 0$), and consider the convolution-type Eq. (3.112). Applying the index Whittaker transform to both sides of the convolution equation, we get

$$(\mathcal{W}_\alpha f)(\tau)\big[1 + (\mathcal{W}_\alpha \theta)(\tau)\big] = (\mathcal{W}_\alpha h)(\tau) \qquad (|\mathrm{Im}\,\tau| \leq \nu). \tag{3.113}$$

Now, Theorem 3.5 shows that the condition

$$1 + (\mathcal{W}_\alpha \theta)(\tau) \neq 0 \quad \text{throughout the strip } |\mathrm{Im}\,\tau| \leq \nu$$

is a necessary and sufficient condition for the existence of a unique $g \in L_{\alpha,\nu}$ satisfying

$$\frac{1}{1 + (\mathcal{W}_\alpha \theta)(\tau)} = 1 + (\mathcal{W}_\alpha g)(\tau) \qquad (|\mathrm{Im}\,\tau| \leq \nu), \tag{3.114}$$

and if this holds then from (3.113) we obtain $(\mathcal{W}_\alpha f)(\tau) = (\mathcal{W}_\alpha h)(\tau)\big[1 + (\mathcal{W}_\alpha g)(\tau)\big]$ $(|\mathrm{Im}\,\tau| \leq \nu)$ or, equivalently,

$$f(x) = h(x) + (h \underset{\alpha}{\diamond} g)(x) = h(x) + \int_0^\infty J_g(x, y)\, h(y)\, dy, \tag{3.115}$$

where $J_g(x, y) = (\mathcal{T}_\alpha^x g)(y)\, r_\alpha(y)$. In summary, we have proved the following:

Theorem 3.6 *Let $J(x, y) = (\mathcal{T}_\alpha^x \theta)(y)\, r_\alpha(y)$ where $\theta \in L_{\alpha,\nu}$ ($\alpha < \frac{1}{2}$, $\nu \geq 0$), and suppose that $1 + (\mathcal{W}_\alpha \theta)(\tau) \neq 0$ for all τ in the strip $|\mathrm{Im}\,\tau| \leq \nu$, including infinity. Then the integral equation* (3.112) *has, for any $h \in L_{\alpha,\nu}$ a unique solution $f \in L_{\alpha,\nu}$ which can be represented in the form* (3.115) *for some $g \in L_{\alpha,\nu}$. Conversely, if $1 + (\mathcal{W}_\alpha \theta)(\tau_0) = 0$ for some τ_0 with $|\mathrm{Im}\,\tau_0| \leq \nu$, then Eq.* (3.112) *is not solvable in the space $L_{\alpha,\nu}$.*

We point out that as long as $\frac{(\mathcal{W}_\alpha \theta)(\tau)}{1+(\mathcal{W}_\alpha \theta)(\tau)} = O(\tau^{-2})$, the representation (3.115) for the solution of the integral equation can be rewritten as

$$f(x) = h(x) - \int_0^\infty \int_0^\infty \frac{(\mathcal{W}_\alpha \theta)(\tau)}{1 + (\mathcal{W}_\alpha \theta)(\tau)} \mathbf{W}_{\alpha,i\tau}(x)\, \mathbf{W}_{\alpha,i\tau}(y)\, \rho_\alpha(\tau) d\tau\, h(y)\, r_\alpha(y) dy \tag{3.116}$$

(here we used (3.54) and Proposition 3.20(a)). In many cases of interest, the index Whittaker transform $(\mathcal{W}_\alpha \theta)(\tau)$ can be computed in closed form using integration formulas for the Whittaker W function (which can be found in published tables of integrals, see e.g. [157, Section 2.19]), so that (3.116) becomes an explicit expression for the solution of the convolution integral equation, which can be evaluated using numerical integration.

The Whittaker translation of the power function $\theta(x) = x^\beta$, whose closed form was computed in Lemma 3.4, yields a large family of Whittaker convolution integral equations to which this theorem can be applied:

Corollary 3.5 *Let* $h \in L_{\alpha,\nu}$ *(*$\alpha < \frac{1}{2}$, $\nu \geq 0$*),* $\lambda \in \mathbb{C}$*, and* $\beta \in \mathbb{C}$ *with* $\operatorname{Re}\beta < -\frac{1}{2} + \alpha - \nu$*. The integral equation*

$$f(x) + \lambda \int_0^\infty (x+y)^\beta \, \mathbf{W}_{\alpha,\alpha-\frac{1}{2}-\beta}\Big(\frac{xy}{x+y}\Big) \, f(y) \, r_\alpha(y) dy = h(x) \tag{3.117}$$

has a unique solution $f \in L_{\alpha,\nu}$ *if and only if the condition*

$$\Gamma(\beta) + \lambda \, \Gamma\big(\beta - \tfrac{1}{2} + \alpha + i\tau\big)\Gamma\big(\beta - \tfrac{1}{2} + \alpha - i\tau\big) \neq 0$$

holds for all $\tau \in \mathbb{C}$ *in the strip* $|\operatorname{Im}\tau| \leq \nu$*, including infinity.*

Proof Let $\theta(x) = \lambda x^\beta$. It is clear from (3.106) that $\theta \in L_{\alpha,\nu}$. We have seen in Lemma 3.4 that

$$(\mathcal{T}_\alpha^x \theta)(y) = \lambda \, (x+y)^\beta \, \mathbf{W}_{\alpha,\alpha-\frac{1}{2}-\beta}\Big(\frac{xy}{x+y}\Big).$$

The index Whittaker transform $\mathcal{W}_\alpha \theta$ is computed using relation 2.19.3.7 in [157], valid for $|\operatorname{Im}\tau| \leq \nu$:

$$(\mathcal{W}_\alpha\theta)(\tau) = \lambda \int_0^\infty x^\beta \, \mathbf{W}_{\alpha,i\tau}(x) \, r_\alpha(x) dx = \frac{\lambda}{\Gamma(-\beta)} \Gamma\big(-\beta - \tfrac{1}{2} + \alpha + i\tau\big)\Gamma\big(-\beta - \tfrac{1}{2} + \alpha - i\tau\big).$$

The corollary is therefore obtained by setting $\theta(x) = \lambda x^\beta$ in Theorem 3.6.

It should be emphasized that Theorem 3.6 is not just an existence and uniqueness theorem for the solution of Whittaker convolution integral equations: under a mild assumption, (3.116) provides an explicit expression for the solution which involves integration with respect to the parameters of the Whittaker function. However, if we are able to determine a closed-form expression for the function $g \in L_{\alpha,\nu}$ which satisfies (3.114), then the representation (3.115) yields a more tractable explicit expression for the solution which does not involve index integrals. This is illustrated in the following corollary:

Corollary 3.6 *If* $h \in L_{-n,\nu}$ *where* $n \in \mathbb{N}_0$ *and* $0 \leq \nu < \frac{1}{2}$*, then the integral equation*

$$f(x) + \frac{n!}{\pi} \int_0^\infty (x+y)^{-n-1} \, \mathbf{W}_{-n,\frac{1}{2}}\Big(\frac{xy}{x+y}\Big) \, f(y) \, y^{2n} e^{-\frac{1}{y}} dy = h(x) \tag{3.118}$$

has a unique solution $f \in L_{-n,\nu}$, which is given by

$$f(x) = h(x) + (h \underset{-n}{\diamond} g_n)(x)$$
$$= h(x) + \int_0^\infty \int_0^\infty k_{-n}(x, y, \xi)\, h(y)\, g_n(\xi)\, (y\xi)^{2n} e^{-\frac{1}{y}-\frac{1}{\xi}}\, dy d\xi,$$

where

$$g_n(x) := -\frac{n!}{\pi^2} \sum_{k=0}^{n} \frac{\Gamma(\frac{1}{2}+k)^2}{k!} x^{-n+k-1} \boldsymbol{W}_{-k,0}(x). \tag{3.119}$$

Proof The integral equation (3.118) is the particular case of (3.117) which is obtained by setting $\alpha = -n$, $\beta = -n-1$ and $\lambda = \frac{n!}{\pi}$. In this case, $(\mathcal{W}_{-n}\theta)(\tau) = \frac{1}{\pi}\Gamma(\frac{1}{2}+i\tau)\Gamma(\frac{1}{2}-i\tau) = \frac{1}{\cosh(\pi\tau)}$. Clearly, if $|\mathrm{Im}\,\tau| < \frac{1}{2}$ then $\mathrm{Re}[\cosh(\pi\tau)] > 0$, hence the solvability condition $1+(\mathcal{W}_{-n}\theta)(\tau) \neq 0$ holds in the strip $|\mathrm{Im}\,\tau| \le \nu < \frac{1}{2}$ and, according to Theorem 3.6, the unique solution of (3.118) is the function $f(x) = h(x) + (h \underset{-n}{\diamond} g)(x)$, where g is the function satisfying

$$(\mathcal{W}_{-n}g)(\tau) = \frac{1}{1+(\mathcal{W}_{-n}\theta)(\tau)} - 1 = -\frac{1}{2\cosh^2(\frac{\pi\tau}{2})}.$$

It remains to show that the function (3.119) satisfies this requirement. Using integral 2.16.48.14 of [156] and recalling the identity (3.16), we find that

$$\int_0^\infty \frac{1}{2\cosh^2(\frac{\pi\tau}{2})} \boldsymbol{W}_{0,i\tau}(x)\, \rho_0(\tau) d\tau = \frac{1}{\pi x} \boldsymbol{W}_{0,0}(x). \tag{3.120}$$

Now, by (3.59) and the recurrence relation for the Gamma function we have

$$\begin{aligned} \left|\Gamma\left(\tfrac{1}{2}+i\tau\right)\right|^2 \frac{d^n}{dx^n} \boldsymbol{W}_{0,i\tau}(x) &= (-1)^n \left|\Gamma\left(\tfrac{1}{2}+i\tau\right)\right|^2 \left|\left(\tfrac{1}{2}+i\tau\right)_n\right|^2 \boldsymbol{W}_{-n,i\tau}(x) \\ &= (-1)^n \left|\Gamma\left(\tfrac{1}{2}+n+i\tau\right)\right|^2 \boldsymbol{W}_{-n,i\tau}(x). \end{aligned} \tag{3.121}$$

Therefore, applying $\frac{d^n}{dx^n}$ to both sides of (3.120) we obtain

$$\begin{aligned} \int_0^\infty \frac{1}{2\cosh^2(\frac{\pi\tau}{2})} \boldsymbol{W}_{-n,i\tau}(x)\, \rho_{-n}(\tau) d\tau &= (-1)^n \frac{d^n}{dx^n}\Big(\frac{1}{\pi x} \boldsymbol{W}_{0,0}(x)\Big) \\ &= \frac{n!}{\pi^2} \sum_{k=0}^{n} \frac{|\Gamma(k+\frac{1}{2})|^2}{k!} x^{-n+k-1} \boldsymbol{W}_{-k,0}(x), \end{aligned}$$

where the possibility of differentiating under the integral is justified as in the proof of Lemma 3.5, and the last equality follows from Leibniz's rule and the identities $\frac{d^{n-k}}{dx^{n-k}}(x^{-1}) = (-1)^{n-k}(n-k)!x^{-n+k-1}$ and (3.121). Recalling (3.54), we see that $\big[\mathcal{W}_{-n}^{-1}\big(\frac{1}{2}\cosh^{-2}(\frac{\pi\tau}{2})\big)\big](x) = -g_n(x)$ and, consequently, $(\mathcal{W}_{-n}g_n)(\tau) = -\frac{1}{2}\cosh^{-2}(\frac{\pi\tau}{2})$, as was to be proved.

The integral equation

$$f(x) + \lambda \int_0^\infty f(y)\frac{e^{-x-y}}{x+y}dy = h(x) \tag{3.122}$$

which has been introduced by Lebedev in [114], can be interpreted as the particular case of (3.117) which is obtained by setting $\alpha = 0$, $\beta = -1$, $\boldsymbol{f}(x) = x^{-1}e^{-x}f(\frac{1}{2x})$ and $\boldsymbol{h}(x) = x^{-1}e^{-x}h(\frac{1}{2x})$. It is shown in [207, Section 17.1] that Lebedev's equation (3.122) is a convolution equation with respect to the Kontorovich–Lebedev convolution, and an explicit representation for the solution has been derived for $\lambda = \frac{1}{\pi}$. Equation (3.118) is therefore a natural generalization of Lebedev's integral equation for which Corollary 3.6 provides an explicit expression for the solution. The existence of explicit solution for the generalized Lebedev equation (3.118) is noteworthy because the Whittaker function $W_{\alpha,\frac{1}{2}}(\cdot)$ is often encountered in problems in physics and chemistry [113].

Remark 3.11 For ease of presentation, throughout the chapter we have introduced the Whittaker convolution as the generalized convolution determined by the Sturm–Liouville operator $\mathcal{A}_\alpha u(y) = y^2u''(y) + (1 + 2(1-\alpha)y)u'(y)$. It is, however, not difficult to extend the construction in order to define the generalized convolution associated with a Sturm–Liouville operator $\mathcal{A} \equiv \mathcal{A}_{\alpha,\gamma,c}$ of the more general form

$$\mathcal{A}u(y) = \gamma y^2u''(y) + \gamma(c+2(1-\alpha)y)u'(y) = \frac{1}{\boldsymbol{r}(y)}(\boldsymbol{p}u')'(y) \qquad (\gamma, c > 0,\ \alpha < \tfrac{1}{2}),$$

where $\boldsymbol{r}(\xi) := \frac{1}{c}r_\alpha(\frac{\xi}{c})$ and $\boldsymbol{p}(\xi) := c\gamma\xi^{2(1-\alpha)}e^{-1/\xi}$. It should be noted that this extension includes the infinitesimal generator of a general (nonstandardized) Shiryaev process (as defined in (3.1)) with drift $\mu > \frac{\sigma^2}{2}$. The crucial observation here is that the solutions of the more general Sturm–Liouville problem $-\mathcal{A}u = \lambda u$ (with Neumann boundary conditions) can also be expressed in terms of the Whittaker W function and, therefore, the corresponding product formula can be obtained by applying elementary changes of variables to the product formula determined in Sect. 3.2. The kernels of the more general product formula also constitute a family of probability densities, so the induced notions of generalized translation, convolution, Lévy processes and moment functions (defined in analogy with those of the previous sections) have essentially the same properties as before.

Table 3.1 Basic results and definitions for the extended Whittaker convolution

Solution of the Sturm–Liouville boundary value problem $[-\mathcal{A}u = \lambda u,$ $u(0) = 1,\ (\boldsymbol{p}u')(0) = 0]$	$\boldsymbol{W}_{\alpha,\Lambda}(\frac{x}{c}) \quad \big[\Lambda := \sqrt{(\frac{1}{2}-\alpha)^2 - \frac{\lambda}{\gamma}}\big]$
Product formula for the Sturm–Liouville solutions	$\boldsymbol{W}_{\alpha,\nu}(\frac{x}{c})\, \boldsymbol{W}_{\alpha,\nu}(\frac{y}{c}) = \int_0^\infty \boldsymbol{W}_{\alpha,\nu}(\frac{\xi}{c})\, \boldsymbol{k}(x,y,\xi)\, \boldsymbol{r}(\xi) d\xi$ $\big[\boldsymbol{k}(x,y,\xi) := k_\alpha(\frac{x}{c}, \frac{y}{c}, \frac{\xi}{c})\big]$
Extended Whittaker translation operator	$(\mathcal{T}^y f)(x) := \int_0^\infty f(\xi)\, \boldsymbol{q}(x,y,\xi)\, \boldsymbol{r}(\xi) d\xi$ $\equiv \big(\mathcal{T}_\alpha^{y/c} f(c\,\cdot)\big)(\frac{x}{c})$
Extended index Whittaker transform of functions and measures	$(\boldsymbol{\mathcal{W}} f)(\tau) := \int_0^\infty f(y)\, \boldsymbol{W}_{\alpha,i\tau}(\frac{y}{c})\, \boldsymbol{r}(y) dy$ $\equiv \big(\mathcal{W}_\alpha f(c\,\cdot)\big)(\tau)$ $(\boldsymbol{\mathcal{W}}\mu)(\lambda) := \int_{\mathbb{R}_0^+} \boldsymbol{W}_{\alpha,\Lambda}(\frac{y}{c})\, \mu(dy) \equiv \widehat{\Theta_{1/c}\mu}(\gamma\lambda)$
Extended Whittaker convolution of functions and measures	$(f \mathbin{\boldsymbol{\diamond}} g)(x) := \int_0^\infty (\mathcal{T}^x f)(\xi)\, g(\xi)\, \boldsymbol{r}(\xi) d\xi$ $= \big(f(c\,\cdot) \underset{\alpha}{\diamond} g(c\,\cdot)\big)(\frac{x}{c})$ $(\mu \mathbin{\boldsymbol{\diamond}} \nu)(d\xi) := \int_{\mathbb{R}_0^+}\int_{\mathbb{R}_0^+} \boldsymbol{q}(x,y,\xi) \boldsymbol{r}(\xi) d\xi\, \mu(dx)\, \nu(dy)$
$\boldsymbol{\diamond}$-*infinitely divisible measures*, $\boldsymbol{\diamond}$-*convolution semigroups* and $\boldsymbol{\diamond}$-*Lévy processes*	Replace $\underset{\alpha}{\diamond}$ by $\boldsymbol{\diamond}$ in the previous definitions
(Canonical) $\boldsymbol{\diamond}$-*moment functions*	$\boldsymbol{\varphi}_k(x) = \varphi_k(\frac{x}{c})$, where $\varphi_k(\cdot)$ are (canonical) $\underset{\alpha}{\diamond}$-moment functions
Extended Whittaker convolution equation	An equation of the form (3.111), with $J(x,y) = (\mathcal{T}^x\theta)(y)\boldsymbol{r}(y)$

Table 3.1 collects the product formula and the definitions of the fundamental objects which underlie the construction of this extension of the Whittaker convolution structure. Using these definitions and the same proofs as before, it is straightforward to check that the main results of the previous sections—such as the Lévy-type martingale characterization for the nonstandardized Shiryaev process or the Wiener-Lévy type theorem for the index Whittaker transform—are also valid for the extended Whittaker convolution structure.

Chapter 4
Generalized Convolutions for Sturm-Liouville Operators

This chapter is dedicated to the problem of constructing *Sturm-Liouville convolutions*, i.e. generalized convolution operators associated with Sturm-Liouville differential expressions. The convolutions constructed here will, in particular, allow us to interpret the diffusion process generated by the Neumann realization $(\mathcal{L}^{(2)}, \mathcal{D}(\mathcal{L}^{(2)}))$ of the Sturm-Liouville operator as a Lévy-like process.

We consider Sturm-Liouville operators of the form (2.10) but without zero order term, that is,

$$\ell = -\frac{1}{r}\frac{d}{dx}\Big(p\,\frac{d}{dx}\Big), \qquad x \in (a,b) \tag{4.1}$$

$(-\infty \leq a < b \leq \infty)$. Throughout the chapter we always assume that the coefficients are such that $p(x), r(x) > 0$ for all $x \in (a,b)$, $p, p', r, r' \in \mathrm{AC}_{\mathrm{loc}}(a,b)$ and $\int_a^c \int_y^c \frac{dx}{p(x)} r(y)dy < \infty$.

Remark 4.1 We shall make extensive use of the fact that the differential expression (4.1) can be transformed into the standard form

$$\widetilde{\ell} = -\frac{1}{A}\frac{d}{d\xi}\Big(A\frac{d}{d\xi}\Big) = -\frac{d^2}{d\xi^2} - \frac{A'}{A}\frac{d}{d\xi}.$$

This is achieved by setting

$$A(\xi) := \sqrt{p(\gamma^{-1}(\xi))\, r(\gamma^{-1}(\xi))}, \tag{4.2}$$

where γ^{-1} is the inverse of the increasing function

$$\gamma(x) = \int_c^x \sqrt{\frac{r(y)}{p(y)}}\, dy,$$

R. Sousa et al., *Convolution-like Structures, Differential Operators and Diffusion Processes*, Lecture Notes in Mathematics 2315,
https://doi.org/10.1007/978-3-031-05296-5_4

$c \in (a, b)$ being a fixed point (if $\sqrt{\frac{r(y)}{p(y)}}$ is integrable near a, we may also take $c = a$). Indeed, we know from Remark 2.1 that a given function $\omega_\lambda : (a, b) \to \mathbb{C}$ satisfies $\ell(\omega_\lambda) = \lambda\omega_\lambda$ if and only if $\widetilde{\omega}_\lambda(\xi) := \omega_\lambda(\gamma^{-1}(\xi))$ satisfies $\widetilde{\ell}(\widetilde{\omega}_\lambda) = \lambda\widetilde{\omega}_\lambda$.

4.1 Known Results and Motivation

As noted in Sect. 2.3, the roots of the problem of constructing Sturm-Liouville convolutions originate in the work of Delsarte and Levitan on generalized translation operators determined by ordinary differential operators [45, 117, 118, 120, 121]. This theory was later developed by Chébli [30], who constructed convolutions satisfying the hypergroup axioms for Sturm-Liouville operators belonging to a family which includes the Bessel operator $\frac{1}{2}\frac{d^2}{dx^2} + \frac{\eta+\frac{1}{2}}{x}\frac{d}{dx}$ $(\eta > -\frac{1}{2})$ and the Jacobi operator (2.42) with $\alpha \geq \beta \geq \frac{1}{2}$. In turn, Zeuner [209, 210] introduced the general notion of a Sturm-Liouville hypergroup and extended the results of Levitan and Chébli to a larger class of differential operators. (Further remarks on the historical development of the topic can be found in [19, pp. 256–257].)

The definition of Sturm-Liouville hypergroup proposed by Zeuner reads as follows:

Definition 4.1 ([209]) A hypergroup $(\mathbb{R}_0^+, *)$ is said to be a *Sturm-Liouville hypergroup* if there exists a function A on $\mathbb{R}_0^+$ satisfying the condition

SL0 $A \in \mathrm{C}(\mathbb{R}_0^+) \cap \mathrm{C}^1(\mathbb{R}^+)$ and $A(x) > 0$ for $x > 0$

such that, for every function $f \in \mathrm{C}_{\mathrm{c,even}}^\infty$, the convolution

$$v_f(x, y) = \int_{\mathbb{R}_0^+} f(\xi)(\delta_x * \delta_y)(d\xi) \tag{4.3}$$

belongs to $\mathrm{C}^2\big((\mathbb{R}_0^+)^2\big)$ and satisfies $(\ell_x v_f)(x, y) = (\ell_y v_f)(x, y)$, $(\partial_y v_f)(x, 0) = 0$ $(x > 0)$, where $\ell_x = -\frac{1}{A(x)}\frac{\partial}{\partial x}(A(x)\frac{\partial}{\partial x})$.

The following fundamental existence theorem for Sturm-Liouville hypergroups was established in [210]:

Theorem 4.1 *Suppose that A satisfies SL0 and is such that*

SL1 *One of the following assertions holds:*

SL1.1 $A(0) = 0$ *and* $\frac{A'(x)}{A(x)} = \frac{\alpha_0}{x} + \alpha_1(x)$ *for x in a neighbourhood of* 0, *where* $\alpha_0 > 0$ *and* $\alpha_1 \in \mathrm{C}^\infty(\mathbb{R})$ *is an odd function;*

SL1.2 $A(0) > 0$ *and* $A \in \mathrm{C}^1(\mathbb{R}_0^+)$.

SL2 *There exists* $\eta \in \mathrm{C}^1(\mathbb{R}_0^+)$ *such that* $\eta \geq 0$, *the functions* $\boldsymbol{\phi}_\eta := \frac{A'}{A} - \eta$, $\boldsymbol{\psi}_\eta := \frac{1}{2}\eta' - \frac{1}{4}\eta^2 + \frac{A'}{2A} \cdot \eta$ *are both decreasing on* $\mathbb{R}^+$ *and* $\lim_{x\to\infty} \boldsymbol{\phi}_\eta(x) = 0$.

Define the convolution $*$ via (4.3) where, for $f \in C^\infty_{c,\text{even}}$, v_f denotes the unique solution of $\ell_x v_f = \ell_y v_f$, $v_f(x,0) = v_f(0,x) = f(x)$, $(\partial_y v_f)(x,0) = (\partial_x v_f)(0,y) = 0$. Then $(\mathbb{R}_0^+, *)$ is a Sturm-Liouville hypergroup.

Remark 4.2 In this chapter we focus on the construction of convolutions defined on noncompact intervals of $\mathbb{R}$. However, it is worth noting that Sturm-Liouville hypergroups have also been studied on compact intervals: a hypergroup $([b_1, b_2], *)$ is said to be a *Sturm-Liouville hypergroup of compact type* [19, Definition 3.5.76] if there exists a function $A \in C[b_1, b_2] \cap C^1(b_1, b_2)$ such that $A(x) > 0$ for $b_1 < x < b_2$ and $\int_{b_1}^{b_2} A(x)dx = 1$ which satisfies the following requirement: for each $f \in C_c^\infty(b_1, b_2)$ the convolution $\int_{[b_1,b_2]} f(\xi)(\delta_x * \delta_y)(d\xi)$ is twice differentiable and satisfies $(\ell_x v_f)(x,y) = (\ell_y v_f)(x,y)$ and $(\partial_y v_f)(x,b_1) = (\partial_y v_f)(x,b_2) = 0$ $(b_1 < x < b_2)$, where $\ell_x = -\frac{1}{A(x)}\frac{\partial}{\partial x}(A(x)\frac{\partial}{\partial x})$ is the associated Sturm-Liouville operator. The simplest example (where A is constant) is the two-point support hypergroup $([0,\beta], \underset{\beta}{\circledcirc})$ defined as $\delta_x \underset{\beta}{\circledcirc} \delta_y = \frac{1}{2}(\delta_{|x-y|} + \delta_{\beta-|\beta-x-y|})$. Another important example (cf. [19, p. 242]) is that of the compact Jacobi hypergroup $([-1,1], \underset{\alpha,\beta}{\circledast})$: for (α,β) such that $-1 < \beta \leq \alpha$ and either $\beta \geq -\frac{1}{2}$ or $\alpha + \beta \geq 0$, this hypergroup is defined as $\delta_x \underset{\alpha,\beta}{\circledast} \delta_y := \boldsymbol{v}_{x,y}^{(\alpha,\beta)}$, where $\boldsymbol{v}_{x,y}^{(\alpha,\beta)}$ is the unique measure such that the product formula

$$R_n^{(\alpha,\beta)}(x)\, R_n^{(\alpha,\beta)}(y) = \int_{[-1,1]} R_n^{(\alpha,\beta)}(\xi)\, \boldsymbol{v}_{x,y}^{(\alpha,\beta)}(d\xi) \qquad (n \in \mathbb{N}_0) \tag{4.4}$$

holds for the Jacobi polynomials $R_n^{(\alpha,\beta)}$ defined in Example 2.2. (It is proved in [70] that for each $x, y \in [-1,1]$ there exists a unique measure $\boldsymbol{v}_{x,y}^{(\alpha,\beta)} \in \mathcal{P}[-1,1]$ such that (4.4) holds.) Further existence theorems for Sturm-Liouville hypergroups of compact type associated with suitable families of differential operators have been established in [19, 210].

Theorem 4.1 is the existence theorem which underlies the general theory of one-dimensional Sturm-Liouville convolution-like operators developed in [19, 163, 184, 210]; it includes, as particular cases, all the concrete examples of hypergroup structures on $\mathbb{R}_0^+$ which have been reported in earlier literature (cf. [19, 68]).

In the previous chapter we studied the problem of existence of a Sturm-Liouville convolution for the particular case of the generator (3.2) of the Shiryaev process, and we established the following result:

Proposition 4.1 *Let $\mathcal{T}_\alpha^y$ and $\underset{\alpha}{\diamond}$ be the Whittaker translation and convolution (Definitions 3.2 and 3.4 respectively), and let $f \in C^\infty_{c,\text{even}}$. Then the function*

$$v_f(x,y) := (\mathcal{T}_\alpha^y f)(x) \equiv \int_{\mathbb{R}_0^+} f(\xi)(\delta_x \underset{\alpha}{\diamond} \delta_y)(d\xi)$$

is a solution of $\mathcal{A}_{\alpha,x} v_f = \mathcal{A}_{\alpha,y} v_f$, $v_f(x,0) = v_f(0,x) = f(x)$, $(\partial_y^{[1]} v_f)(x,0) = (\partial_x^{[1]} v_f)(0,y) = 0$. *(Here* $\mathcal{A}_{\alpha,x}$ *and* $\mathcal{A}_{\alpha,y}$ *denote the differential operator* (3.2) *acting on the variable x and y respectively, and* $\partial_\xi^{[1]} := p_\alpha(\xi)\frac{\partial}{\partial \xi}$.*)*

Proof The fact that v_f is a solution of $\mathcal{A}_{\alpha,x} v_f = \mathcal{A}_{\alpha,y} v_f$ follows from the proof of Lemma 3.5, and the claimed boundary conditions at the axes $x = 0$ and $y = 0$ were proved in Lemma 3.10.

This result is not a particular case of Theorem 4.1 because the Sturm-Liouville operator $\mathcal{A}_\alpha$ does not belong to the family of operators satisfying assumptions SL1–SL2. (Note that $\mathcal{A}_\alpha$ is transformed, via the change of variables $z = \log x$, into the operator $\frac{d^2}{dz^2} + (1 - 2\alpha + e^{-z})\frac{d}{dz}$ defined on the interval $(a,b) = (-\infty,\infty)$.) Moreover, it was observed in Sect. 3.5 that, unlike the convolutions of Theorem 4.1, the Whittaker convolution does not satisfy the compactness axiom H6 of hypergroups. However, many of the properties of the Whittaker convolution established in Chap. 3 are remarkably similar to those of Sturm-Liouville hypergroups. This leads to natural questions, namely whether one can construct other Sturm-Liouville convolutions which do not satisfy the compactness axiom and, more specifically, whether it is possible to achieve this by extending the PDE approach of [210] to the Sturm-Liouville operator $\mathcal{A}_\alpha$ and other operators of a similar sort. A positive answer to these questions is given within this chapter.

4.2 Laplace-Type Representation

The possibility of constructing a generalized convolution associated with the Sturm-Liouville expression (4.1) is strongly connected with the positivity-preservingness of solutions of the hyperbolic Cauchy problem $\ell_x v = \ell_y v$, $v(x,0) = v(0,x) = f(x)$, $(\partial_y v)(x,0) = (\partial_x v)(0,y) = 0$. We now introduce an assumption which will be seen to be sufficient for the Cauchy problem to be positivity preserving. Recall that the function A, defined in (4.2), is the coefficient associated with the transformation of ℓ into the standard form (Remark 4.1).

Assumption MP *We have* $\gamma(b) = \int_c^b \sqrt{\frac{r(y)}{p(y)}}\,dy = \infty$, *and there exists* $\eta \in \mathrm{C}^1(\gamma(a),\infty)$ *such that* $\eta \geq 0$, *the functions* $\boldsymbol{\phi}_\eta := \frac{A'}{A} - \eta$, $\boldsymbol{\psi}_\eta := \frac{1}{2}\eta' - \frac{1}{4}\eta^2 + \frac{A'}{2A}\cdot\eta$ *are both decreasing on* $(\gamma(a),\infty)$ *and* $\boldsymbol{\phi}_\eta$ *satisfies* $\lim_{\xi\to\infty}\boldsymbol{\phi}_\eta(\xi) = 0$.

The reader will notice that this assumption is similar to condition SL2 in the existence theorem for Sturm-Liouville hypergroups stated above (Theorem 4.1), but it is more general as it does not require the function η to be C^1 at the left endpoint of the interval.

As in Sect. 2.2, in the sequel we denote by $w_\lambda(\cdot)$ the unique solution of (2.12), and $\{a_m\}_{m\in\mathbb{N}}$ will denote a sequence $b > a_1 > a_2 > \ldots$ with $\lim a_m = a$. Having in mind the product formula that we shall establish for Sturm-Liouville expressions

(4.1) which satisfy Assumption MP, in this section we prove the related fact that the solution $w_\lambda(x)$ of the initial value problem (2.12) admits a representation as the Fourier transform of a subprobability measure. To this end, we need a few lemmas. We start by stating some important properties which hold for all Sturm-Liouville operators of the form (4.1) which satisfy Assumption MP.

Lemma 4.1 *If Assumption MP holds, then*

(a) *The function* $\frac{A'}{A}$ *is nonnegative, and there exists a finite limit* $\sigma := \lim_{\xi\to\infty} \frac{A'(\xi)}{2A(\xi)} \in \mathbb{R}_0^+$.

(b) *If* $\lambda \le \sigma^2$*, then* $w_\lambda(x) > 0$ *for all* $x \in [a, b)$.

(c) *If* $\lambda > \sigma^2$*, then* $w_\lambda(\cdot)$ *has infinitely many zeros on* $[a, b)$.

(d) *b is a natural endpoint for the Sturm-Liouville operator* ℓ.

Proof The proofs of **(a)** and **(b)** are rather technical and rely on a careful study of (the coefficients of) the differential operator $\widetilde{\ell}$; see, respectively, Section 2 and Proposition 4.2 of [210].

Concerning part **(c)**, we first apply the Liouville transformation (Remark 2.1) to deduce that that the function $\sqrt{A(\xi)}\, w_\lambda(\gamma^{-1}(\xi))$ is a solution of $-v''+(\mathfrak{q}-\lambda)v = 0$, where

$$\mathfrak{q}(\xi) = \Bigl(\frac{A'(\xi)}{2A(\xi)}\Bigr)^2 + \Bigl(\frac{A'(\xi)}{2A(\xi)}\Bigr)' = \frac{1}{4}\boldsymbol{\phi}_\eta^2(\xi) + \boldsymbol{\psi}_\eta(\xi) + \frac{1}{2}\boldsymbol{\phi}_\eta'(\xi), \qquad \xi \in (\gamma(a), \infty). \tag{4.5}$$

We know from Assumption MP and [210, Lemma 2.9] that $\lim_{\xi\to\infty} \boldsymbol{\phi}_\eta(\xi) = 0$ and $\lim_{\xi\to\infty} \eta'(\xi) = 0$. In turn, the fact that $\boldsymbol{\phi}_\eta$ is positive and decreasing clearly implies that $\boldsymbol{\phi}_\eta' \in L^1([c,\infty), d\xi)$ for $c > \gamma(a)$ and, therefore, $\lim_{\xi\to\infty} \boldsymbol{\phi}_\eta'(\xi) = 0$. We thus have $\lim_{\xi\to\infty} \mathfrak{q}(\xi) = \sigma^2$. Using a basic oscillation criterion for second order ordinary differential equations [48, XIII.7.37], we conclude that $\sqrt{A(\xi)}\, w_\lambda(\gamma^{-1}(\xi))$ has infinitely many zeros on $[\gamma(a), \infty)$ whenever $\lambda > \sigma^2$, so that (c) holds.

Part **(d)** follows from the general fact that the existence of oscillatory solutions for the Sturm-Liouville equation (that is, solutions with infinitely many zeros) implies that the essential spectrum of any self-adjoint realization of the Sturm-Liouville expression is nonempty [48, XIII.7.39], which in turn implies that, in the Feller boundary classification, at least one endpoint must be natural [137, Theorem 3.1].

Our second lemma states that the family of Sturm-Liouville expressions satisfying Assumption MP is closed under changes of variable determined by the multiplication of the coefficients by (squared) strictly positive solutions of the Sturm-Liouville problem. It is based on a known result on changes of spectral functions for Sturm-Liouville operators and Krein strings ([110], see also [51, Section 6.9]).

Lemma 4.2 *Let $\ell = -\frac{1}{r}\frac{d}{dx}\bigl(p\,\frac{d}{dx}\bigr)$ be a Sturm-Liouville expression satisfying Assumption MP. For $-\infty < \kappa \leq \sigma^2$, consider the modified differential expression*

$$\ell^{\langle\kappa\rangle} = -\frac{1}{r^{\langle\kappa\rangle}}\frac{d}{dx}\Bigl(p^{\langle\kappa\rangle}\frac{d}{dx}\Bigr), \qquad x \in (a,b),$$

where $p^{\langle\kappa\rangle} = w_\kappa^2 \cdot p$ and $r^{\langle\kappa\rangle} = w_\kappa^2 \cdot r$. Then Assumption MP also holds for $\ell^{\langle\kappa\rangle}$, and the function

$$w_\lambda^{\langle\kappa\rangle}(x) := \frac{w_{\kappa+\lambda}(x)}{w_\kappa(x)} \tag{4.6}$$

is, for each $\lambda \in \mathbb{C}$, the unique solution of $\ell^{\langle\kappa\rangle}(w) = \lambda w$, $w(a) = 1$ and $(p^{\langle\kappa\rangle}w')(a) = 0$. Moreover, the spectral measure associated with $\ell^{\langle\kappa\rangle}$ (Theorem 2.5) is given by

$$\boldsymbol{\rho}_{\mathcal{L}}^{\langle\kappa\rangle}(\lambda_1,\lambda_2] = \boldsymbol{\rho}_{\mathcal{L}}(\lambda_1+\kappa,\lambda_2+\kappa] \qquad (-\infty < \lambda_1 \leq \lambda_2 < \infty).$$

Proof Fix $-\infty < \kappa \leq \sigma^2$. The functions A and $A^{\langle\kappa\rangle}$ associated to the operators ℓ and $\ell^{\langle\kappa\rangle}$ respectively (defined as in (4.2)) are connected by $A^{\langle\kappa\rangle} = \widetilde{w}_\kappa^2 \cdot A$, where $\widetilde{w}_\kappa(\xi) = w_\kappa(\gamma^{-1}(\xi))$.

In order to show that Assumption MP holds for $\ell^{\langle\kappa\rangle}$, write $\tilde{a}_m = \gamma(a_m)$ and consider the function $A^{\langle\kappa,m\rangle}(\xi) := \widetilde{w}_{\kappa,m}^2(\xi) \cdot A(\xi)$, where $\tilde{a}_m \leq \xi < \infty$ and $\widetilde{w}_{\lambda,m}(\xi) = w_{\lambda,m}(\gamma^{-1}(\xi))$. Let $\eta^{\langle\kappa,m\rangle} := \eta + 2\frac{\widetilde{w}_{\kappa,m}'}{\widetilde{w}_{\kappa,m}}$, where η satisfies the conditions of Assumption MP. By Lemma 4.1(b) we have $\eta^{\langle\kappa,m\rangle} \in \mathrm{C}^1[\tilde{a}_m,\infty)$, and it is easily seen (cf. [210, Example 4.6]) that

$$\boldsymbol{\phi}_{\eta^{\langle\kappa,m\rangle}} := \frac{(A^{\langle\kappa,m\rangle})'}{A^{\langle\kappa,m\rangle}} - \eta^{\langle\kappa,m\rangle} = \boldsymbol{\phi}_\eta, \qquad \boldsymbol{\psi}_{\eta^{\langle\kappa,m\rangle}} = \boldsymbol{\psi}_\eta - \kappa, \qquad \eta^{\langle\kappa,m\rangle}(\tilde{a}_m) = \eta(\tilde{a}_m) \geq 0,$$

and then one can show that $\eta^{\langle\kappa,m\rangle} \geq 0$ (see [210, Remark 2.12]), hence Assumption MP holds for the function $A^{\langle\kappa,m\rangle}$. If we now let $\eta^{\langle\kappa\rangle}(\xi) := \eta(\xi) + 2\frac{\widetilde{w}_\kappa'(\xi)}{\widetilde{w}_\kappa(\xi)} = \lim_{m\to\infty}\eta^{\langle\kappa,m\rangle}(\xi)$ (where $\gamma(a) < \xi < \infty$; the second equality is due to (2.18)), then it is clear that the limit function $\eta^{\langle\kappa\rangle}$ satisfies Assumption MP for the function $A^{\langle\kappa\rangle}$ associated with the operator $\ell^{\langle\kappa\rangle}$.

A simple computation gives

$$\begin{aligned}
-\frac{1}{r^{\langle\kappa\rangle}}\Bigl[p^{\langle\kappa\rangle}\Bigl(\frac{w_{\kappa+\lambda}}{w_\kappa}\Bigr)'\Bigr]' &= -\frac{1}{w_\kappa^2\cdot r}\bigl[p\,w_{\kappa+\lambda}'\,w_\kappa - p\,w_{\kappa+\lambda}w_\kappa'\bigr]' \\
&= -\frac{1}{w_\kappa^2}\bigl[\ell(w_{\kappa+\lambda})\,w_\kappa - w_{\kappa+\lambda}\,\ell(w_\kappa)\bigr] = \lambda\frac{w_{\kappa+\lambda}(x)}{w_\kappa(x)},
\end{aligned}$$

so that $\ell^{\langle\kappa\rangle}(w_\lambda^{\langle\kappa\rangle}) = \lambda w_\lambda^{\langle\kappa\rangle}$. The boundary conditions at a are also straightforwardly checked. To prove the last assertion, notice that the eigenfunction expansions associated with ℓ and $\ell^{\langle\kappa\rangle}$ are related through the identity

$$\Bigl(\mathcal{F}^{\langle\kappa\rangle}\frac{f}{w_\kappa}\Bigr)(\lambda) = (\mathcal{F}f)(\kappa+\lambda), \qquad f \in L^2(r)$$

(where, as in Sect. 2.2, we write $L^p(r) := L^p\bigl((a,b); r(x)dx\bigr)$), and therefore

$$\|\mathcal{F}f\|_{L^2(\mathbb{R},\boldsymbol{\rho}_{\mathcal{L}})} = \|f\|_{L^2(r)} = \biggl\|\frac{f}{w_\kappa}\biggr\|_{L^2(r^{\langle\kappa\rangle})} = \bigl\|(\mathcal{F}f)(\kappa+\cdot)\bigr\|_{L^2(\mathbb{R},\boldsymbol{\rho}_{\mathcal{L}}^{\langle\kappa\rangle})}.$$

Recalling the uniqueness of the spectral measure for which the isometric property in Theorem 2.5 holds, we deduce that $\boldsymbol{\rho}_{\mathcal{L}}^{\langle\kappa\rangle}(\lambda_1,\lambda_2] = \boldsymbol{\rho}_{\mathcal{L}}(\lambda_1+\kappa,\lambda_2+\kappa]$.

Corollary 4.1 *If $0 < \lambda \leq \sigma^2$, then $w_\lambda(\cdot)$ is strictly decreasing and such that* $\lim_{x\uparrow b} w_\lambda(x) = 0$.

Proof By the previous lemma, $w_\lambda(x) = \bigl[w_{-\lambda}^{\langle\lambda\rangle}(x)\bigr]^{-1}$. By Corollary 2.1, $w_{-\lambda}^{\langle\lambda\rangle}(x)$ is strictly increasing and unbounded, yielding the result.

The remaining ingredient for the proof of the Laplace representation is the weak maximum principle for the hyperbolic PDE $\partial_x^2 u = \partial_y^2 u + \boldsymbol{\phi}_\eta(y)\,\partial_y u - \boldsymbol{\psi}_\eta(y)\,u$. (This equation is equivalent, up to a change of variables, to the PDE $\partial_x^2 u = -\ell_y u$.) In the following lemma and corollary we state and prove this maximum principle in a general form which also serves as a preparation for our study of the hyperbolic PDE $\ell_x u = \ell_y u$ (Sect. 4.3).

Lemma 4.3 *Let the functions $\boldsymbol{\phi}_1, \boldsymbol{\phi}_2, \boldsymbol{\psi}_1, \boldsymbol{\psi}_2 : (\gamma(a),\infty) \longrightarrow \mathbb{R}$ be such that*

$$\boldsymbol{\phi}_2, \boldsymbol{\psi}_2 \text{ are decreasing}, \qquad 0 \leq \boldsymbol{\phi}_1 \leq \boldsymbol{\phi}_2, \qquad 0 \leq \boldsymbol{\psi}_1 \leq \boldsymbol{\psi}_2, \qquad \lim_{\xi\to\infty} \boldsymbol{\phi}_2(\xi) = 0. \tag{4.7}$$

Denote by $\wp_j$ $(j = 1,2)$ the differential expression

$$\wp_j(v) := -v'' - \boldsymbol{\phi}_j v' + \boldsymbol{\psi}_j v = -\frac{1}{A_{\boldsymbol{\phi}_j}}(A_{\boldsymbol{\phi}_j} v')' + \boldsymbol{\psi}_j v,$$

where $A_{\boldsymbol{\phi}_j}(x) = \exp(\int_\beta^x \boldsymbol{\phi}_j(\xi)d\xi)$ (with $\beta > \gamma(a)$ arbitrary). For $\gamma(a) < c \leq y \leq x$, consider the triangle $\Delta_{c,x,y} := \{(\xi,\zeta) \in \mathbb{R}^2 \mid \zeta \geq c,\ \xi+\zeta \leq x+y,\ \xi-\zeta \geq x-y\}$, and let $v \in \mathrm{C}^2(\Delta_{c,x,y})$. Then the following integral equation holds:

$$A_{\boldsymbol{\phi}_1}(x) A_{\boldsymbol{\phi}_2}(y)\, v(x,y) = H + I_0 + I_1 + I_2 + I_3 - I_4, \tag{4.8}$$

where

$$H := \tfrac{1}{2} A_{\phi_2}(c)\big[A_{\phi_1}(x-y+c)\, v(x-y+c,c) + A_{\phi_1}(x+y-c)\, v(x+y-c,c)\big], \tag{4.9}$$

$$I_0 := \tfrac{1}{2} A_{\phi_2}(c) \int_{x-y+c}^{x+y-c} A_{\phi_1}(s)(\partial_y v)(s,c)\, ds, \tag{4.10}$$

$$I_1 := \tfrac{1}{2} \int_c^y A_{\phi_1}(x-y+s) A_{\phi_2}(s)\big[\phi_2(s) + \phi_1(x-y+s)\big] v(x-y+s,s)\, ds, \tag{4.11}$$

$$I_2 := \tfrac{1}{2} \int_c^y A_{\phi_1}(x+y-s) A_{\phi_2}(s)\big[\phi_2(s) - \phi_1(x+y-s)\big] v(x+y-s,s)\, ds, \tag{4.12}$$

$$I_3 := \tfrac{1}{2} \int_{\Delta_{c,x,y}} A_{\phi_1}(\xi) A_{\phi_2}(\zeta)\big[\psi_2(\zeta) - \psi_1(\xi)\big] v(\xi,\zeta)\, d\xi\, d\zeta, \tag{4.13}$$

$$I_4 := \tfrac{1}{2} \int_{\Delta_{c,x,y}} A_{\phi_1}(\xi) A_{\phi_2}(\zeta)\, (\wp_{2,\zeta} v - \wp_{1,\xi} v)(\xi,\zeta)\, d\xi\, d\zeta, \tag{4.14}$$

and $\wp_{j,z}$ *denotes the differential expression* $\wp_j$ *acting on the variable* z.

Proof Just compute

$$\begin{aligned} I_4 - I_3 &= \tfrac{1}{2} \int_{\Delta_{c,x,y}} \bigg(\frac{\partial}{\partial \xi}\big[A_{\phi_1}(\xi) A_{\phi_2}(\zeta)\, (\partial_\xi v)(\xi,\zeta)\big] \\ &\qquad\qquad - \frac{\partial}{\partial \zeta}\big[A_{\phi_1}(\xi) A_{\phi_2}(\zeta)\, (\partial_\zeta v)(\xi,\zeta)\big]\bigg) d\xi\, d\zeta \\ &= I_0 - \tfrac{1}{2} \int_c^y A_{\phi_1}(x-y+s) A_{\phi_2}(s)\, (\partial_\zeta v + \partial_\xi v)(x-y+s,s)\, ds \\ &\quad - \tfrac{1}{2} \int_c^y A_{\phi_1}(x+y-s) A_{\phi_2}(s)\, (\partial_\zeta v - \partial_\xi v)(x+y-s,s)\, ds \\ &= I_0 + I_1 - \int_c^y \frac{d}{ds}\big[A_{\phi_1}(x-y+s) A_{\phi_2}(s)\, v(x-y+s,s)\big] ds \\ &\quad + I_2 - \int_c^y \frac{d}{ds}\big[A_{\phi_1}(x+y-s) A_{\phi_2}(s)\, v(x+y-s,s)\big] ds, \end{aligned}$$

where in the second equality we used Green's theorem, and the third equality follows easily from the fact that $(A_{\phi_j})' = \phi_j A_{\phi_j}$.

Corollary 4.2 (Weak Maximum Principle) *In the conditions of Lemma 4.3, let* $\gamma(a) < c \leq y_0 \leq x_0$. *If* $u \in \mathrm{C}^2(\Delta_{c,x_0,y_0})$ *satisfies*

$$\begin{aligned} (\wp_{2,y}u - \wp_{1,x}u)(x,y) &\leq 0, \qquad && (x,y) \in \Delta_{c,x_0,y_0}, \\ u(x,c) &\geq 0, && x \in [x_0 - y_0 + c, x_0 + y_0 - c], \\ (\partial_y u)(x,c) &\geq 0, && x \in [x_0 - y_0 + c, x_0 + y_0 - c], \end{aligned} \tag{4.15}$$

then $u \geq 0$ *in* Δ_{c,x_0,y_0}.

Proof Pick a function $\omega \in \mathrm{C}^2[c,\infty)$ such that $\wp_2\omega < 0$, $\omega(c) > 0$ and $\omega'(c) \geq 0$. Clearly, it is enough to show that for all $\varepsilon > 0$ we have $v(x,y) := u(x,y) + \varepsilon\omega(y) > 0$ for $(x,y) \in \Delta_{c,x_0,y_0}$.

By Lemma 4.3, the integral equation (4.8) holds for the function v. Assume by contradiction that there exist $\varepsilon > 0$, $(x,y) \in \Delta_{c,x_0,y_0}$ for which we have $v(x,y) = 0$ and $v(\xi,\zeta) \geq 0$ for all $(\xi,\zeta) \in \Delta_{c,x,y} \subset \Delta_{c,x_0,y_0}$. It is clear from the choice of ω that $v(\cdot,c) > 0$, thus we have $H \geq 0$ in the right hand side of (4.8). Similarly, $(\partial_y v)(\cdot,c) = (\partial_y u)(\cdot,c) + \varepsilon\omega'(c) \geq 0$, hence $I_0 \geq 0$. Since the functions ϕ_1, ϕ_2, ψ_1, ψ_2 satisfy (4.7) and we are assuming that $u \geq 0$ on $\Delta_{c,x,y}$, we have $I_1 \geq 0$, $I_2 \geq 0$ and $I_3 \geq 0$. In addition, $I_4 < 0$ because $(\wp_{2,\zeta}v - \wp_{1,\xi}v)(\xi,\zeta) = (\wp_{2,\zeta}u - \wp_{1,\xi}u)(\xi,\zeta) + (\wp_2\omega)(\zeta) < 0$. Consequently, (4.8) yields $0 = A_{\phi_1}(x)A_{\phi_2}(y)v(x,y) \geq -I_4 > 0$. This contradiction shows that $v(x,y) > 0$ for all $(x,y) \in \Delta_{c,x_0,y_0}$.

Finally, we state the announced Laplace-type representation for the solutions of the Sturm-Liouville initial value problem.

Theorem 4.1 (Laplace-Type Representation) *Let* ℓ *be a Sturm-Liouville expression of the form* (4.1), *and suppose that Assumption MP holds. Let* w_λ *be the solution of the initial value problem* (2.12). *For each* $x \in [a,b)$ *there exists a subprobability measure* π_x *on* $\mathbb{R}$ *such that*

$$w_{\tau^2+\sigma^2}(x) = \int_{\mathbb{R}} e^{i\tau s}\,\pi_x(ds) = \int_{\mathbb{R}} \cos(\tau s)\,\pi_x(ds) \qquad (\tau \in \mathbb{C}), \tag{4.16}$$

where $\sigma = \lim_{\xi\to\infty} \frac{A'(\xi)}{2A(\xi)}$. *In particular, the boundedness property* (2.19) *extends to*

$$|w_{\tau^2+\sigma^2}(x)| \leq 1 \quad \text{on the strip } |\mathrm{Im}(\tau)| \leq \sigma \quad (a \leq x < b). \tag{4.17}$$

We first show that a similar representation holds for the solutions $w_{\lambda,m}$ of the initial value problem on the approximating intervals (a_m, b) (Lemma 2.2); the result of Theorem 4.1 will then be deduced by a limiting argument.

Proposition 4.2 *Let* ℓ *be a Sturm-Liouville expression of the form* (4.1), *and suppose that Assumption MP holds. Let* $w_{\lambda,m}$ *be defined as in Lemma 2.2. For each*

$m \in \mathbb{N}$ and $x \in [a_m, b)$ there exists a subprobability measure $\pi_{x,m}$ on $\mathbb{R}$ such that

$$w_{\tau^2+\sigma^2,m}(x) = \int_{\mathbb{R}} e^{i\tau s} \pi_{x,m}(ds) = \int_{\mathbb{R}} \cos(\tau s)\, \pi_{x,m}(ds) \qquad (\tau \in \mathbb{C}). \tag{4.18}$$

Proof Throughout the proof we assume, without loss of generality, that we have chosen $c = a_m$ in the definition of the function γ introduced in Remark 4.1, so that $\gamma(a_m) = 0$.

We begin by proving that the result holds when $\sigma = 0$. Let η, $\boldsymbol{\phi}_\eta$, $\boldsymbol{\psi}_\eta$ be defined as in Assumption MP. The function $\boldsymbol{\vartheta}_{\lambda,m}(y) := \exp(\frac{1}{2}\int_0^y \eta(\xi)d\xi)\, w_{\lambda,m}(\gamma^{-1}(y))$ is the solution of

$$\wp(u) = \lambda u \quad (0 < \xi < \infty), \qquad u(0) = 1, \qquad u'(0) = 0,$$

where $\wp(v) := -v'' - \boldsymbol{\phi}_\eta v' + \boldsymbol{\psi}_\eta v$. From this it follows that the function $u_\tau(x, y) := \cos(\tau x)\, \boldsymbol{\vartheta}_{\tau^2,m}(y)$ $(x, y \in \mathbb{R}_0^+)$ is, for each $\tau \in \mathbb{C}$, a solution of the hyperbolic PDE $\partial_x^2 u = -\wp_y u$. It follows from Corollary 4.2 that the Cauchy problem

$$(\partial_x^2 + \wp_y)u = 0, \qquad u(x, 0) = f(x), \qquad (\partial_y u)(x, 0) = 0$$

has the property that if $f \in \mathrm{C}_\mathrm{c}^\infty(\mathbb{R})$, $f \geq 0$ then the solution u_f is such that $u_f(x, y) \geq 0$ for all $x \geq y \geq 0$. Thus $f \mapsto u_f(x, y)$ is a positive linear functional on $\mathrm{C}_\mathrm{c}^\infty(\mathbb{R})$ and, consequently , $u_f(x, y) = \int_{\mathbb{R}} f\, d\mu_{x,y,m}$ for all $f \in \mathrm{C}_\mathrm{c}^\infty(\mathbb{R})$, where $\mu_{x,y,m}$ is, for each $x \geq y \geq 0$, a finite positive Borel measure; moreover, it follows from the domain of dependence for the Cauchy problem that $\mu_{x,y,m}$ has compact support. In particular we can write

$$\cos(\tau x)\, \boldsymbol{\vartheta}_{\tau^2,m}(y) = \int_{\mathbb{R}} \cos(\tau s)\, \mu_{x,y,m}(ds), \qquad x \geq y \geq 0. \tag{4.19}$$

Assume that each measure $\mu_{x,y,m}$ is symmetric (if not, replace it by its symmetrization), and let $\varkappa_{y,m} = \mu_{y,y,m} \underset{\mathbb{R}}{*} \frac{1}{2}(\delta_y + \delta_{-y}) - \mu_{2y,y,m}$. We then have

$$\begin{aligned}\int_{\mathbb{R}} \cos(\tau s)\, \varkappa_{y,m}(ds) &= \int_{\mathbb{R}} \cos(\tau s)\, \mu_{y,y,m}(ds) \int_{\mathbb{R}} \cos(\tau s) \left(\tfrac{1}{2}(\delta_y + \delta_{-y})\right)(ds) \\ &\quad - \int_{\mathbb{R}} \cos(\tau s)\, \mu_{2y,y,m}(ds) \\ &= \left(\cos^2(\tau y) - 2\cos(\tau y)\right) \boldsymbol{\vartheta}_{\tau^2,m}(y) \\ &= \boldsymbol{\vartheta}_{\tau^2,m}(y).\end{aligned}$$

We claim that $\varkappa_{y,m}$ is a positive measure. Indeed, we have

$$\cos(\tau x)\, \boldsymbol{\vartheta}_{\tau^2,m}(y) = \int_{\mathbb{R}} \cos(\tau s)\bigl(\varkappa_{y,m} \underset{\mathbb{R}}{*} \tfrac{1}{2}(\delta_x + \delta_{-x})\bigr)(ds),$$

where the right-hand side is, by (4.19), a positive-definite function of $\tau \in \mathbb{R}$; therefore, the convolution $\varkappa_{y,m} \underset{\mathbb{R}}{*} \frac{1}{2}(\delta_x + \delta_{-x})$ is, for all $x \geq y \geq 0$, a positive Borel measure. Since the support of $\varkappa_{y,m}$ is compact, the supports of $\varkappa_{y,m} \underset{\mathbb{R}}{*} \delta_x$ and $\varkappa_{y,m} \underset{\mathbb{R}}{*} \delta_{-x}$ are disjoint for x sufficiently large, and this implies that the measures $\varkappa_{y,m} \underset{\mathbb{R}}{*} \delta_x$ and (consequently) $\varkappa_{y,m}$ are both positive. Setting $\pi_{x,m} := \exp\bigl(-\frac{1}{2}\int_0^{\gamma(x)} \eta(\xi)d\xi\bigr)\varkappa_{\gamma(x),m}$, we conclude that (4.18) holds for all $\tau \in \mathbb{C}$. Since $w_{0,m}(x) \equiv 1$, we have $\pi_{x,m} \in \mathcal{P}(\mathbb{R})$ for all $x \in [a_m, b)$.

Suppose now that $\sigma > 0$. Then the result for the case $\sigma = 0$ can be applied to the operator $\ell^{\langle\sigma^2\rangle}$ defined in Lemma 4.2 and the corresponding eigenfunctions $w_{\lambda,m}^{\langle\sigma^2\rangle}(x) := \frac{w_{\lambda+\sigma^2,m}(x)}{w_{\sigma^2,m}(x)}$. (Indeed, it follows from Lemma 4.1 that the function $A^{\langle\sigma^2\rangle}$ associated to the operator $\ell^{\langle\sigma^2\rangle}$, defined as in (4.2), is such that $\lim_{\xi\to\infty} \frac{(A^{\langle\sigma^2\rangle})'(\xi)}{2A^{\langle\sigma^2\rangle}(\xi)} = 0$.) Hence

$$\frac{w_{\tau^2+\sigma^2,m}(x)}{w_{\sigma^2,m}(x)} = \int_{\mathbb{R}} \cos(\tau s)\, \boldsymbol{\pi}^{x,m}(ds),$$

where $\boldsymbol{\pi}^{x,m}$ is, for each $x \in [a, b)$, a symmetric probability measure. Setting $\pi_{x,m} := w_{\sigma^2,m}(x)\, \boldsymbol{\pi}^{x,m}$, we obtain (4.18). By Lemma 2.3 we have $w_{\sigma^2,m}(\cdot) \leq 1$, hence each $\pi_{x,m}$ is a subprobability measure.

Proof of Theorem 4.1 We proved in Proposition 4.2 that for each $m \in \mathbb{N}$ there exists a symmetric subprobability measure $\pi_{x,m}$ whose Fourier transform is the function $\tau \mapsto w_{\tau^2+\sigma^2,m}(x)$ $(\tau \in \mathbb{R})$. We also know (from Lemmas 2.1–2.2) that $w_{\tau^2+\sigma^2,m}(x) \longrightarrow w_{\tau^2+\sigma^2}(x)$ pointwise as $m \to \infty$, the limit function being continuous in τ. Applying the Lévy continuity theorem (e.g. [9, Theorem 23.8]), we conclude that $w_{\tau^2+\sigma^2}(x)$ is the Fourier transform of a symmetric subprobability measure π_x and, in addition, the measures $\pi_{x,m}$ converge weakly to π_x as $m \to \infty$. Therefore, for $x > a$ we have

$$w_{\tau^2+\sigma^2}(x) = \int_{\mathbb{R}} \cos(\tau s)\, \pi_x(ds) \qquad (\tau \in \mathbb{R}). \tag{4.20}$$

In order to extend (4.20) to $\tau \in \mathbb{C}$, we let $0 \leq \phi_1 \leq \phi_2 \leq \ldots$ be functions with compact support such that $\phi_n \uparrow 1$ pointwise, and for fixed $x > a$, $\kappa > 0$ we compute

$$\begin{aligned}
&\int_{\mathbb{R}} \cosh(\kappa s)\, \pi_x(ds) \\
&\quad = \lim_{n\to\infty} \int_{\mathbb{R}} \phi_n(s) \cosh(\kappa s)\, \pi_x(ds) \\
&\quad = \lim_{n\to\infty} \lim_{m\to\infty} \int_{\mathbb{R}} \phi_n(s) \cosh(\kappa s)\, \pi_{x,m}(ds) \\
&\quad \leq \lim_{m\to\infty} \int_{\mathbb{R}} \cosh(\kappa s)\, \pi_{x,m}(ds) \;=\; \lim_{m\to\infty} w_{\sigma^2-\kappa^2,m}(x) \;=\; w_{\sigma^2-\kappa^2}(x) < \infty.
\end{aligned}$$

From this estimate we easily see that the right-hand side of (4.20) is an entire function of τ; therefore, by analytic continuation, (4.20) holds for all $\tau \in \mathbb{C}$.

Finally, if $|\mathrm{Im}(\tau)| \leq \sigma$ then

$$|w_{\tau^2+\sigma^2}(x)| \leq \int_{\mathbb{R}} |\cos(\tau s)| \pi_x(ds) \leq \int_{\mathbb{R}} \cosh(\sigma s)\, \pi_x(ds) = w_0(x) = 1,$$

and therefore (4.17) is true.

We finish this section by presenting a description of the spectrum of the Neumann realization of the Sturm-Liouville operator ℓ which will later be useful, and whose proof relies on the Laplace representation. Recall that the Neumann realization $(\mathcal{L}^{(2)}, \mathcal{D}(\mathcal{L}^{(2)}))$ was defined in Theorem 2.5 as the self-adjoint operator obtained by restricting the Sturm-Liouville operator ℓ to the domain which (considering that, by Lemma 4.1(d), the endpoint b is limit point) was defined in (2.20) as

$$\mathcal{D}(\mathcal{L}^{(2)}) = \bigl\{u \in L^2(r) \bigm| u, u' \in \mathrm{AC}_{\mathrm{loc}}(a,b),\ \ell(u) \in L^2(r),\ (pu')(a) = 0\bigr\}.$$

Proposition 4.3 *Let ℓ be a Sturm-Liouville expression of the form* (4.1), *and suppose that Assumption MP holds. The spectral measure $\boldsymbol{\rho}_{\mathcal{L}}$ of Proposition 2.5 is such that* $\mathrm{supp}(\boldsymbol{\rho}_{\mathcal{L}}) = [\sigma^2, \infty)$. *In addition, $\mathcal{L}$ has purely absolutely continuous spectrum in (σ^2, ∞).*

Proof It follows from the proof of Lemma 4.1 that the operator $\mathcal{L}$ is unitarily equivalent to a self-adjoint realization of the differential expression $-\frac{d^2}{d\xi^2} + \mathfrak{q}$ $(\gamma(a) < \xi < \infty)$, where $\mathfrak{q}$ is defined by (4.5) and satisfies $\mathfrak{q} = \mathfrak{q}_1 + \mathfrak{q}_2$, with $\lim_{\xi\to\infty} \mathfrak{q}_1(\xi) = \sigma^2$ and $\mathfrak{q}_2 \in L^1([c,\infty), d\xi)$ for $c > \gamma(a)$. Using a general result on the spectral properties of Sturm-Liouville operators stated in [201, Theorem 15.3], we conclude that the spectrum of $\mathcal{L}$ is purely absolutely continuous on (σ^2, ∞) and the essential spectrum equals $[\sigma^2, \infty)$. (The result of [201] is stated for Sturm-Liouville operators whose left endpoint is regular, but we can apply it here because a well-known result [185, Theorem 9.11] ensures that the essential spectrum of $\mathcal{L}$ is the union of the essential spectrums of self-adjoint realizations of ℓ restricted to the intervals (a, c) and (c, b), $a < c < b$. Recall also that, as noted in the proof of Lemma 4.1, Sturm-Liouville operators with no natural endpoints have a purely discrete spectrum.)

It remains to show that $\mathcal{L}$ has no eigenvalues on $[0, \sigma^2]$. Indeed, if we assume that $0 \leq \lambda_0 \leq \sigma^2$ is an eigenvalue of $\mathcal{L}$, then w_{λ_0} belongs to $\mathcal{D}(\mathcal{L}^{(2)})$ and therefore, by the Laplace representation (4.16), w_λ belongs to $\mathcal{D}(\mathcal{L}^{(2)})$ for all $\lambda \geq \sigma^2$; since the eigenvalues are discrete, this is a contradiction.

4.3 The Existence Theorem for Sturm-Liouville Product Formulas

As in the particular cases of the Kingman and the Whittaker convolutions, the product formula for the solutions w_λ of the Sturm-Liouville problem (2.12) is the tool which will allow us to introduce a generalized convolution associated with the operator ℓ. The probabilistic property of the product formula (i.e. the property that the kernel of the product formula is composed of probability measures) is the requirement which will ensure that the convolution preserves the space of probability measures.

The aim of this section is to show that Assumption MP is a sufficient condition for the existence of such a probabilistic product formula. Namely, we will prove the following result:

Theorem 4.2 (Product Formula for w_λ) *Let ℓ be a Sturm-Liouville expression of the form* (4.1)*, and suppose that Assumption MP holds. For each $x, y \in [a, b)$ there exists a measure $\boldsymbol{\nu}_{x,y} \in \mathcal{P}[a, b)$ such that the product $w_\lambda(x)\, w_\lambda(y)$ admits the integral representation*

$$w_\lambda(x)\, w_\lambda(y) = \int_{[a,b)} w_\lambda(\xi)\, \boldsymbol{\nu}_{x,y}(d\xi), \qquad x, y \in [a, b),\ \lambda \in \mathbb{C}. \tag{4.21}$$

4.3.1 The Associated Hyperbolic Cauchy Problem

The proof of Theorem 4.2 relies crucially on the basic properties (existence, uniqueness and positivity-preservingness of solution) of the hyperbolic Cauchy problem associated with ℓ, i.e., of the boundary value problem defined by

$$\begin{aligned} (\ell_x h)(x, y) &= (\ell_y h)(x, y) \qquad (x, y \in (a, b)), \\ h(x, a) &= f(x), \\ (\partial_y^{[1]} h)(x, a) &= 0, \end{aligned} \tag{4.22}$$

where $\partial_y^{[1]} = p(y) \frac{\partial}{\partial y}$.

Since $\ell_y - \ell_x = \frac{p(x)}{r(x)}\frac{\partial^2}{\partial x^2} - \frac{p(y)}{r(y)}\frac{\partial^2}{\partial y^2} +$ lower order terms, the equation $\ell_x h = \ell_y h$ is hyperbolic at the line $y = a$ if $\frac{p(a)}{r(a)} > 0$; otherwise, the initial conditions of the Cauchy problem are given at a line of parabolic degeneracy. If $\gamma(a) = -\int_a^c \sqrt{\frac{r(y)}{p(y)}}dy > -\infty$, then we can remove the degeneracy via the change of variables $x = \gamma(\xi)$, $y = \gamma(\zeta)$ (cf. Remark 4.1), through which the partial differential equation is transformed to the standard form $\widetilde{\ell}_\xi u = \widetilde{\ell}_\zeta u$, with initial condition at the line $\zeta = \gamma(a)$. In the case $\gamma(a) = -\infty$, the standard form of the equation is also parabolically degenerate in the sense that its initial line is $\zeta = -\infty$.

Theorem 4.3 (Existence of Solution) *Let ℓ be a Sturm-Liouville expression of the form* (4.1), *and suppose that $x \mapsto p(x)r(x)$ is an increasing function. If $f \in \mathcal{D}(\mathcal{L}^{(2)})$ and $\ell(f) \in \mathcal{D}(\mathcal{L}^{(2)})$, then the function*

$$h_f(x, y) := \int_{[\sigma^2,\infty)} w_\lambda(x)\, w_\lambda(y)\, (\mathcal{F}f)(\lambda)\, \boldsymbol{\rho}_{\mathcal{L}}(d\lambda) \tag{4.23}$$

solves the Cauchy problem (4.22).

For ease of notation, unless necessary we drop the dependence in h and denote (4.23) by $h(x, y)$.

Proof Let us begin by justifying that $\ell_x h$ can be computed via differentiation under the integral sign. Since w_λ is a solution of the initial value problem (2.12), we have $(pw_\lambda')(x) = -\lambda \int_a^x w_\lambda(\xi)\, r(\xi) d\xi$ and therefore (by Lemma 2.3) $|(pw_\lambda')(x)| \leq \lambda \int_a^x r(\xi) d\xi$. Hence

$$\begin{aligned}&\int_{[\sigma^2,\infty)} \big|(\mathcal{F}f)(\lambda)\, (pw_\lambda')(x)\, w_\lambda(y)\big|\, \boldsymbol{\rho}_{\mathcal{L}}(d\lambda)\\ &\qquad \leq \int_a^x r(\xi)d\xi \cdot \int_{[\sigma^2,\infty)} \lambda \big|(\mathcal{F}f)(\lambda)\, w_\lambda(y)\big|\, \boldsymbol{\rho}_{\mathcal{L}}(d\lambda) < \infty,\end{aligned} \tag{4.24}$$

where the convergence (which is uniform in compacts) follows from (2.24) and Lemma 2.4. The convergence of the differentiated integral yields that $(\partial_x^{[1]} h)(x, y) = \int_{[\sigma^2,\infty)} (\mathcal{F}f)(\lambda)\, (pw_\lambda')(x)\, w_\lambda(y)\, \boldsymbol{\rho}_{\mathcal{L}}(d\lambda)$. Since $(\ell w_\lambda)(x) = \lambda w_\lambda(x)$, in the same way we check that $\int_{[\sigma^2,\infty)} (\mathcal{F}f)(\lambda)\, (\ell w_\lambda)(x)\, w_\lambda(y)\, \boldsymbol{\rho}_{\mathcal{L}}(d\lambda)$ converges absolutely and uniformly on compacts and is therefore equal to $(\ell_x h)(x, y)$. Consequently,

$$(\ell_x h)(x, y) = (\ell_y h)(x, y) = \int_{[\sigma^2,\infty)} \lambda\, (\mathcal{F}f)(\lambda)\, w_\lambda(x)\, w_\lambda(y)\, \boldsymbol{\rho}_{\mathcal{L}}(d\lambda). \tag{4.25}$$

Concerning the boundary conditions, Lemma 2.4(b) together with the fact that $w_\lambda(a) = 1$ imply that $h(x, a) = f(x)$, and from (4.24) we easily see that

$\lim_{y\downarrow a}(\partial_y^{[1]}h)(x,y) = 0$. This shows that h is a solution of the Cauchy problem (4.22).

Under the assumptions of the theorem, the solution (4.23) of the hyperbolic Cauchy problem satisfies the following conditions:

($\boldsymbol{\alpha}$) $h(\cdot, y) \in \mathcal{D}(\mathcal{L}^{(2)})$ for all $a < y < b$;
($\boldsymbol{\beta}$) There exists a zero $\boldsymbol{\rho}_{\mathcal{L}}$-measure set $\Lambda_0 \subset [\sigma^2, \infty)$ such that for each $\lambda \in [\sigma^2, \infty) \setminus \Lambda_0$ we have

$$\mathcal{F}[\ell_y h(\cdot, y)](\lambda) = \ell_y[\mathcal{F}h(\cdot, y)](\lambda) \quad \text{for all } a < y < b, \tag{4.26}$$

$$\lim_{y\downarrow a}[\mathcal{F}h(\cdot, y)](\lambda) = (\mathcal{F}f)(\lambda), \qquad \lim_{y\downarrow a}\partial_y^{[1]}\mathcal{F}[h(\cdot, y)](\lambda) = 0. \tag{4.27}$$

Indeed, by Theorem 2.5 we have $[\mathcal{F}h(\cdot, y)](\lambda) = (\mathcal{F}f)(\lambda)\, w_\lambda(y)$ for all $\lambda \in \operatorname{supp}(\boldsymbol{\rho}_{\mathcal{L}})$ and $a < y < b$. Since $f \in \mathcal{D}(\mathcal{L}^{(2)})$ and $|w_\lambda(\cdot)| \leq 1$ (Lemma 2.3), it is clear from (2.23) that $h(x, y)$ satisfies ($\boldsymbol{\alpha}$). Moreover, it follows from (4.25) that $\mathcal{F}[\ell_y h(\cdot, y)](\lambda) = \lambda\,(\mathcal{F}f)(\lambda)\, w_\lambda(y) = \ell_y[\mathcal{F}h(\cdot, y)](\lambda)$, hence (4.26) holds. The properties (4.27) follow immediately from Lemma 2.1.

Next we show that the solution from the above existence theorem is the unique solution satisfying conditions ($\boldsymbol{\alpha}$)–($\boldsymbol{\beta}$):

Theorem 4.4 (Uniqueness) *Let ℓ be a Sturm-Liouville expression of the form* (4.1), *and suppose that $x \mapsto p(x)r(x)$ is an increasing function. Let $f \in \mathcal{D}(\mathcal{L}^{(2)})$ and let $h_1, h_2 \in \mathrm{C}^2\big((a,b)^2\big)$ be two solutions of $(\ell_x h)(x, y) = (\ell_y h)(x, y)$. Suppose that both h_1 and h_2 satisfy conditions ($\boldsymbol{\alpha}$)–($\boldsymbol{\beta}$). Then*

$$h_1(x, y) \equiv h_2(x, y) \qquad \textit{for all } x, y \in (a, b). \tag{4.28}$$

Proof Fix $\lambda \in \mathbb{R}_0^+ \setminus \Lambda_0$ and let $\Psi_j(y, \lambda) := [\mathcal{F}h_j(\cdot, y)](\lambda)$. We have

$$\ell_y \Psi_j(y, \lambda) = \mathcal{F}[\ell_y h_j(\cdot, y)](\lambda) = \mathcal{F}[\ell_x h_j(\cdot, y)](\lambda) = \lambda \Psi_j(y, \lambda), \qquad a < y < b$$

where the first equality is due to (4.26) and the last step follows from (2.24). Moreover,

$$\lim_{y\downarrow a} \Psi_j(y, \lambda) = (\mathcal{F}f)(\lambda) \quad \text{and} \quad \lim_{y\downarrow a}(\partial_y^{[1]}\Psi_j)(y, \lambda) = 0$$

by (4.27). It thus follows from Lemma 2.1 that

$$[\mathcal{F}h_j(\cdot, y)](\lambda) = \Psi_j(y, \lambda) = (\mathcal{F}f)(\lambda)\, w_\lambda(y), \qquad a < y < b.$$

This equality holds for $\boldsymbol{\rho}_{\mathcal{L}}$-almost every λ, so the isometric property of $\mathcal{F}$ gives $h_1(\cdot, y) = h_2(\cdot, y)$ Lebesgue-a.e.; since the h_j are continuous, we conclude that (4.28) holds.

If the hyperbolic equation $\ell_x h = \ell_y h$ (or the transformed equation $\widetilde{\ell}_\xi u = \widetilde{\ell}_y u$) is uniformly hyperbolic, the existence and uniqueness of solution for this Cauchy problem is a standard result which follows from the classical theory of hyperbolic problems in two variables (see e.g. [40, Chapter V]); in fact, the existence and uniqueness holds under much weaker restrictions on the initial condition. However, the existence and uniqueness theorems above become nontrivial in the presence of a (non-removable) parabolic degeneracy at the initial line.

Indeed, even though many authors have addressed Cauchy problems for degenerate hyperbolic equations in two variables, most studies are restricted to equations where the $\frac{\partial^2}{\partial x^2}$ term vanishes at an initial line $y = y_0$ (we refer to [18, §2.3], [159, Section 5.4] and references therein). Much less is known for hyperbolic equations whose $\frac{\partial^2}{\partial y^2}$ term vanishes at the same initial line: it is known that the Cauchy problem is, in general, not well-posed, and the relevance of determining conditions for its well-posedness has long been pointed out [18, §2.4], but as far as we are aware little progress has been made on this problem (for related work see [130]). The application of spectral techniques to hyperbolic Cauchy problems associated with Sturm-Liouville operators is by no means new, see e.g. [29, 30] and references therein; applying such techniques to degenerate cases is the crucial new idea in the above approach.

An existence theorem analogous to Theorem 4.3 also holds when the initial line is shifted away from the degeneracy, and this has the important consequence that the solution of the degenerate Cauchy problem is the pointwise limit of solutions of nondegenerate problems. These facts are proved in the following proposition.

Proposition 4.4 (Pointwise Approximation by Solutions of Problems with Shifted Boundary) *Let ℓ be a Sturm-Liouville expression of the form* (4.1), *and suppose that $x \mapsto p(x)r(x)$ is an increasing function. If $f \in \mathcal{D}(\mathcal{L}^{(2)})$ and $\ell(f) \in \mathcal{D}(\mathcal{L}^{(2)})$, then for each $m \in \mathbb{N}$ the function*

$$h_m(x, y) = \int_{[\sigma^2,\infty)} w_\lambda(x)\, w_{\lambda,m}(y)\, (\mathcal{F}f)(\lambda)\, \boldsymbol{\rho}_{\mathcal{L}}(d\lambda) \qquad \big(x \in (a, b),\ y \in (a_m, b)\big) \tag{4.29}$$

is a solution of the Cauchy problem

$$(\ell_x h_m)(x, y) = (\ell_y h_m)(x, y), \qquad h_m(x, a_m) = f(x), \qquad (\partial_y^{[1]} h_m)(x, a_m) = 0. \tag{4.30}$$

Moreover, we have

$$\lim_{m\to\infty} h_m(x, y) = h(x, y) \qquad \textit{pointwise for each } x, y \in (a, b), \tag{4.31}$$

where $h(x, y)$ is the solution (4.23) *of the Cauchy problem* (4.22).

Proof Let us begin by justifying that $(\partial_x^{[1]} h_m)(x, y)$ and $(\ell_x h_m)(x, y)$ can be computed via differentiation under the integral sign. The differentiated integrals are given by

$$\int_{[\sigma^2,\infty)} (pw_\lambda')(x)\, w_{\lambda,m}(y)\, (\mathcal{F}f)(\lambda)\, \boldsymbol{\rho}_{\mathcal{L}}(d\lambda), \tag{4.32}$$

$$\int_{[\sigma^2,\infty)} w_\lambda(x)\, w_{\lambda,m}(y)\, [\mathcal{F}(\ell(f))](\lambda)\, \boldsymbol{\rho}_{\mathcal{L}}(d\lambda) \tag{4.33}$$

(for the latter, we used the identities $(\ell w_\lambda)(x) = \lambda w_\lambda(x)$ and (2.24)), and their absolute and uniform convergence on compacts follows from the fact that $f, \ell(f) \in \mathcal{D}(\mathcal{L}^{(2)})$, together with Lemma 2.4(b) and the inequality $|w_{\lambda,m}(\cdot)| \leq 1$ (which follows from Lemma 2.3 if we replace a by a_m). This justifies that $(\partial_x^{[1]} h_m)(x, y)$ and $(\ell_x h_m)(x, y)$ are given by (4.32), (4.33) respectively.

We also need to ensure that $(\partial_y^{[1]} h_m)(x, y)$ and $(\ell_y h_m)(x, y)$ are given by the corresponding differentiated integrals, and to that end we must check that

$$\int_{[\sigma^2,\infty)} w_\lambda(x)\, (pw_{\lambda,m}')(y)\, (\mathcal{F}f)(\lambda)\, \boldsymbol{\rho}_{\mathcal{L}}(d\lambda)$$

converges absolutely and uniformly. Indeed, it follows from (2.17) that for $y \geq a_m$ we have $(pw_{\lambda,m}')(y) = \lambda \int_{a_m}^y w_{\lambda,m}(\xi)\, r(\xi) d\xi$ and consequently $|(pw_{\lambda,m}')(y)| \leq \lambda \int_{a_m}^y r(\xi) d\xi$; hence

$$\begin{aligned} &\int_{[\sigma^2,\infty)} \bigl| w_\lambda(x)\, (pw_{\lambda,m}')(y)\, (\mathcal{F}f)(\lambda) \bigr| \boldsymbol{\rho}_{\mathcal{L}}(d\lambda) \\ &\qquad \leq \int_{a_m}^y r(\xi) d\xi \cdot \int_{[\sigma^2,\infty)} \lambda \bigl| w_\lambda(x) (\mathcal{F}f)(\lambda) \bigr| \boldsymbol{\rho}_{\mathcal{L}}(d\lambda), \end{aligned} \tag{4.34}$$

and the uniform convergence in compacts follows from (2.24) and Lemma 2.4(b).

The verification of the boundary conditions is straightforward: Lemma 2.4(b) together with the fact that $w_{\lambda,m}(a_m) = 1$ imply that $h_m(x, a_m) = f(x)$, and from (4.34) we easily see that $(\partial_y^{[1]} h_m)(x, a_m) = 0$. This shows that the function h_m defined by (4.29) is a solution of the Cauchy problem (4.30).

Since $w_{\lambda,m}(y) \to w_\lambda(y)$ as $m \to \infty$ (Lemma 2.2), the pointwise convergence $h_m(x, y) \to h(x, y)$ follows from the dominated convergence theorem (which is applicable due to Lemmas 2.3 and 2.4(b)).

It should be noted that the above existence and uniqueness theorems hold for all Sturm-Liouville operators of the form (4.1) and such that the function $x \mapsto p(x)r(x)$ is increasing (and thus they are applicable to many operators which do not satisfy Assumption MP).

The role of Assumption MP is to ensure that the solution of the Cauchy problem has the positivity-preservingness property stated in the next proposition (and corollary), whose proof relies on the weak maximum principle of Corollary 4.2.

Proposition 4.5 (Positivity of Solution for the Problem with Shifted Boundary) *Let ℓ be a Sturm-Liouville expression of the form* (4.1)*, and suppose that Assumption MP holds. Let $m \in \mathbb{N}$. If $f \in \mathcal{D}(\mathcal{L}^{(2)})$, $\ell(f) \in \mathcal{D}(\mathcal{L}^{(2)})$ and $f \geq 0$, then the function h_m given by* (4.29) *is such that*

$$h_m(x, y) \geq 0 \qquad \text{for } x \geq y > a_m. \tag{4.35}$$

If, in addition, $f \leq C$ (where C is a constant), then $h_m(x, y) \leq C$ for $x \geq y > a_m$.

Proof Let $\tilde{a}_m := \gamma(a_m)$ and $B(x) := \exp(\frac{1}{2}\int_{\tilde{a}_m}^x \eta(\xi)d\xi)$. It follows from Proposition 4.4 that the function $u_m(x, y) := B(x)B(y)h_m(\gamma^{-1}(x), \gamma^{-1}(y))$ is a solution of the Cauchy problem

$$(\wp_x u_m)(x, y) = (\wp_y u_m)(x, y), \qquad x, y > \tilde{a}_m, \tag{4.36}$$

$$u_m(x, \tilde{a}_m) = B(x)\, f(\gamma^{-1}(x)), \qquad x > \tilde{a}_m, \tag{4.37}$$

$$(\partial_y u_m)(x, \tilde{a}_m) = \tfrac{1}{2}\eta(\tilde{a}_m)\, B(x)\, f(\gamma^{-1}(x)), \qquad x > \tilde{a}_m, \tag{4.38}$$

where $\wp_x := -\frac{\partial^2}{\partial x^2} - \boldsymbol{\phi}_\eta(x)\frac{\partial}{\partial x} + \boldsymbol{\psi}_\eta(x)$. Clearly, u_m satisfies the inequalities (4.15) for arbitrary $x_0 \geq y_0 \geq \tilde{a}_m$ (here $\wp_1 = \wp_2$ and $c = \tilde{a}_m$). By Corollary 4.2, $u_m(x_0, y_0) \geq 0$ for all $x_0 \geq y_0 > \tilde{a}_m$; consequently, (4.35) holds.

The proof that $f \leq C$ implies $h_m \leq C$ is straightforward: if we have $f \leq C$, then $\widetilde{u}_m(x, y) = B(x)B(y)\big(C - h_m(\gamma^{-1}(x), \gamma^{-1}(y))\big)$ is a solution of (4.36) with initial conditions

$$\widetilde{u}_m(x, \tilde{a}_m) = B(x)\big(C - f(\gamma^{-1}(x))\big) \geq 0,$$

$$(\partial_y \widetilde{u}_m)(x, \tilde{a}_m) = \tfrac{1}{2}\eta(\tilde{a}_m)\, B(x)\big(C - f(\gamma^{-1}(x))\big) \geq 0,$$

thus the reasoning of the previous paragraph yields that $C - h_m \geq 0$ for $x \geq y > \tilde{a}_m$.

Corollary 4.3 (Positivity of Solution for the Cauchy Problem (4.22)**)** *Let ℓ be a Sturm-Liouville expression of the form* (4.1)*, and suppose that Assumption MP holds. If $f \in \mathcal{D}(\mathcal{L}^{(2)})$, $\ell(f) \in \mathcal{D}(\mathcal{L}^{(2)})$ and $f \geq 0$, then the function h given by* (4.23) *is such that*

$$h(x, y) \geq 0 \qquad \text{for } x, y \in (a, b).$$

If, in addition, $f \leq C$, then $h(x, y) \leq C$ for $x, y \in (a, b)$.

Proof This is an immediate consequence of Proposition 4.5 together with the pointwise convergence property (4.31). (By (4.23) we have $f(x, y) = f(y, x)$, thus the conclusion holds for all $x, y \in (a, b)$.)

4.3.2 *The Time-Shifted Product Formula*

Before proving that there exists a product formula of the form (4.21) for the Sturm-Liouville solutions $\{w_\lambda(\cdot)\}_{\lambda\in\mathbb{C}}$, we will show that a similar product formula holds for the family of functions $\{e^{-t\lambda}w_\lambda(\cdot)\}_{\lambda\in\mathbb{C}}$. This auxiliary result will be called the *time-shifted product formula* because the latter family is obtained by applying the diffusion semigroup generated by ℓ to the solutions $w_\lambda(\cdot)$. Indeed, we saw in Sect. 2.2.3 that

$$e^{-t\lambda}w_\lambda(x) = (T_t w_\lambda)(x) = [\mathcal{F}p(t, x, \cdot)](\lambda),$$

where $\{T_t\}_{t\geq 0}$ denotes the Feller semigroup generated by the Neumann realization of ℓ and $p(t, x, y)$ denotes the Feller transition density (2.33).

By the inversion formula (2.22) for the $\mathcal{L}$-transform, a natural candidate for the measure of the product formula for $\{w_\lambda(\cdot)\}_{\lambda\in\mathbb{C}}$ is

$$\nu_{x,y}(d\xi) = \int_{[\sigma^2,\infty)} w_\lambda(x)\, w_\lambda(y)\, w_\lambda(\xi)\, \boldsymbol{\rho}_{\mathcal{L}}(d\lambda)\, r(\xi)d\xi.$$

This is only a formal solution, because in general the integral does not converge. But it suffices to include the regularization term $e^{-t\lambda}$ in order to obtain an integral which (under the assumptions of the existence and uniqueness theorems above) always converges absolutely:

Lemma 4.4 *Let ℓ be a Sturm-Liouville expression of the form* (4.1), *and suppose that $x \mapsto p(x)r(x)$ is an increasing function. Let $t_0 > 0$ and K_1, K_2 compact subsets of* (a, b). *The integral*

$$\int_{[\sigma^2,\infty)} e^{-t\lambda}\, w_\lambda(x)\, w_\lambda(y)\, w_\lambda(\xi)\, \boldsymbol{\rho}_{\mathcal{L}}(d\lambda)$$

converges absolutely and uniformly on $(t, x, y, \xi) \in [t_0, \infty) \times K_1 \times K_2 \times [a, b)$.

Proof This follows from Lemma 2.3 and the uniform convergence property of the integral representation of the transition density of the Feller semigroup $\{T_t\}_{t\geq 0}$ (Proposition 2.14).

In what follows we write

$$q_t(x, y, \xi) := \int_{[\sigma^2,\infty)} e^{-t\lambda}\, w_\lambda(x)\, w_\lambda(y)\, w_\lambda(\xi)\, \boldsymbol{\rho}_{\mathcal{L}}(d\lambda). \tag{4.39}$$

This function, which is (at least formally) the density of the measure of the time-shifted product formula, is for fixed t, x, y the density (with respect to $r(\xi)d\xi$) of a subprobability measure:

Lemma 4.5 *Let ℓ be a Sturm-Liouville expression of the form* (4.1), *and suppose that Assumption MP holds. The function $q_t(x, y, \xi)$ is nonnegative and such that $\int_a^b q_t(x, y, \xi)\, r(\xi)d\xi \leq 1$ for all $(t, x, y) \in \mathbb{R}^+ \times (a, b) \times (a, b)$.*

Throughout the proof (and in the sequel) we write $\mathcal{D}^{(2,0)} := \mathcal{D}(\mathcal{L}^{(2)}) \cap \mathcal{D}(\mathcal{L}^{(0)})$, where

$$\mathcal{D}(\mathcal{L}^{(0)}) = \bigl\{u \in \mathrm{C}_0[a, b) \bigm| u, u' \in \mathrm{AC}_{\mathrm{loc}}(a, b),\ \ell(u) \in \mathrm{C}_0[a, b),\ (pu')(a) = 0\bigr\}$$

is the domain of the Feller semigroup $\{T_t\}_{t\geq 0}$ (cf. Sect. 2.2.3). Note that if $g \in \mathrm{C}_\mathrm{c}^2[a, b)$ with $g' \in \mathrm{C}_\mathrm{c}(a, b)$, then $g \in \mathcal{D}^{(2,0)}$; consequently, any indicator function of an interval $I \subset [a, b)$ is the pointwise limit of functions $g_n \in \mathcal{D}^{(2,0)}$.

Proof Since $q_t(x, y, \cdot) \in \mathrm{C}_\mathrm{b}[a, b)$, it suffices to show that for all $g \in \mathcal{D}^{(2,0)}$ with $0 \leq g \leq 1$ we have

$$0 \leq \mathcal{Q}_{t,g}(x, y) \leq 1 \qquad \bigl(t > 0,\ x, y \in (a, b)\bigr),$$

where $\mathcal{Q}_{t,g}(x, y) := \int_a^b g(\xi)\, q_t(x, y, \xi)\, r(\xi)d\xi$.

Fix $t > 0$ and $g \in \mathcal{D}^{(2,0)}$ with $0 \leq g \leq 1$. Since $[\mathcal{F}q_t(x, y, \cdot)](\lambda) = e^{-t\lambda}\, w_\lambda(x)\, w_\lambda(y)$, it follows from the isometric property of the $\mathcal{L}$-transform (Theorem 2.5) that

$$\mathcal{Q}_{t,g}(x, y) = \int_{[\sigma^2,\infty)} e^{-t\lambda}\, w_\lambda(x)\, w_\lambda(y)\, (\mathcal{F}g)(\lambda)\, \boldsymbol{\rho}_{\mathcal{L}}(d\lambda).$$

Differentiating under the integral sign we easily check (by dominated convergence and using Lemma 2.4(b)) that $\ell_x \mathcal{Q}_{t,g} = \ell_y \mathcal{Q}_{t,g}$, $(\partial_y^{[1]} \mathcal{Q}_{t,g})(x, a) = 0$ and

$$\mathcal{Q}_{t,g}(x, a) = \int_{[\sigma^2,\infty)} e^{-t\lambda}\, w_\lambda(x)\, (\mathcal{F}g)(\lambda)\, \boldsymbol{\rho}_{\mathcal{L}}(d\lambda) = (T_t g)(x),$$

where the last equality follows from (2.31). The fact that $0 \leq g \leq 1$ clearly implies that $0 \leq (T_t g)(x) \leq 1$ for $x \in (a, b)$. One can verify via (2.23) that the function $f(x) = (T_t g)(x)$ is such that $f \in \mathcal{D}(\mathcal{L}^{(2)})$ and $\ell(f) \in \mathcal{D}(\mathcal{L}^{(2)})$. It then follows from the positivity property of the hyperbolic Cauchy problem (Corollary 4.3) that $0 \leq \mathcal{Q}_{t,g}(x, y) \leq 1$ for all $x, y \in (a, b)$, as claimed.

Proposition 4.6 (Time-Shifted Product Formula) *Let ℓ be a Sturm-Liouville expression of the form* (4.1), *and suppose that Assumption MP holds. The product*

$e^{-t\lambda} w_\lambda(x)\, w_\lambda(y)$ admits the integral representation

$$e^{-t\lambda} w_\lambda(x)\, w_\lambda(y) = \int_a^b w_\lambda(\xi)\, q_t(x, y, \xi)\, r(\xi) d\xi, \qquad t > 0,\ x, y \in (a, b),\ \lambda \geq 0, \tag{4.40}$$

where the integral in the right hand side is absolutely convergent.

In particular, $\int_a^b q_t(x, y, \xi)\, r(\xi) d\xi = 1$ for all $t > 0$, $x, y \in (a, b)$.

Proof The absolute convergence of the integral in the right hand side is immediate from Lemmas 2.3 and 4.5.

By Theorem 2.5, the equality in (4.40) holds $\boldsymbol{\rho}_{\mathcal{L}}$-almost everywhere. Since $\text{supp}(\boldsymbol{\rho}_{\mathcal{L}}) = [\sigma^2, \infty)$ (Lemma 4.3), the fact that both sides of (4.40) are continuous functions of $\lambda \geq 0$ allows us to extend by continuity the equality (4.40) to all $\lambda \geq \sigma^2$. If $\sigma = 0$, we are done.

Suppose that $\sigma > 0$. By (4.17) and Lemma 4.5, together with standard results on the analyticity of parameter-dependent integrals, the function $\tau \mapsto \int_a^b w_{\tau^2+\sigma^2}(\xi)\, q_t(x, y, \xi)\, r(\xi) d\xi$ is an analytic function of τ in the strip $|\text{Im}(\tau)| < \sigma$. It is also clear that $\tau \mapsto e^{-t(\tau^2+\sigma^2)} w_{\tau^2+\sigma^2}(x)\, w_{\tau^2+\sigma^2}(y)$ is an entire function. By analytic continuation we see that these two functions are equal for all τ in the strip $|\text{Im}(\tau)| < \sigma$; consequently, (4.40) holds.

The last statement is obtained by setting $\lambda = 0$.

4.3.3 The Product Formula for w_λ as the Limit Case

Unsurprisingly, the product formula (4.21) is deduced by taking the limit as $t \downarrow 0$ in the time-shifted product formula (4.40). If the functions $w_\lambda(\cdot)$ belong to $\text{C}_0[a, b)$, the limit can be straightforwardly taken in the vague topology of measures. As shown below, the class of modified Sturm-Liouville operators described in Lemma 4.2 can then be used to extend the product formula to the case where the functions $w_\lambda(\cdot)$ do not belong to $\text{C}_0[a, b)$

Theorem 4.5 (Product Formula for w_λ) *Let ℓ be a Sturm-Liouville expression of the form (4.1), and suppose that Assumption MP holds. For $x, y \in (a, b)$ and $t > 0$, let $\boldsymbol{\nu}_{t,x,y} \in \mathcal{P}[a, b)$ be the measure defined by $\boldsymbol{\nu}_{t,x,y}(d\xi) = q_t(x, y, \xi)\, r(\xi) d\xi$. Then for each $x, y \in (a, b)$ there exists a measure $\boldsymbol{\nu}_{x,y} \in \mathcal{P}[a, b)$ such that $\boldsymbol{\nu}_{t,x,y} \xrightarrow{w} \boldsymbol{\nu}_{x,y}$ as $t \downarrow 0$. Moreover, the product $w_\lambda(x)\, w_\lambda(y)$ admits the integral representation*

$$w_\lambda(x)\, w_\lambda(y) = \int_{[a,b)} w_\lambda(\xi)\, \boldsymbol{\nu}_{x,y}(d\xi), \qquad x, y \in (a, b),\ \lambda \in \mathbb{C}. \tag{4.41}$$

In particular, Theorem 4.2 holds.

Proof Let $\{t_n\}_{n\in\mathbb{N}}$ be an arbitrary decreasing sequence with $t_n \downarrow 0$. It is a basic fact that any sequence of probability measures contains a vaguely convergent subsequence (e.g. [10, p. 213]), thus there exists a subsequence $\{t_{n_k}\}$ and a measure $\boldsymbol{\nu}_{x,y} \in \mathcal{M}_+[a,b)$ such that $\boldsymbol{\nu}_{t_{n_k},x,y} \overset{v}{\longrightarrow} \boldsymbol{\nu}_{x,y}$ as $k \to \infty$. Let us show that all such subsequences $\{\boldsymbol{\nu}_{t_{n_k},x,y}\}$ have the same vague limit. Suppose that t_k^1, t_k^2 are two different sequences with $t_k^j \downarrow 0$ and that $\boldsymbol{\nu}_{t_k^j,x,y} \overset{v}{\longrightarrow} \boldsymbol{\nu}_{x,y}^j$ as $k \to \infty$ $(j = 1, 2)$. For $g \in \mathcal{D}^{(2,0)}$ we have

$$\begin{aligned}\int_{[a,b)} g(\xi)\, \boldsymbol{\nu}_{x,y}^j(d\xi) &= \lim_{k\to\infty} \int_{[a,b)} g(\xi)\, \boldsymbol{\nu}_{t_k^j,x,y}(d\xi) \\ &= \lim_{k\to\infty} \int_{[\sigma^2,\infty)} e^{-t_k^j\lambda}\, w_\lambda(x)\, w_\lambda(y)\, (\mathcal{F}g)(\lambda)\, \boldsymbol{\rho}_{\mathcal{L}}(d\lambda) \\ &= \int_{[\sigma^2,\infty)} w_\lambda(x)\, w_\lambda(y)\, (\mathcal{F}g)(\lambda)\, \boldsymbol{\rho}_{\mathcal{L}}(d\lambda)\end{aligned}$$

(the second equality was justified in the proof of Lemma 4.5, and dominated convergence yields the last equality). In particular, $\int_{[a,b)} g(\xi)\, \boldsymbol{\nu}_{x,y}^1(d\xi) = \int_{[a,b)} g(\xi)\, \boldsymbol{\nu}_{x,y}^2(d\xi)$ for all $g \in \mathcal{D}^{(2,0)}$, and this implies that $\boldsymbol{\nu}_{x,y}^1 = \boldsymbol{\nu}_{x,y}^2$. Since all subsequences have the same vague limit, we conclude that $\boldsymbol{\nu}_{t,x,y} \overset{v}{\longrightarrow} \boldsymbol{\nu}_{x,y}$ as $t \downarrow 0$.

Suppose first that $\sigma := \lim_{\xi\to\infty} \frac{A'(\xi)}{2A(\xi)} > 0$. Then Corollary 4.1 ensures that $\lim_{x\uparrow b} w_\lambda(x) = 0$ for $0 < \lambda \leq \sigma^2$, and by the Laplace-type representation (4.16) we have $w_\lambda(\cdot) \leq w_{\sigma^2}(\cdot)$ for $\lambda > \sigma^2$, hence $w_\lambda \in \mathrm{C}_0[a,b)$ for all $\lambda > 0$. Accordingly, by taking the limit as $t \downarrow 0$ of both sides of (4.40) we deduce that the product formula (4.41) holds for all $\lambda > 0$.

To prove that (4.41) is valid in the general case, let $\kappa < 0$ be arbitrary. We know that the operator $\ell^{\langle\kappa\rangle}$ defined in Lemma 4.2 satisfies Assumption MP; by Lemma 4.1 we have $\lim_{\xi\to\infty} \frac{(A^{\langle\kappa\rangle})'(\xi)}{2A^{\langle\kappa\rangle}(\xi)} > 0$ and consequently (by the reasoning in the previous paragraph) the corresponding Sturm-Liouville solutions (4.6) belong to $\mathrm{C}_0[a,b)$ for all $\lambda > 0$. From the previous part of the proof,

$$w_\lambda^{\langle\kappa\rangle}(x)\, w_\lambda^{\langle\kappa\rangle}(y) = \int_a^b w_\lambda^{\langle\kappa\rangle}(\xi)\, \boldsymbol{\nu}_{x,y}^{\langle\kappa\rangle}(d\xi), \qquad x, y \in (a,b),\ \lambda > 0, \tag{4.42}$$

with $\boldsymbol{\nu}_{x,y}^{\langle\kappa\rangle}$ constructed as before. We easily verify that $q_t^{\langle\kappa\rangle}(x,y,\xi) r^{\langle\kappa\rangle}(\xi) = \frac{e^{t\kappa} w_\kappa(\xi)}{w_\kappa(x) w_\kappa(y)} q_t(x,y,\xi) r(\xi)$ and, consequently, $\boldsymbol{\nu}_{x,y}^{\langle\kappa\rangle}(d\xi) = \frac{w_\kappa(\xi)}{w_\kappa(x) w_\kappa(y)} \boldsymbol{\nu}_{x,y}(d\xi)$. It thus follows from (4.42) that

$$w_{\kappa+\lambda}(x)\, w_{\kappa+\lambda}(y) = \int_a^b w_{\kappa+\lambda}(\xi)\, \boldsymbol{\nu}_{x,y}(d\xi), \qquad x, y \in (a,b),\ \lambda > 0,$$

where $\kappa < 0$ is arbitrary; hence (4.41) holds for all $\lambda \in \mathbb{R}$. If we then set $\lambda = \tau^2 + \sigma^2$ in (4.41), we straightforwardly verify that both sides are entire functions of τ (for the right hand side, this follows from the Laplace-type representation (4.16) and the fact that the integral converges for all $\lambda < 0$), so by analytic continuation the product formula holds for all $\lambda \in \mathbb{C}$.

Given that $w_0(x) \equiv 1$, setting $\lambda = 0$ in (4.41) shows that $\boldsymbol{\nu}_{x,y} \in \mathcal{P}[a, b)$; consequently, the measures $\boldsymbol{\nu}_{t,x,y}$ converge to $\boldsymbol{\nu}_{x,y}$ in the weak topology (cf. [10, Theorem 30.8]). Clearly, the product formula (4.41) can be extended to $x, y \in [a, b)$ by setting $\boldsymbol{\nu}_{x,a} := \delta_x$ and $\boldsymbol{\nu}_{a,y} := \delta_y$, hence Theorem 4.2 holds.

It is worth commenting that the reasoning used in this proof also allows us to justify that the time-shifted product formula (4.40) is valid for all $\lambda \in \mathbb{C}$.

As shown in the proof above, the measure $\boldsymbol{\nu}_{x,y}$ of the product formula (4.41) is characterized by the identity

$$\int_{[a,b)} f(\xi)\, \boldsymbol{\nu}_{x,y}(d\xi) = \int_{[\sigma^2,\infty)} w_\lambda(x)\, w_\lambda(y)\, (\mathcal{F}f)(\lambda)\, \boldsymbol{\rho}_{\mathcal{L}}(d\lambda), \qquad f \in \mathcal{D}^{(2,0)}. \tag{4.43}$$

Furthermore, the relation between this measure and the measure $\boldsymbol{\nu}_{t,x,y}(d\xi) = q_t(x, y, \xi)\, r(\xi) d\xi$ of the time-shifted product formula (4.40) can be written explicitly:

Corollary 4.4 *The measure $\boldsymbol{\nu}_{t,x,y}$ can be written in terms of the measure $\boldsymbol{\nu}_{x,y}$ and the transition kernel $p(t, x, y)$ of the Feller semigroup generated by the Sturm-Liouville operator ℓ as*

$$\boldsymbol{\nu}_{t,x,y}(d\xi) = \int_a^b \boldsymbol{\nu}_{z,y}(d\xi)\, p(t, x, z)\, r(z) dz \qquad \big(t > 0,\ x, y \in (a, b)\big).$$

Proof Recalling (2.31) and the proof of the previous proposition, we find that for $g \in \mathcal{D}^{(2,0)}$ we have

$$\begin{aligned}
&\int_a^b \int_{[a,b)} g(\xi)\, \boldsymbol{\nu}_{z,y}(d\xi)\, p(t, x, z)\, r(z) dz \\
&= \int_a^b \int_{[\sigma^2,\infty)} w_\lambda(z)\, w_\lambda(y)\, (\mathcal{F}g)(\lambda)\, \boldsymbol{\rho}_{\mathcal{L}}(d\lambda)\, p(t, x, z)\, r(z) dz \\
&= \int_{[\sigma^2,\infty)} e^{-t\lambda}\, w_\lambda(x)\, w_\lambda(y)\, (\mathcal{F}g)(\lambda)\, \boldsymbol{\rho}_{\mathcal{L}}(d\lambda) \\
&= \int_a^b g(\xi)\, q_t(x, y, \xi)\, r(\xi) d\xi,
\end{aligned}$$

hence the measures $\boldsymbol{\nu}_{t,x,y}(d\xi)$ and $\int_a^b \boldsymbol{\nu}_{z,y}(d\xi)\, p(t, x, z)\, r(z) dz$ are the same.

4.4 Sturm-Liouville Transform of Measures

In analogy with the definition of the index Whittaker transform of measures (Definition 3.3), it is natural to define the $\mathcal{L}$-transform of finite complex measures so that (2.21) is the $\mathcal{L}$-transform of an absolutely continuous measure with density $f(\cdot)\,r(\cdot)$:

Definition 4.3 Let $\mu \in \mathcal{M}_{\mathbb{C}}[a,b)$. The *$\mathcal{L}$-transform* of the measure μ is the function defined by the integral

$$\widehat{\mu}(\lambda) = \int_{[a,b)} w_\lambda(x)\, \mu(dx), \qquad \lambda \geq 0.$$

It is immediate from Lemma 2.3 that $|\widehat{\mu}(\lambda)| \leq \widehat{\mu}(0) = \|\mu\|$ for all $\mu \in \mathcal{M}_+[a,b)$. In addition, this definition leads to various properties which, as in the case of the Whittaker transform (cf. Proposition 3.4), resemble those of the Fourier transform of complex measures:

Proposition 4.7 *Let ℓ be a Sturm-Liouville expression of the form* (4.1)*, and suppose that Assumption MP holds. Let $\widehat{\mu}$ be the $\mathcal{L}$-transform of $\mu \in \mathcal{M}_{\mathbb{C}}[a,b)$. The following properties hold:*

(i) *$\widehat{\mu}$ is continuous on $\mathbb{R}_0^+$. Moreover, if a family of measures $\{\mu_j\} \subset \mathcal{M}_{\mathbb{C}}[a,b)$ is tight and uniformly bounded, then $\{\widehat{\mu_j}\}$ is equicontinuous on $\mathbb{R}_0^+$.*

(ii) *Each measure $\mu \in \mathcal{M}_{\mathbb{C}}[a,b)$ is uniquely determined by $\widehat{\mu}|_{[\sigma^2,\infty)}$.*

(iii) *If $\{\mu_n\}$ is a sequence of measures belonging to $\mathcal{M}_+[a,b)$, $\mu \in \mathcal{M}_+[a,b)$, and $\mu_n \xrightarrow{w} \mu$, then*

$$\widehat{\mu_n} \xrightarrow[n\to\infty]{} \widehat{\mu} \qquad \textit{uniformly for } \lambda \textit{ in compact sets.}$$

(iv) *Suppose that $\lim_{x\uparrow b} w_\lambda(x) = 0$ for all $\lambda > 0$. If $\{\mu_n\}$ is a sequence of measures belonging to $\mathcal{M}_+[a,b)$ whose $\mathcal{L}$-transforms are such that*

$$\widehat{\mu_n}(\lambda) \xrightarrow[n\to\infty]{} f(\lambda) \qquad \textit{pointwise in } \lambda \geq 0$$

for some real-valued function f which is continuous at a neighborhood of zero, then $\mu_n \xrightarrow{w} \mu$ for some measure $\mu \in \mathcal{M}_+[a,b)$ such that $\widehat{\mu} \equiv f$.

Proof

(i) It suffices to prove the second statement. Set $C = \sup_j \|\mu_j\|$. Fix $\lambda_0 \geq 0$ and $\varepsilon > 0$. By the tightness assumption, we can choose $\beta \in (a,b)$ such that $|\mu_j|(\beta,b) < \varepsilon$ for all j. Since the family of derivatives $\{\partial_\lambda w_{(\cdot)}(x)\}_{x\in(a,\beta]}$ is locally bounded on $\mathbb{R}_0^+$ (to verify this, differentiate the series (2.14) term by term and then compute an upper bound as in (2.15)), we can choose $\delta > 0$ such that

$$|\lambda - \lambda_0| < \delta \quad \implies \quad |w_\lambda(x) - w_{\lambda_0}(x)| < \varepsilon \ \text{ for all } a < x \leq \beta.$$

Consequently,

$$\begin{aligned}\big|\widehat{\mu_j}(\lambda) - \widehat{\mu_j}(\lambda_0)\big| &= \bigg|\int_{(a,b)} (w_\lambda(x) - w_{\lambda_0}(x))\mu_j(dx)\bigg| \\ &\leq \int_{(\beta,b)} |w_\lambda(x) - w_{\lambda_0}(x)||\mu_j|(dx) + \int_{(a,\beta]} |w_\lambda(x) - w_{\lambda_0}(x)||\mu_j|(dx) \leq (2+C)\varepsilon\end{aligned}$$

for all j, provided that $|\lambda - \lambda_0| < \delta$, which means that $\{\widehat{\mu_j}\}$ is equicontinuous at λ_0.

(ii) Let $\mu \in \mathcal{M}_{\mathbb{C}}[a,b)$ be such that $\widehat{\mu}(\lambda) = 0$ for all $\lambda \geq \sigma^2$. We need to show that μ is the zero measure. For each $g \in \mathcal{D}^{(2,0)}$ we have for $a < x < b$

$$\begin{aligned}0 &= \int_{[\sigma^2,\infty)} (\mathcal{F}g)(\lambda)\, w_\lambda(x)\, \widehat{\mu}(\lambda)\, \boldsymbol{\rho}_{\mathcal{L}}(d\lambda) \\ &= \int_{[a,b)} \int_{[\sigma^2,\infty)} (\mathcal{F}g)(\lambda)\, w_\lambda(x)\, w_\lambda(y)\, \boldsymbol{\rho}_{\mathcal{L}}(d\lambda)\, \mu(dy),\end{aligned}$$

where the change of order of integration is valid because, by Lemmas 2.3 and 2.4(b), the double integral converges absolutely; therefore

$$\begin{aligned}\int_{[a,b)} g(y)\, \mu(dy) &= \int_{[a,b)} \int_{[\sigma^2,\infty)} (\mathcal{F}g)(\lambda)\, w_\lambda(y)\, \boldsymbol{\rho}_{\mathcal{L}}(d\lambda) \mu(dy) \\ &= \int_{[a,b)} \lim_{x\downarrow a} \int_{[\sigma^2,\infty)} (\mathcal{F}g)(\lambda)\, w_\lambda(x)\, w_\lambda(y)\, \boldsymbol{\rho}_{\mathcal{L}}(d\lambda)\mu(dy) \\ &= \lim_{x\downarrow a} \int_{[a,b)} \int_{[\sigma^2,\infty)} (\mathcal{F}g)(\lambda)\, w_\lambda(x)\, w_\lambda(y)\, \boldsymbol{\rho}_{\mathcal{L}}(d\lambda)\mu(dy) \\ &= 0,\end{aligned}$$

using Lemma 2.4, the identity (4.43) and dominated convergence. This shows that $\int_{[a,b)} g(y)\, \mu(dy) = 0$ for all $g \in \mathcal{D}^{(2,0)}$ and, consequently, μ is the zero measure.

(iii) Since $w_\lambda(\cdot)$ is continuous and bounded, the pointwise convergence $\widehat{\mu_n}(\lambda) \to \widehat{\mu}(\lambda)$ follows from the definition of weak convergence of measures. By Prokhorov's theorem $\{\mu_n\}$ is tight and uniformly bounded, thus (by part (i)) $\{\widehat{\mu_n}\}$ is equicontinuous on $\mathbb{R}_0^+$. The same argument from the proof of Proposition 3.4(c) yields that the convergence $\widehat{\mu_n} \to \widehat{\mu}$ is uniform on compact sets.

(iv) The proof follows the same argument as that of Proposition 3.4(iv), replacing the interval $\mathbb{R}_0^+$ and the function $\boldsymbol{W}_{\alpha,\Delta_\lambda}(\cdot)$ by $[a, b)$ and $w_\lambda(\cdot)$ respectively.

Remark 4.3

I. If $\lim_{x\uparrow b} w_\lambda(x) = 0$ for all $\lambda > 0$, then as in Remark 3.4 we obtain the following analogue of the Lévy continuity theorem: *the $\mathcal{L}$-transform is a topological homeomorphism between $\mathcal{P}[a, b)$ with the weak topology and the set $\widehat{\mathcal{P}}$ of $\mathcal{L}$-transforms of probability measures with the topology of uniform convergence in compact sets.*

II. Much like weak convergence, vague convergence of measures can be formulated via the $\mathcal{L}$-transform, provided that $\lim_{x\uparrow b} w_\lambda(x) = 0$ for all $\lambda > 0$. Indeed, we can state:

II.1 If $\{\mu_n\} \subset \mathcal{M}_+[a, b)$, $\mu \in \mathcal{M}_+[a, b)$, and $\mu_n \overset{v}{\longrightarrow} \mu$, then $\lim \widehat{\mu_n}(\lambda) = \widehat{\mu}(\lambda)$ pointwise for each $\lambda > 0$;

II.2 If $\{\mu_n\} \subset \mathcal{M}_+[a, b)$, $\{\mu_n\}$ is uniformly bounded and $\lim \widehat{\mu_n}(\lambda) = f(\lambda)$ pointwise in $\lambda > 0$ for some function $f \in \mathrm{B_b}(\mathbb{R}^+)$, then $\mu_n \overset{v}{\longrightarrow} \mu$ for some measure $\mu \in \mathcal{M}_+[a, b)$ such that $\widehat{\mu} \equiv f$.

(The first part is trivial, and the second part is proved as follows: since any uniformly bounded sequence of positive measures contains a vaguely convergent subsequence, for any subsequence $\{\mu_{n_k}\}$ there exists a further subsequence $\{\mu_{n_{k_j}}\}$ and a measure μ such that $\mu_{n_{k_j}} \overset{v}{\longrightarrow} \mu$; then II.1 implies that $\widehat{\mu}(\lambda) = f(\lambda)$ for $\lambda > 0$, so the vague limit of such a subsequence is unique and, consequently, $\mu_n \overset{v}{\longrightarrow} \mu$.)

Consider the following stronger version of Assumption MP:

Assumption MP$_\infty$ *The operator $\ell = -\frac{1}{r}\frac{d}{dx}\bigl(p\,\frac{d}{dx}\bigr)$ satisfies Assumption MP and its coefficients satisfy $\lim_{x\uparrow b} p(x)r(x) = \infty$.*

This assumption will play an important role in the subsequent sections, mostly because it ensures that the properties stated in the Remark 4.3 hold. Indeed, one can state:

Lemma 4.6 *Let $\ell = -\frac{1}{r}\frac{d}{dx}\bigl(p\,\frac{d}{dx}\bigr)$ be a Sturm-Liouville operator satisfying Assumption MP. Then Assumption MP$_\infty$ holds if and only if $\lim_{x\uparrow b} w_\lambda(x) = 0$ for all $\lambda > 0$.*

Proof This follows from known results on the asymptotic behaviour of solutions of the Sturm-Liouville equation $-u'' - \frac{A'}{A}u' = \lambda u$, see [64, proof of Lemma 3.7].

We note that, in particular, the lemma states that the condition $\lim_{x\uparrow b} w_\lambda(x) = 0$ ($\lambda > 0$) holds whenever $\sigma > 0$. This particular case had already been pointed out in the proof of Theorem 4.5.

4.5 Sturm-Liouville Convolution of Measures

In what follows we always assume that the Sturm-Liouville expression ℓ satisfies Assumption MP. (In general we allow for operators such that $\lim_{x\uparrow b} p(x)r(x) < \infty$; whenever this is not the case, we will explicitly state that Assumption MP$_\infty$ is required to hold.)

As usual (cf. Definitions 2.6 and 3.4, Proposition 2.16), we define the convolution $* : \mathcal{M}_\mathbb{C}[a,b) \times \mathcal{M}_\mathbb{C}[a,b) \longrightarrow \mathcal{M}_\mathbb{C}[a,b)$ as the natural extension of the mapping $(x, y) \mapsto \delta_x * \delta_y := \boldsymbol{\nu}_{x,y}$ (where $\boldsymbol{\nu}_{x,y}$ is the measure of the product formula (4.41)), and we define the translation of functions as the integral with respect to the convolution of Dirac measures:

Definition 4.5 Let $\mu, \nu \in \mathcal{M}_\mathbb{C}[a,b)$. The complex measure

$$(\mu * \nu)(d\xi) = \int_{[a,b)} \int_{[a,b)} \boldsymbol{\nu}_{x,y}(d\xi)\, \mu(dx)\, \nu(dy)$$

is called the *$\mathcal{L}$-convolution* of the measures μ and ν. The *$\mathcal{L}$-translation* of a Borel measurable function $f : [a,b) \longrightarrow \mathbb{C}$ is defined as

$$(\mathcal{T}^y f)(x) := \int_{[a,b)} f(\xi)\, \boldsymbol{\nu}_{x,y}(d\xi) \equiv \int_{[a,b)} f(\xi)\, (\delta_x * \delta_y)(d\xi), \qquad x, y \in [a,b).$$

More generally, the *$\mathcal{L}$-translation by* $\mu \in \mathcal{M}_+[a,b)$ is defined as $(\mathcal{T}^\mu f)(x) := \int_{[a,b)} f(\xi)\, (\delta_x * \mu)(d\xi)$.

We will see that, in the same spirit of Sects. 3.5–3.8, analogues of many basic notions of (generalized) probabilistic harmonic analysis can be developed on the measure algebra determined by the $\mathcal{L}$-convolution. Our first proposition states the unsurprising fact that the $\mathcal{L}$-convolution is trivialized by the Sturm-Liouville transform of measures:

Proposition 4.8 *Let $\mu, \nu, \pi \in \mathcal{M}_\mathbb{C}[a,b)$. We have $\pi = \mu * \nu$ if and only if*

$$\widehat{\pi}(\lambda) = \widehat{\mu}(\lambda)\, \widehat{\nu}(\lambda) \qquad \textit{for all } \lambda \geq 0.$$

Proof Identical to that of Proposition 3.6 (replacing $\boldsymbol{W}_{\alpha,\Delta_\lambda}(\cdot)$ by $w_\lambda(\cdot)$, etc.).

The following result collects some basic properties of the measure algebra determined by the $\mathcal{L}$-convolution.

Proposition 4.9 *The space $(\mathcal{M}_\mathbb{C}[a,b), *)$, equipped with the total variation norm, is a commutative Banach algebra over $\mathbb{C}$ whose identity element is the Dirac measure δ_a. The subset $\mathcal{P}[a,b)$ is closed under the $\mathcal{L}$-convolution. Moreover, the map $(\mu, \nu) \mapsto \mu * \nu$ is continuous (in the weak topology) from $\mathcal{M}_\mathbb{C}[a,b) \times \mathcal{M}_\mathbb{C}[a,b)$ to $\mathcal{M}_\mathbb{C}[a,b)$.*

Proof Since $\widehat{\mu * \nu} = \widehat{\mu} \cdot \widehat{\nu}$ (Proposition 4.8), the commutativity, associativity and bilinearity of the $\mathcal{L}$-convolution follow at once from the uniqueness property of the $\mathcal{L}$-transform (Proposition 4.7(ii)). One can verify directly from the definition of the $\mathcal{L}$-convolution that the submultiplicativity property $\|\mu * \nu\| \leq \|\mu\| \cdot \|\nu\|$ holds, and that equality holds whenever $\mu, \nu \in \mathcal{M}_+[a, b)$; it is also clear that the convolution of positive measures is a positive measure. We conclude that the Banach algebra property holds and that $\mathcal{P}[a, b)$ is closed under convolution.

If $\lim_{x \uparrow b} w_\lambda(x) = 0$ for all $\lambda > 0$, the identity $\widehat{\boldsymbol{\nu}_{x,y}}(\lambda) = w_\lambda(x) w_\lambda(y)$ implies (by Proposition 4.7(iv)) that $(x, y) \mapsto \boldsymbol{\nu}_{x,y}$ is continuous in the weak topology. If the functions $w_\lambda(x)$ do not vanish at the limit $x \uparrow b$, let $\kappa < 0$ be arbitrary and let $h \in \mathrm{C_b}[a, b)$. Since w_κ is increasing and unbounded (Corollary 2.1), $\frac{h}{w_\kappa} \in \mathrm{C_0}[a, b)$. If we let $\boldsymbol{\nu}_{x,y}^{\langle\kappa\rangle}$ be the measure defined in the proof of Theorem 4.5, then by Remark 4.3.III the map $(x, y) \mapsto \boldsymbol{\nu}_{x,y}^{\langle\kappa\rangle}$ is continuous, and thus

$$(x, y) \longmapsto \int_{[a,b)} \frac{f(\xi)}{w_\kappa(\xi)} \boldsymbol{\nu}_{x,y}^{\langle\kappa\rangle}(d\xi) = \frac{1}{w_\kappa(x) w_\kappa(y)} \int_{[a,b)} f(\xi)\, \boldsymbol{\nu}_{x,y}(d\xi)$$

is continuous. This shows that $(x, y) \mapsto \int_{[a,b)} f(\xi)\, \boldsymbol{\nu}_{x,y}(d\xi)$ is continuous for all $f \in \mathrm{C_b}[a, b)$ and therefore $(x, y) \mapsto \boldsymbol{\nu}_{x,y}$ is continuous in the weak topology. Finally, for $f \in \mathrm{C_b}[a, b)$ and $\mu_n, \nu_n \in \mathcal{M}_\mathbb{C}[a, b)$ with $\mu_n \xrightarrow{w} \mu$ and $\nu_n \xrightarrow{w} \nu$ we have

$$\begin{aligned}
\lim_n \int_{[a,b)} f(\xi)(\mu_n * \nu_n)(d\xi) &= \lim_n \int_{[a,b)} \int_{[a,b)} \biggl(\int_{[a,b)} f\, d\boldsymbol{\nu}_{x,y} \biggr) \mu_n(dx) \nu_n(dy) \\
&= \int_{[a,b)} \int_{[a,b)} \biggl(\int_{[a,b)} f\, d\boldsymbol{\nu}_{x,y} \biggr) \mu(dx) \nu(dy) \\
&= \int_{[a,b)} f(\xi)(\mu * \nu)(d\xi),
\end{aligned}$$

due to the continuity of the function in parenthesis; this proves that $(\mu, \nu) \mapsto \mu * \nu$ is continuous.

Next we summarize some useful facts about the generalized translation introduced in Definition 4.5. For simplicity we write $\|\cdot\|_p \equiv \|\cdot\|_{L^p(r)}$ $(1 \leq p \leq \infty)$.

Proposition 4.10 *Let $\mu \in \mathcal{M}_+[a, b)$. The $\mathcal{L}$-translation operator $\mathcal{T}^\mu$ has the following properties:*

(i) *Let $1 \leq p \leq \infty$. If $f \in L^p(r)$, then $(\mathcal{T}^\mu f)(x)$ is a Borel measurable function of $x \in [a, b)$ and satisfies $\|\mathcal{T}^\mu f\|_p \leq \|\mu\| \cdot \|f\|_p$.*
(ii) *If $f \in L^2(r)$, then $\mathcal{F}(\mathcal{T}^\mu f)(\lambda) = \widehat{\mu}(\lambda)\, (\mathcal{F} f)(\lambda)$ for $\boldsymbol{\rho}_\mathcal{L}$-a.e. λ.*
(iii) *If $f \in \mathrm{C_b}[a, b)$, then $\mathcal{T}^\mu f \in \mathrm{C_b}[a, b)$.*
(iv) *Suppose that Assumption MP$_\infty$ holds. If $f \in \mathrm{C_0}[a, b)$, then $\mathcal{T}^\mu f \in \mathrm{C_0}[a, b)$.*

Proof

(i) It suffices to prove the result for nonnegative f. The map $\nu \mapsto \mu * \nu$ is weakly continuous (Proposition 4.9) and takes $\mathcal{M}_+[a,b)$ into itself. By a technical result proved in [94, Section 2.3], this implies that, for each Borel measurable $h \geq 0$, the function $x \mapsto (\mathcal{T}^\mu f)(x)$ is Borel measurable. It follows that $\int_{[a,b)} g(x)(\mu * r)(dx) := \int_a^b (\mathcal{T}^\mu g)(x) r(x) dx$ ($g \in \mathrm{C_c}[a,b)$) defines a positive Borel measure. For $a \leq c_1 < c_2 < b$, let $\mathbb{1}_{[c_1,c_2)}$ be the indicator function of $[c_1,c_2)$, let $f_n \in \mathcal{D}^{(2,0)}$ be a sequence of nonnegative functions such that $f_n \to \mathbb{1}_{[c_1,c_2)}$ pointwise, and write $\mathfrak{C} = \{g \in \mathrm{C}_\mathrm{c}^\infty(a,b) \mid 0 \leq g \leq 1\}$. We compute

$$
\begin{aligned}
(\mu * r)[c_1,c_2) &= \lim_n \int_{[a,b)} f_n(x)(\mu * r)(dx) \\
&= \lim_n \sup_{g \in \mathfrak{C}} \int_a^b (\mathcal{T}^\mu f_n)(x)\, g(x)\, r(x) dx \\
&= \lim_n \sup_{g \in \mathfrak{C}} \int_{[\sigma^2,\infty)} (\mathcal{F} f_n)(\lambda)\, (\mathcal{F} g)(\lambda)\, \widehat{\mu}(\lambda)\, \rho_{\mathcal{L}}(d\lambda) \\
&= \lim_n \sup_{g \in \mathfrak{C}} \int_a^b f_n(x)\, (\mathcal{T}^\mu g)(x)\, r(x) dx \\
&\leq \|\mu\| \cdot \lim_n \int_a^b f_n(x)\, r(x) dx \\
&= \|\mu\| \cdot \int_{c_1}^{c_2} r(x) dx,
\end{aligned}
$$

where the third and fourth equalities follow from (4.43) and the isometric property of the $\mathcal{L}$-transform (Theorem 2.5), and the inequality holds because $\|\mathcal{T}^\mu g\|_\infty \leq \|\mu\| \cdot \|g\|_\infty \leq \|\mu\|$. Therefore, $\|\mathcal{T}^\mu f\|_1 = \|f\|_{L^1([a,b),\mu * r)} \leq \|\mu\| \cdot \|f\|_1$ for each Borel measurable $f \geq 0$. Since $\delta_x * \mu \in \mathcal{M}_+[a,b)$, Hölder's inequality yields that $\|\mathcal{T}^\mu f\|_p \leq \|\mu\|^{1/q} \cdot \|\mathcal{T}^\mu |f|^p\|_1^{1/p} \leq \|\mu\| \cdot \|f\|_p$ for $1 < p < \infty$.

Finally, if $f \in L_\infty(r)$, $f \geq 0$ then $f = f_\mathbf{b} + f_\mathbf{0}$, where $0 \leq f_\mathbf{b} \leq \|f\|_\infty$ and $f_\mathbf{0} = 0$ Lebesgue-almost everywhere. Since $\|\mathcal{T}^\mu f_\mathbf{0}\|_1 \leq \|\mu\| \cdot \|f_\mathbf{0}\|_1 = 0$, we have $\mathcal{T}^\mu f_\mathbf{0} = 0$ Lebesgue-a.e., and therefore $\|\mathcal{T}^\mu f\|_\infty = \|\mathcal{T}^\mu f_\mathbf{b}\|_\infty \leq \|\mu\| \cdot \|f\|_\infty$.

(ii) For $f \in \mathcal{D}^{(2,0)}$, this identity follows at once from (4.43). The property extends to all $f \in L_2(r)$ by the standard continuity argument.

(iii) This follows immediately from the fact that $(\mu,\nu) \mapsto \mu * \nu$ is weakly continuous (Proposition 4.9).

(iv) It remains to show that $(\mathcal{T}^\mu h)(x) \to 0$ as $x \uparrow b$. Since $w_\lambda(x)\widehat{\mu}(\lambda) \to 0$ as $x \uparrow b$ ($\lambda > 0$), it follows from Remark 4.3.II that $\delta_x * \mu \xrightarrow{v} \mathbf{0}$ as $x \uparrow b$, where

$\mathbf{0}$ denotes the zero measure; this means that for each $f \in \mathrm{C}_0[a,b)$ we have

$$(\mathcal{T}^\mu f)(x) = \int_{[a,b)} f(\xi)(\delta_x * \mu)(d\xi) \longrightarrow \int_{[a,b)} f(\xi)\,\mathbf{0}(d\xi) = 0 \qquad \text{as } x \uparrow b,$$

showing that $\mathcal{T}^\mu f \in \mathrm{C}_0[a,b)$.

4.5.1 Infinite Divisibility and Lévy-Khintchine Type Representation

The set $\mathcal{P}_{\mathrm{id}}$ of *$\mathcal{L}$-infinitely divisible distributions* is defined in the usual way:

$$\mathcal{P}_{\mathrm{id}} = \bigl\{\mu \in \mathcal{P}[a,b) \bigm| \text{for all } n \in \mathbb{N} \text{ there exists } \nu_n \in \mathcal{P}[a,b) \text{ such that } \mu = \nu_n^{*n}\bigr\},$$

where ν_n^{*n} denotes the n-fold $\mathcal{L}$-convolution of ν_n with itself.

Lemma 4.7 *Suppose that Assumption MP$_\infty$ holds. If $\mu \in \mathcal{P}_{\mathrm{id}}$, then*

$$\widehat{\mu}(\lambda) = e^{-\psi_\mu(\lambda)},$$

where $\psi_\mu(\lambda)$ ($\lambda \geq 0$) is a positive continuous function such that $\psi_\mu(0) = 0$. Moreover, measures $\mu \in \mathcal{P}_{\mathrm{id}}$ have no nontrivial idempotent divisors, i.e., if $\mu = \vartheta * \nu$ (with $\vartheta, \nu \in \mathcal{P}[a,b)$) where ϑ is idempotent with respect to the $\mathcal{L}$-convolution (that is, it satisfies $\vartheta = \vartheta * \vartheta$), then $\vartheta = \delta_0$.

Proof Same as that of Lemma 3.6.

The function $\psi_\mu(\lambda)$ described in the lemma will be called the *log $\mathcal{L}$-transform* of μ. As in the case of the log-Whittaker transform (cf. Proposition 3.7), its growth is at most linear:

Proposition 4.11 *Suppose that Assumption MP$_\infty$ holds, and let $\mu \in \mathcal{P}_{\mathrm{id}}$. Then*

$$\psi_\mu(\lambda) \leq C_\mu(1+\lambda) \qquad \textit{for all } \lambda \geq 0$$

for some constant $C_\mu > 0$ which is independent of λ.

Proof Let $\nu_n \in \mathcal{P}[a,b)$ be the measure such that $\widehat{\nu_n}(\lambda) \equiv \exp(-\frac{1}{n}\psi_\mu(\lambda))$. The inequality $n\bigl(1-\widehat{\nu_n}(\lambda)\bigr) \leq \psi_\mu(\lambda)$ ($n \in \mathbb{N}$) and the limit $\lim_{n\to\infty} n\bigl(1-\widehat{\nu_n}(\lambda)\bigr) = \psi_\mu(\lambda)$ are justified as in the proof of Proposition 3.7.

Pick $\lambda_1 > 0$. We know that $\lim_{x\uparrow b} w_{\lambda_1}(x) = 0$ (Lemma 4.6), hence there exists $\beta \in (a,b)$ such that $|w_{\lambda_1}(x)| \leq \frac{1}{2}$ for all $\beta \leq x < b$. Combining this with (2.19), we deduce that for all $\lambda \geq 0$ we have

$$\begin{aligned} n \int_{[\beta,b)} \bigl(1 - w_\lambda(x)\bigr)\nu_n(dx) &\leq 2n \int_{[\beta,b)} \nu_n(dx) \\ &\leq 4n \int_{[\beta,b)} \bigl(1 - w_{\lambda_1}(x)\bigr)\nu_n(dx) \\ &\leq 4n\bigl(1 - \widehat{\nu_n}(\lambda_1)\bigr) \leq 4\psi_\mu(\lambda_1). \end{aligned} \tag{4.44}$$

Next, it follows from the proof of Proposition 4.7(i) that we can choose $\lambda_2 > 0$ such that $1 - w_\lambda(x) < \frac{1}{2}$ for all $0 \leq \lambda \leq \lambda_2$ and all $a < x \leq \beta$. Define $\eta_1(x) := \int_a^x \frac{1}{p(y)} \int_a^y r(\xi)d\xi\, dy \equiv \eta_1(x; -1)$, where $\eta_1(\cdot, \cdot)$ is the function defined in (2.13). Recalling (2.16), we obtain

$$1 - w_{\lambda_2}(x) = \lambda_2 \int_a^x \frac{1}{p(y)} \int_a^y w_{\lambda_2}(\xi) r(\xi) d\xi\, dy \geq \frac{\lambda_2}{2}\eta_1(x) \qquad \text{for all } a \leq x \leq \beta.$$

On the other hand, by (2.19) we have $1 - w_\lambda(x) \leq \lambda \int_a^x \frac{1}{p(y)} \int_a^y |w_\lambda(\xi)| r(\xi) d\xi\, dy \leq \lambda\eta_1(x)$ for all $x \in [a, b)$ and $\lambda \geq 0$. Consequently,

$$\begin{aligned} n \int_{[a,\beta)} \bigl(1 - w_\lambda(x)\bigr)\nu_n(dx) &\leq \lambda n \int_{[a,\beta)} \eta_1(x)\, \nu_n(dx) \\ &\leq \frac{2\lambda n}{\lambda_2} \int_{[a,\beta)} \bigl(1 - w_{\lambda_2}(x)\bigr)\nu_n(dx) \\ &\leq \frac{2\lambda n}{\lambda_2}\bigl(1 - \widehat{\nu_n}(\lambda_2)\bigr) \leq \frac{2\lambda}{\lambda_2}\psi_\mu(\lambda_2). \end{aligned} \tag{4.45}$$

Combining (4.44) and (4.45), one sees that for all $n \in \mathbb{N}$ and $\lambda \geq 0$ we have $n(1 - \widehat{\nu_n}(\lambda)) \leq C_\mu(1 + \lambda)$, where $C_\mu = \max\bigl\{4\psi_\mu(\lambda_1), \frac{2}{\lambda_2}\psi_\mu(\lambda_2)\bigr\}$. The conclusion follows by taking the limit as $n \to \infty$.

The log $\mathcal{L}$-transforms of $\mathcal{L}$-infinitely divisible distributions also admit an analogue of the classical Lévy-Khintchine representation. The relevant notions of compound Poisson and Gaussian measures are similar to those for the Whittaker convolution:

Definition 4.6 Let $\mu \in \mathcal{P}[a, b)$ and $c > 0$. The measure $\mathbf{e}(c\mu) \in \mathcal{P}[a, b)$ defined by

$$\mathbf{e}(c\mu) = e^{-c} \sum_{n=0}^{\infty} \frac{c^n}{n!} \mu^{*n}$$

(the infinite sum converging in the weak topology) is said to be the *$\mathcal{L}$-compound Poisson measure* associated with $c\mu$.

It is immediate that $\mathbf{e}(c\mu) \in \mathcal{P}_{\text{id}}$ and that its log $\mathcal{L}$-transform is $\psi_{\mathbf{e}(c\mu)}(\lambda) = c(1 - \widehat{\mu}(\lambda))$.

Definition 4.7 A measure $\mu \in \mathcal{P}[a, b)$ is called an *$\mathcal{L}$-Gaussian measure* if $\mu \in \mathcal{P}_{\text{id}}$ and

$$\mu = \mathbf{e}(c\nu) * \vartheta \quad \big(c > 0,\ \nu \in \mathcal{P}[a, b),\ \vartheta \in \mathcal{P}_{\text{id}}\big) \qquad \Longrightarrow \qquad \nu = \delta_a.$$

Theorem 4.5 (Lévy-Khintchine Type Formula) *Suppose that Assumption MP$_\infty$ holds. The log $\mathcal{L}$-transform of a measure $\mu \in \mathcal{P}_{\text{id}}$ can be represented in the form*

$$\psi_\mu(\lambda) = \psi_\alpha(\lambda) + \int_{(a,b)} \big(1 - w_\lambda(x)\big)\nu(dx), \tag{4.46}$$

where ν is a σ-finite measure on (a, b) which is finite on the complement of any neighbourhood of a and such that

$$\int_{(a,b)} \big(1 - w_\lambda(x)\big)\nu(dx) < \infty,$$

and α is an $\mathcal{L}$-Gaussian measure with log $\mathcal{L}$-transform $\psi_\alpha(\lambda)$. Conversely, each function of the form (4.46) *is a log $\mathcal{L}$-transform of some $\mu \in \mathcal{P}_{\text{id}}$.*

This Lévy-Khintchine type representation, together with its counterpart for the Whittaker convolution (Theorem 3.3), are both particular cases of the general Lévy-Khintchine formula for stochastic convolutions stated in Proposition 2.18 and whose proof was sketched in Sect. 3.5.2. (The fact that the $\mathcal{L}$-convolution satisfies axiom V6 of Definition 2.5 is argued in the same way as in Sect. 3.5.2.)

4.5.2 Convolution Semigroups

Definition 4.8 A family $\{\mu_t\}_{t \geq 0} \subset \mathcal{P}[a, b)$ is called an *$\mathcal{L}$-convolution semigroup* if it satisfies the conditions

- $\mu_s * \mu_t = \mu_{s+t}$ for all $s, t \geq 0$;
- $\mu_0 = \delta_a$;
- $\mu_t \overset{w}{\longrightarrow} \delta_a$ as $t \downarrow 0$.

If the Sturm-Liouville operator satisfies Assumption MP$_\infty$, then there exists a one-to-one correspondence $\{\mu_t\}_{t \geq 0} \mapsto \mu_1 \in \mathcal{P}_{\text{id}}$ between the set of $\mathcal{L}$-convolution semigroups and the set of $\mathcal{L}$-infinitely divisible distributions. (This can be justified exactly as in Remark 3.8.) Consequently, any convolution semigroup $\{\mu_t\}$ has an $\mathcal{L}$-transform of the form $\widehat{\mu_t}(\lambda) = \exp\big(-t\,\psi_{\mu_1}(\lambda)\big)$, where $\psi_{\mu_1}(\cdot)$ is a function of the form (4.46).

The family of generalized translation operators determined by a given $\mathcal{L}$-convolution semigroup has the expected Feller-type properties:

Proposition 4.12 *Suppose that Assumption* MP_∞ *holds, and let* $\{\mu_t\}_{t\geq 0}$ *be an* $\mathcal{L}$*-convolution semigroup. Then the family* $\{T_t\}_{t\geq 0}$ *defined by*

$$T_t : \mathrm{C_b}(\mathbb{R}_0^+) \longrightarrow \mathrm{C_b}(\mathbb{R}_0^+), \qquad T_t f := \mathcal{T}^{\mu_t} f$$

is a conservative Feller semigroup such that the identity $T_t \mathcal{T}^\nu f = \mathcal{T}^\nu T_t f$ *holds for all* $t \geq 0$ *and* $\nu \in \mathcal{M}_\mathbb{C}[a,b)$*. The restriction* $\big\{T_t|_{\mathrm{C_c}[a,b)}\big\}$ *can be extended to a strongly continuous contraction semigroup* $\{T_t^{(p)}\}$ *on the space* $L^p(r)$ $(1 \leq p < \infty)$*. Moreover, the operators* $T_t^{(p)}$ *are given by* $T_t^{(p)} f = \mathcal{T}^{\mu_t} f$ $(f \in L^p(r))$*.*

Proof Similar to that of Propositions 3.8–3.9.

Proposition 4.13 *Suppose that Assumption* MP_∞ *holds. Let* $\{\mu_t\}$ *be an* $\mathcal{L}$*-convolution semigroup with log* $\mathcal{L}$*-transform* ψ *and let* $\{T_t^{(2)}\}$ *be the associated Markovian semigroup on* $L^2(r)$*. Then the infinitesimal generator* $(\mathcal{G}^{(2)}, \mathcal{D}(\mathcal{G}^{(2)}))$ *of the semigroup* $\{T_t^{(2)}\}$ *is the self-adjoint operator given by*

$$\big(\mathcal{F}(\mathcal{G}^{(2)} f)\big)(\lambda) = -\psi \cdot (\mathcal{F} f), \qquad f \in \mathcal{D}(\mathcal{G}^{(2)}),$$

where

$$\mathcal{D}(\mathcal{G}^{(2)}) = \biggl\{ f \in L^2(r) \biggm| \int_0^\infty |\psi(\lambda)|^2 \big|(\mathcal{F} f)(\lambda)\big|^2 \rho_{\mathcal{L}}(\lambda) d\lambda < \infty \biggr\}.$$

Proof Similar to that of Proposition 3.10.

4.5.3 Additive and Lévy Processes

Definition 4.9 An $[a,b)$-valued Markov chain $\{S_n\}_{n\in\mathbb{N}_0}$ is said to be *$\mathcal{L}$-additive* if there exist measures $\mu_n \in \mathcal{P}[a,b)$ such that

$$P[S_n \in B | S_{n-1} = x] = (\mu_n * \delta_x)(B), \qquad n \in \mathbb{N},\ x \in [a,b),\ B \text{ a Borel subset of } [a,b). \tag{4.47}$$

If $\mu_n = \mu$ for all n, then $\{S_n\}$ is said to be an *$\mathcal{L}$-random walk*.

An explicit construction can be given for $\mathcal{L}$-additive Markov chains, based on the following lemma:

Lemma 4.8 *There exists a Borel measurable* $\Phi : [a,b) \times [a,b) \times [0,1] \longrightarrow [a,b)$ *such that*

$$(\delta_x * \delta_y)(B) = \mathfrak{m}\{\Phi(x,y,\cdot) \in B\}, \qquad x, y \in [a,b),\ B \text{ a Borel subset of } [a,b),$$

where $\mathfrak{m}$ *denotes Lebesgue measure on* $[0,1]$.

Proof Let $\Phi(x,y,\xi) = \max\bigl(a, \sup\{z \in [a,b) : (\delta_x * \delta_y)[a,z] < \xi\}\bigr)$. Using the continuity of the $\mathcal{L}$-convolution, one can show that Φ is Borel measurable, see [19, Theorem 7.1.3]. It is straightforward that $\mathfrak{m}\{\Phi(x,y,\cdot) \in [a,c]\} = \mathfrak{m}\{(\delta_x * \delta_y)[a,c] \geq \xi\} = (\delta_x * \delta_y)[a,c]$.

Let $X_1, U_1, X_2, U_2, \ldots$ be a sequence of independent random variables (on a given probability space $(\Omega, \mathfrak{A}, \pi)$) where the X_n have distribution $P_{X_n} = \mu_n \in \mathcal{P}[a,b)$ and each of the (auxiliary) random variables U_n has the uniform distribution on $[0,1]$. Set

$$S_0 = 0, \qquad S_n = S_{n-1} \oplus_{U_n} X_n, \tag{4.48}$$

where $X \oplus_U Y := \Phi(X,Y,U)$. Then the distributions P_{S_n} of the random variables S_n are such that $P_{S_n} = P_{S_{n-1}} * \mu_n$ $(n \in \mathbb{N}_0)$ and, consequently, $\{S_n\}_{n\in\mathbb{N}_0}$ is an $\mathcal{L}$-additive Markov chain satisfying (4.47). The identity $P_{S_n} = P_{S_{n-1}} * \mu_n$ is easily checked:

$$\begin{aligned} P_{S_n}(B) &= P\bigl[\Phi(S_{n-1}, X_n, U_n) \in B\bigr] \\ &= \int_{[a,b)} \int_{[a,b)} \mathfrak{m}\{\Phi(x,y,\cdot) \in B\} P_{S_{n-1}}(dx) P_{X_n}(dy) \\ &= \int_{[a,b)} \int_{[a,b)} (\delta_x * \delta_y)(B) P_{S_{n-1}}(dx) P_{X_n}(dy) \\ &= (P_{S_{n-1}} * \mu_n)(B). \end{aligned}$$

The continuous-time analogue of $\mathcal{L}$-random walks are the $\mathcal{L}$-Lévy processes, defined in analogy with Definition 3.8:

Definition 4.10 An $[a,b)$-valued Markov process $Y = \{Y_t\}_{t\geq 0}$ is said to be an *$\mathcal{L}$-Lévy process* if there exists an $\mathcal{L}$-convolution semigroup $\{\mu_t\}_{t\geq 0}$ such that for $0 \leq s \leq t$ the transition probabilities of Y are given by

$$P[Y_t \in B | Y_s = x] = (\mu_{t-s} * \delta_x)(B), \qquad x \in [a,b),\ B \text{ a Borel subset of } [a,b). \tag{4.49}$$

If we let $\nu \in \mathcal{P}[a,b)$ be a given measure and $\{\mu_t\}_{t\geq 0}$ a given $\mathcal{L}$-convolution semigroup, then in the same manner as in the previous chapter one can construct a $\mathcal{L}$-Lévy process satisfying (4.49) and such that $P[X_0 \in \cdot] = \nu$. Like in Corollary 2.3 and Proposition 3.11, the class of Lévy processes includes the diffusion generated by the associated Sturm-Liouville operator (as defined in Sect. 2.2.3):

Proposition 4.14 *Suppose that Assumption MP_∞ holds. The diffusion process X generated by the Neumann realization $(\mathcal{L}^{(2)}, \mathcal{D}(\mathcal{L}^{(2)}))$ of ℓ is a $\mathcal{L}$-Lévy process.*

Proof Similar to that of Proposition 3.11.

An analogue of the well-known theorem on approximation of Lévy processes by triangular arrays holds for $\mathcal{L}$-Lévy processes (below the notation $\xrightarrow{d}$ stands for convergence in distribution):

Proposition 4.15 *Suppose that Assumption MP_∞ holds, and let X be an $[a, b)$-valued random variable. The following assertions are equivalent:*

(i) $X = Y_1$ *for some $\mathcal{L}$-Lévy process* $Y = \{Y_t\}_{t\geq 0}$;
(ii) *The distribution of X is $\mathcal{L}$-infinitely divisible;*
(iii) $S_{m_n}^n \xrightarrow{d} X$ *for some sequence of $\mathcal{L}$-random walks $S^1, S^2, \ldots$ (with $S_0^j = a$) and some integers $m_n \to \infty$.*

Proof The equivalence between (i) and (ii) is a restatement of the one to one correspondence $\{\mu_t\}_{t\geq 0} \longleftrightarrow \mu_1$ between $\mathcal{L}$-infinitely divisible measures and $\mathcal{L}$-convolution semigroups. It is obvious that (i) implies (iii): simply let $m_n = n$ and S^n the random walk whose step distribution is the law of $Y_{1/n}$.

Suppose that (iii) holds and let π_n, μ be the distributions of S_j^n, X respectively. Choose $\varepsilon > 0$ small enough so that $\widehat{\mu}(\lambda) > C_\varepsilon > 0$ for $\lambda \in [0, \varepsilon]$, where $C_\varepsilon > 0$ is a constant. By (iii) and Proposition 4.7(iii), $\widehat{\pi_n}(\lambda)^{m_n} \to \widehat{\mu}(\lambda)$ uniformly on compacts, which implies that $\widehat{\pi_n}(\lambda) \to 1$ for all $\lambda \in [0, \varepsilon]$ and, therefore, by Proposition 4.7(iv) $\pi_n \xrightarrow{w} \delta_a$. Now let $k \in \mathbb{N}$ be arbitrary. Since $\pi_n \xrightarrow{w} \delta_a$, we can assume that each m_n is a multiple of k. Write $\nu_n = \pi_n^{*(m_n/k)}$, so that $\nu_n^{*k} \xrightarrow{w} \mu$. By relative compactness of $\mathrm{D}(\{\pi_n^{*m_n}\})$ (see [198, Corollary 1]), the sequence $\{\nu_n\}_{n\in\mathbb{N}}$ has a weakly convergent subsequence, say $\nu_{n_j} \xrightarrow{w} \mu_k$ as $j \to \infty$, and from this it clearly follows that $\mu_k^{*k} = \mu$. Consequently, (ii) holds.

An $\mathcal{L}$-convolution semigroup $\{\mu_t\}_{t\geq 0}$ such that μ_1 is an $\mathcal{L}$-Gaussian measure is called an *$\mathcal{L}$-Gaussian convolution semigroup*, and an $\mathcal{L}$-Lévy process associated with an $\mathcal{L}$-Gaussian convolution semigroup is called an *$\mathcal{L}$-Gaussian process.*

Proposition 4.16 (Alternative Characterizations of $\mathcal{L}$-Gaussian Convolution Semigroups) *Suppose that Assumption MP_∞ holds and that $A \in \mathrm{C}^3(a, b)$ (where A is the function defined in* (4.2)*). Let $Y = \{Y_t\}_{t\geq 0}$ be an $\mathcal{L}$-Lévy process, let $\{\mu_t\}_{t\geq 0}$ be the associated $\mathcal{L}$-convolution semigroup and let $(\mathcal{G}^{(0)}, \mathcal{D}(\mathcal{G}^{(0)}))$ be the infinitesimal generator of the Feller semigroup associated with Y. The following conditions are equivalent:*

(i) μ_1 *is a Gaussian measure;*
(ii) $\lim_{t\downarrow 0} \frac{1}{t}\mu_t\bigl([a, b) \setminus \mathcal{V}_a\bigr) = 0$ *for every neighbourhood $\mathcal{V}_a$ of the point a;*

(iii) $\lim_{t\downarrow 0}\frac{1}{t}(\mu_t * \delta_x)\bigl([a,b)\setminus \mathcal{V}_x\bigr) = 0$ *for every* $x \in [a,b)$ *and every neighbourhood* $\mathcal{V}_x$ *of the point* x*;*

(iv) Y *has a modification whose paths are a.s. continuous.*

If any of these conditions hold then the infinitesimal generator of Y *is a local operator, i.e.,* $(\mathcal{G}^{(0)} f)(x) = (\mathcal{G}^{(0)} g)(x)$ *whenever* $f, g \in \mathcal{D}(\mathcal{G}^{(0)})$ *and* $h = g$ *on some neighbourhood of* $x \in [a,b)$*.*

Proof

(i) $\Longleftrightarrow$ ***(ii):*** This can be proved as in Lemma 3.8 (with the obvious adaptations).

(ii) $\Longleftrightarrow$ ***(iii):*** To prove the nontrivial direction, assume that (ii) holds, and fix $x \in (a,b)$. Let $\mathcal{V}_x$ be a neighbourhood of the point x not containing a and write $E_x = [a,b)\setminus\mathcal{V}_x$. Pick a function $f \in \mathrm{C}^4$ such that $0 \le f \le 1$, $f = 0$ on E_x and $f = 1$ on some smaller neighbourhood $\mathcal{U}_x \subset \mathcal{V}_x$ of the point x. (Using the assumption $A \in \mathrm{C}^3(a,b)$, it is easy to check that $f, \ell(f) \in \mathcal{D}^{(2,0)}$.)

We begin by showing that

$$\lim_{y\downarrow a}\frac{1-(\mathcal{T}^x f)(y)}{1-w_\lambda(y)} = 0 \qquad \text{for each } \lambda > 0. \tag{4.50}$$

Indeed, it follows from (4.43) that $\lim_{y\downarrow a}(\mathcal{T}^x f)(y) = 1$, $\lim_{y\downarrow a}\partial_y^{[1]}(\mathcal{T}^x f)(y) = 0$ and

$$\begin{aligned}\ell_y(\mathcal{T}^x f)(y) &= \int_{\mathbb{R}_0^+}\lambda\,(\mathcal{F}f)(\lambda)\,w_\lambda(x)\,w_\lambda(y)\,\boldsymbol{\rho}_{\mathcal{L}}(d\lambda)\\ &= \bigl(\mathcal{T}^x\ell(f)\bigr)(y) \xrightarrow[y\downarrow a]{} \ell(f)(x) = 0,\end{aligned}$$

hence using L'Hôpital's rule twice we find that for $\lambda > 0$ we have $\lim_{y\downarrow a}\frac{1-(\mathcal{T}^x f)(y)}{1-w_\lambda(y)} = \lim_{y\downarrow a}\frac{\ell_y(\mathcal{T}^x f)(y)}{\lambda\, w_\lambda(y)} = 0$.

By (4.50), for each $\lambda > 0$ there exists $a_\lambda > a$ such that $(\mathcal{T}^x\mathbb{1}_{E_x})(y) \le \bigl(\mathcal{T}^x(\mathbb{1} - f)\bigr)(y) \le 1 - w_\lambda(y)$ for all $y \in [a, a_\lambda)$. We then estimate

$$\begin{aligned}\frac{1}{t}(\mu_t * \delta_x)(E_x) &= \frac{1}{t}\int_{[a,b)}(\mathcal{T}^x\mathbb{1}_{E_x})(y)\mu_t(dy)\\ &\le \frac{1}{t}\int_{[a,a_\lambda)}\bigl(1-w_\lambda(y)\bigr)\mu_t(dy) + \frac{1}{t}\mu_t[a_\lambda,b)\\ &\le \frac{1}{t}\int_{[a,b)}\bigl(1-w_\lambda(y)\bigr)\mu_t(dy) + \frac{1}{t}\mu_t[a_\lambda,b)\\ &= \frac{1}{t}\bigl(1-\widehat{\mu_t}(\lambda)\bigr) + \frac{1}{t}\mu_t[a_\lambda,b).\end{aligned}$$

Given that we are assuming that (ii) holds and, by the $\mathcal{L}$-semigroup property, $\lim_{t\downarrow 0}\frac{1}{t}\bigl(1-\widehat{\mu_t}(\lambda)\bigr) = \lim_{t\downarrow 0}\frac{1}{t}\bigl(1-\widehat{\mu_1}(\lambda)^t\bigr) = -\log\widehat{\mu_1}(\lambda)$, the above inequality gives

$$\limsup_{t\downarrow 0}\frac{1}{t}(\mu_t * \delta_x)(E_x) \leq -\log\widehat{\mu_1}(\lambda).$$

This holds for arbitrary $\lambda > 0$. Since the right-hand side is continuous and vanishes for $\lambda = 0$, we conclude that $\lim_{t\downarrow 0}\frac{1}{t}(\mu_t * \delta_x)(E_x) = 0$, as desired.

(iii)⟹(iv): This follows from Proposition 2.4.

(iv) ⟹ (iii): This is a general fact which is known as Ray's theorem on one-dimensional diffusion processes. The proof can be found in [88, Theorem 5.2.1].

The final assertion follows from Proposition 2.5.

Remark 4.4 As in the context of the Whittaker convolution, one can introduce the notion of a moment sequence associated with the $\mathcal{L}$-convolution.

The *canonical $\mathcal{L}$-moment functions* $\widetilde{\varphi}_k$ $(k \in \mathbb{N})$ can be defined recursively as the solution of the initial value problem

$$\ell(\widetilde{\varphi}_k)(x) = -2\sigma k\widetilde{\varphi}_{k-1}(x) - k(k-1)\widetilde{\varphi}_{k-2}(x), \qquad \widetilde{\varphi}_k(a) = 0, \qquad (p\widetilde{\varphi}_k')(a) = 0$$

(where $\widetilde{\varphi}_{-1}(x) := 0$ and $\widetilde{\varphi}_0(x) := 1$). Equivalently, we can write

$$\begin{aligned}\widetilde{\varphi}_k(x) &= k\int_0^x \frac{1}{p(y)}\int_0^y r(\xi)\bigl[2\sigma\widetilde{\varphi}_{k-1}(\xi) + (k-1)\widetilde{\varphi}_{k-2}(\xi)\bigr]\,d\xi\,dy \\ &= \frac{\partial^k}{\partial\tau^k}\Big|_{\tau=\sigma} w_{\sigma^2-\tau^2}(x) \\ &= \int_{-\infty}^{\infty} s^k e^{\sigma s}\pi_x(ds).\end{aligned}$$

Using the product formula (4.41), one can check that the canonical $\mathcal{L}$-moment functions are a solution of the functional equation $(\mathcal{T}^y\varphi_k)(x) = \sum_{j=0}^k \binom{k}{j}\varphi_j(x)\varphi_{k-j}(y)$, meaning that the $\widetilde{\varphi}_k$ play a role similar to that of the monomials under the classical convolution.

The canonical $\mathcal{L}$-moment functions are a tool for establishing strong laws of large numbers for $\mathcal{L}$-additive Markov chains. In particular, the following results hold for a given $\mathcal{L}$-additive Markov chain $\{S_n\}$ constructed as in (4.48):

I. *If* $\{r_n\}_{n\in\mathbb{N}}$ *is a sequence of positive numbers such that* $\lim_n r_n = \infty$ *and* $\sum_{n=1}^{\infty}\frac{1}{r_n}\bigl(\mathbb{E}[\varphi_2(X_n)] - \mathbb{E}[\varphi_1(X_n)]^2\bigr) < \infty$, *then*

$$\lim_n \frac{1}{\sqrt{r_n}}\bigl(\varphi_1(S_n) - \mathbb{E}[\varphi_1(S_n)]\bigr) = 0 \qquad \pi\text{-a.s.}$$

II. *If $\{S_n\}$ is an $\mathcal{L}$-random walk such that $\mathbb{E}[\varphi_2(X_1)^{\theta/2}] < \infty$ for some $1 \leq \theta < 2$, then $\mathbb{E}[\varphi_1(X_1)] < \infty$ and*

$$\lim_n \frac{1}{n^{1/\theta}} \bigl(\varphi_1(S_n) - n\mathbb{E}[\varphi_1(X_1)]\bigr) = 0 \qquad \pi\text{-a.s.}$$

III. *Suppose that $\varphi_1 \equiv 0$. If $\{r_n\}_{n\in\mathbb{N}}$ is a sequence of positive numbers such that $\lim_n r_n = \infty$ and $\sum_{n=1}^{\infty} \frac{1}{r_n}\mathbb{E}[\varphi_2(X_n)] < \infty$, then*

$$\lim_n \frac{1}{r_n} \varphi_2(S_n) = 0 \qquad \pi\text{-a.s.}$$

IV. *Suppose that $\varphi_1 \equiv 0$. If $\{S_n\}$ is an $\mathcal{L}$-random walk such that $\mathbb{E}[\varphi_2(X_1)^{\theta}] < \infty$ for some $0 < \theta < 1$, then*

$$\lim_n \frac{1}{n^{1/\theta}} \varphi_2(S_n) = 0 \qquad \pi\text{-a.s.}$$

The above statements can be proved exactly as in the hypergroup framework, see [19, Section 7.3]. In addition, one can show that the modified moments $\mathbb{E}[\widetilde{\varphi}_k(X)]$ can be computed via the $\mathcal{L}$-transform of measures and that a martingale property holds for $\mathcal{L}$-moment functions applied to $\mathcal{L}$-Lévy processes (these results are proved as in Propositions 3.16 and 3.17).

Under additional assumptions on the coefficients of ℓ, one can also establish a Lévy-type characterization similar to that of Theorem 3.4 for the diffusion process associated with the Sturm-Liouville operator. In particular, an adaptation of the proof of Theorem 3.4 yields the following result: *Set $\eta_1(x) := \int_a^x \frac{1}{p(y)} \int_a^y r(\xi)d\xi\, dy$ and $\eta_2(x) := \int_a^x \frac{1}{p(y)} \int_a^y \eta_1(\xi) r(\xi)d\xi\, dy$. Suppose that $a > -\infty$, Assumption MP$_\infty$ holds and one of the following conditions is satisfied:*

- $\eta_1(x) = c_1(x-a)+c_2(x-a)^2+o((x-a)^2)$ *and* $\eta_2(x) = c_3(x-a)^2+o((x-a)^2)$ *as $x \downarrow a$, with $c_1, c_3 > 0$;*
- $\eta_1(x) = c_1(x-a)^2+c_2(x-a)^4+o((x-a)^4)$ *and* $\eta_2(x) = c_3(x-a)^4+o((x-a)^4)$ *as $x \downarrow a$, with $c_1, c_3 > 0$.*

Let $X = \{X_t\}_{t\geq 0}$ be an $[a,b)$-valued Markov process with a.s. continuous paths. Then the following assertions are equivalent:

(i) *X is the diffusion process generated by the Neumann realization $(\mathcal{L}^{(2)}, \mathcal{D}(\mathcal{L}^{(2)}))$ of ℓ;*
(ii) *$\{\eta_1(X_t) - t\}_{t\geq 0}$ and $\{\eta_2(X_t) - t\,\eta_1(X_t) + \frac{t^2}{2}\}_{t\geq 0}$ are martingales (or local martingales);*
(iii) *$\{\eta_1(X_t) - t\}_{t\geq 0}$ is a local martingale with $[\eta_1(X)]_t = 2\int_0^t \frac{p(X_s)}{r(X_s)}(\eta_1'(X_s))^2\, ds$.*

4.6 Sturm-Liouville Hypergroups

Our purpose here is to discuss whether the convolution algebra structure constructed in the previous section satisfies the hypergroup axioms H1–H8 introduced in Definition 2.4. We will determine a sufficient condition that leads to an existence theorem for Sturm-Liouville hypergroups which is more general than that of Zeuner (stated above in Theorem 4.1). In addition to this, we will introduce a notion of degenerate hypergroup which includes the Whittaker convolution and many other Sturm-Liouville convolutions whose associated hyperbolic Cauchy problems are also parabolically degenerate.

4.6.1 The Nondegenerate Case

We saw in Proposition 4.9 that the $\mathcal{L}$-convolution satisfies the hypergroup axioms H1–H5 (with $K = [a, b)$ and $\mathrm{e} = a$ as the identity element; H5 holds for the identity involution $\check{x} = x$). In order to verify axioms H6–H8, one needs to determine the support of $\boldsymbol{\nu}_{x,y} = \delta_x * \delta_y$.

A detailed study of $\mathrm{supp}(\boldsymbol{\nu}_{x,y})$ was carried out by Zeuner in [210]. The next proposition shows that the results of Zeuner can be applied to the $\mathcal{L}$-convolution, provided that the differential operator (4.1) has coefficients $p = r = A$ defined on $\mathbb{R}^+$, and there exists $\eta \in \mathrm{C}^1(\mathbb{R}_0^+)$ satisfying the conditions given in Assumption MP.

Proposition 4.17 *Let*

$$\ell = -\frac{1}{A}\frac{d}{dx}\Big(A\frac{d}{dx}\Big), \qquad x \in \mathbb{R}^+,$$

where $A(x) > 0$ for all $x \geq 0$. Suppose that there exists $\eta \in \mathrm{C}^1(\mathbb{R}_0^+)$ such that $\eta \geq 0$, the functions $\boldsymbol{\phi}_\eta$, $\boldsymbol{\psi}_\eta$ are both decreasing on $\mathbb{R}^+$ and $\lim_{x\to\infty} \boldsymbol{\phi}_\eta(x) = 0$. Let $x_0 = \sup\{x \geq 0 \mid \boldsymbol{\psi}_\eta(x) = \boldsymbol{\psi}_\eta(0)\}$ and $x_1 = \inf\{x > 0 \mid \boldsymbol{\phi}_\eta(x) = 0\}$. Then:

(a) *If $x_0 = \infty$, $x_1 = 0$ and $\eta(0) = 0$ then $\mathrm{supp}(\delta_x * \delta_y) = \{|x - y|, x + y\}$ for all $x, y \geq 0$.*

(b) *If $0 < x_0 < \infty$, $x_1 = 0$ and $\eta(0) = 0$ then*

$$\mathrm{supp}(\delta_x * \delta_y) = \begin{cases} \{|x-y|, x+y\}, & x+y \leq x_0, \\ \{|x-y|\} \cup [2x_0 - x - y, x+y], & x, y < x_0 < x+y, \\ [|x-y|, x+y], & \max\{x, y\} \geq x_0. \end{cases}$$

(c) *If* $x_0 = \infty$, $0 < x_1 < \infty$ *and* $\eta(0) = 0$ *then*

$$\mathrm{supp}(\delta_x * \delta_y) = \begin{cases} [|x-y|, x+y], & \min\{x,y\} \leq 2x_1, \\ [|x-y|, 2x_1 + |x-y|] \cup [x+y-2x_1, x+y], & \min\{x,y\} > 2x_1. \end{cases}$$

(d) *If* $0 < 3x_1 < x_0 < \infty$ *and* $\eta(0) = 0$ *then*

$$\mathrm{supp}(\delta_x * \delta_y) = \\ = \begin{cases} [|x-y|, x+y], & \min\{x,y\} \leq 2x_1 \text{ or } \max\{x,y\} \geq x_0 - x_1, \\ [|x-y|, 2x_1 + |x-y|] \\ \qquad \cup [x+y-2x_1, x+y], & \min\{x,y\} > 2x_1 \text{ and } \max\{x,y\} < x_0 - x_1. \end{cases}$$

(e) *If* $x_0 \leq 3x_1$ *or* $\eta(0) > 0$ *then* $\mathrm{supp}(\delta_x * \delta_y) = [|x-y|, x+y]$ *for all* $x, y \geq 0$.

The proof depends on the following lemma which ensures that the existence and uniqueness theorems for the associated hyperbolic Cauchy problem (Theorems 4.3–4.4) are also valid for initial conditions $f \in \mathcal{D}^{(2,0)}$.

Lemma 4.9 *If* $f \in \mathcal{D}^{(2,0)}$*, then there exists a unique solution* $h \in \mathrm{C}^2\big((a,b)^2\big)$ *of the Cauchy problem* (4.22) *satisfying conditions* **(α)**–**(β)** *of Sect. 4.3.1, and this unique solution is given by* (4.23)*.*

Proof The fact that there exists at most one solution of (4.22) satisfying the given requirements is proved in the same way.

Let $f \in \mathcal{D}^{(2,0)}$ and consider the function $h(x,y)$ defined by (4.23). The limit $\lim_{y\downarrow a} h(x,y) = f(x)$ follows from Lemma 2.4(b) and dominated convergence. Similarly, we have

$$\lim_{y\downarrow a} (\partial_y^{[1]} h)(x,y) = \lim_{y\downarrow a} \int_{\mathbb{R}_0^+} (\mathcal{F}f)(\lambda)\, w_\lambda(x)\, w_\lambda^{[1]}(y)\, \boldsymbol{\rho}_{\mathcal{L}}(d\lambda) = 0$$

(the absolute and uniform convergence of the differentiated integral justifies the differentiation under the integral sign). Now fix $y \in (a,b)$. By (4.43), we have $h(\cdot, y) = \mathcal{T}^y f$. Using (2.24) and Lemma 4.10(ii), we obtain

$$\begin{aligned} \mathcal{F}(\ell_x(\mathcal{T}^y f))(\lambda) &= \lambda\, \mathcal{F}(\mathcal{T}^y f)(\lambda) = \lambda\, w_\lambda(y)\, (\mathcal{F}f)(\lambda) \\ &= w_\lambda(y)\, \mathcal{F}(\ell(f))(\lambda) = \mathcal{F}(\mathcal{T}^y \ell(f))(\lambda), \end{aligned}$$

hence $\ell_x(\mathcal{T}^y f)(x) = (\mathcal{T}^y \ell(f))(x)$ for almost every x. Since (by the weak continuity of $(x,y) \mapsto \boldsymbol{\nu}_{x,y}$, see Proposition 4.9) $(x,y) \mapsto (\mathcal{T}^y \ell(f))(x)$ is continuous, it follows that

$$\ell_x h(x,y) = (\mathcal{T}^y \ell(f))(x), \qquad \text{for all } x, y \in (a,b).$$

Exactly the same reasoning shows that $\ell_y h(x,y) = (\mathcal{T}^y \ell(f))(x)$, hence $h \in \mathrm{C}^2\big((a,b)^2\big)$ is a solution of $\ell_x u = \ell_y u$.

It remains to check that the function (4.23) satisfies conditions $(\boldsymbol{\alpha})$–$(\boldsymbol{\beta})$. As seen above we have $\mathcal{F}(h(\cdot,y))(\lambda) = w_\lambda(y)\,(\mathcal{F}f)(\lambda)$ and $\mathcal{F}[\ell_y h(\cdot,y)](\lambda) = \mathcal{F}[\mathcal{T}^y \ell(f)](\lambda) = \lambda\, w_\lambda(y)\,(\mathcal{F}f)(\lambda)$, hence (4.26)–(4.27) hold. Moreover, it is immediate from (2.23) that $h(\cdot,y) \in \mathcal{D}(\mathcal{L}^{(2)})$, and therefore $(\boldsymbol{\alpha})$ holds.

Proof of Proposition 4.17 Fix $z \geq 0$, and let $\{f_\varepsilon\} \subset \mathcal{D}^{(2,0)}$ be a family of functions such that

$$\begin{aligned} f_\varepsilon(\xi) > 0 &\quad \text{for } z-\varepsilon < \xi < z+\varepsilon, \\ f_\varepsilon(\xi) = 0 &\quad \text{for } \xi \leq z-\varepsilon \text{ and } \xi \geq z+\varepsilon. \end{aligned} \tag{4.51}$$

Observe that $z \in \operatorname{supp}(\delta_x * \delta_y)$ if and only if $\int_{\mathbb{R}_0^+} f_\varepsilon\, d(\delta_x * \delta_y) > 0$ for all $\varepsilon > 0$. Now, we know from Lemma 4.9 that the function

$$h_{f_\varepsilon}(x,y) := \int_{\mathbb{R}_0^+} f_\varepsilon\, d(\delta_x * \delta_y) = \int_{[\sigma^2,\infty)} w_\lambda(x)\, w_\lambda(y)\,(\mathcal{F}f_\varepsilon)(\lambda)\, \boldsymbol{\rho}_{\mathcal{L}}(d\lambda)$$

(the second equality is due to (4.43)) is a nonnegative solution of the Cauchy problem (4.22) with $f \equiv f_\varepsilon$; writing $B(x) := \exp(\frac{1}{2}\int_0^x \eta(\xi)d\xi)$, it follows that $u_{f_\varepsilon}(x,y) = B(x)B(y)h_{f_\varepsilon}(x,y)$ is a solution of $\wp_x u - \wp_y u = 0$, where $\wp_x := -\frac{\partial^2}{\partial x^2} - \boldsymbol{\phi}_\eta(x)\frac{\partial}{\partial x} + \boldsymbol{\psi}_\eta(x)$. Applying Lemma 4.3 with $c > 0$ and $\wp_1(v) = \wp_2(v) = -v'' - \boldsymbol{\phi}_\eta v' + \boldsymbol{\psi}_\eta v$ and then letting $c \downarrow 0$, we deduce that the following integral equation holds:

$$\frac{A(x)A(y)}{B(x)^2 B(y)^2}u_{f_\varepsilon}(x,y) = H + I_0 + I_1 + I_2 + I_3, \tag{4.52}$$

where $H = \frac{1}{2}A(0)\big[\frac{A(x-y)}{B(x-y)}f_\varepsilon(x-y) + \frac{A(x+y)}{B(x+y)}f_\varepsilon(x+y)\big]$, $I_0 = \frac{\eta(0)}{4}\int_{x-y}^{x+y}\frac{A(s)}{B(s)}f_\varepsilon(s)\, ds$ and I_1, I_2, I_3 are given by (4.11)–(4.13) with $c = 0$ and $v = u_{f_\varepsilon}$. Since f_ε and h_{f_ε} are nonnegative, all the terms in the right-hand side of (4.52) are nonnegative; consequently, we have $z \in \operatorname{supp}(\delta_x * \delta_y)$ if and only if at least one of the terms in the right-hand side of (4.52) is strictly positive for all $\varepsilon > 0$. In order to ascertain whether this holds or not, one needs to perform a thorough analysis of the integrals I_0, I_1, I_2 and I_3. This has been done by Zeuner in [210, Proposition 3.9]; his results lead to the conclusion stated in the proposition.

Theorem 4.6 (Existence Theorem for Sturm-Liouville Hypergroups) *Let ℓ be a differential expression of the form* (4.1). *Suppose that $\gamma(a) > -\infty$ and that there exists $\eta \in \mathrm{C}^1[\gamma(a),\infty)$ satisfying the conditions given in Assumption MP. Then $\big([a,b),*\big)$ is a hypergroup.*

Proof We need to check that the $\mathcal{L}$-convolution satisfies axioms H6–H8. Assume first that ℓ satisfies the assumptions of Proposition 4.17. Then the explicit expressions for $\mathrm{supp}(\delta_x * \delta_y)$ show that (in each of the cases *(a)*–*(e)*) $\mathrm{supp}(\delta_x * \delta_y)$ is compact, depends continuously on (x, y) and contains e $= 0$ if and only if $x = y$, so that H6–H8 hold. (Verifying the continuity is easy after noting that the topology in the space of compact subsets can be metrized by the Hausdorff metric, cf. [106, Subsection 4.1].)

In the general case of an operator ℓ of the form (4.1), the hypothesis that $\gamma(a) > -\infty$ means that $\sqrt{\frac{r(y)}{p(y)}}$ is integrable near a, and thus we may assume that $\gamma(a) = 0$ (otherwise, replace the interior point c by the endpoint a in the definition of the function γ). By assumption the transformed operator $\widetilde{\ell} = -\frac{1}{A}\frac{d}{d\xi}(A\frac{d}{d\xi})$ defined via (4.2) satisfies the assumptions of Proposition 4.17; by the above, the associated convolution, which we denote by $\widetilde{*}$, satisfies H6–H8. From the product formulas for the solutions $w_\lambda(x)$ and $\widetilde{w}_\lambda(\xi) = w_\lambda(\gamma^{-1}(\xi))$ we deduce that

$$\begin{aligned}\int_{[a,b)} w_\lambda \, d(\delta_x * \delta_y) &= w_\lambda(x) w_\lambda(y) \\ &= \widetilde{w}_\lambda(\gamma(x))\widetilde{w}_\lambda(\gamma(y)) \\ &= \int_{\mathbb{R}_0^+} w_\lambda(\gamma^{-1}(z))\big(\delta_{\gamma(x)} \widetilde{*} \delta_{\gamma(y)}\big)(dz)\end{aligned}$$

and, consequently, $\delta_x * \delta_y = \gamma^{-1}(\delta_{\gamma(x)} \widetilde{*} \delta_{\gamma(y)})$. In particular, $\mathrm{supp}(\delta_x * \delta_y) = \gamma^{-1}\big(\mathrm{supp}(\delta_{\gamma(x)} \widetilde{*} \delta_{\gamma(y)})\big)$; since γ is a continuous bijection, we immediately conclude that the convolution $*$ also satisfies axioms H6–H8.

Recalling the definition of hypergroup isomorphism given in Sect. 2.3, we see that the hypergroups $\big([a, b), *\big)$ and $\big(\mathbb{R}_0^+, \widetilde{*}\big)$ considered above (associated with the differential operators ℓ and $\widetilde{\ell}$ respectively) are isomorphic.

For Sturm-Liouville operators on $\mathbb{R}^+$ of the form $\ell(u) = -u'' - \frac{A'}{A}u'$, the assumption of the theorem above can be re-expressed in terms of conditions SL0–SL2 introduced in Sect. 4.1:

Corollary 4.5 *Suppose that A satisfies SL0 and SL2. For $f \in \mathcal{D}^{(2,0)}$, denote by v_f the unique solution of $\ell_x v_f = \ell_y v_f$, $v_f(x, 0) = v_f(0, x) = f(x)$, $(\partial_y^{[1]} v_f)(x, 0) = (\partial_x^{[1]} v_f)(0, y) = 0$ such that conditions ($\boldsymbol{\alpha}$)–($\boldsymbol{\beta}$) of Sect. 4.3.1 hold for $h = v_f$. Define the convolution $*$ via* (4.3). *Then $\big(\mathbb{R}_0^+, *\big)$ is a hypergroup.*

Proof Just notice that, by (4.43) and Lemma 4.9, the definition of convolution given in the statement of the corollary is equivalent to Definition 4.5.

The statement of Corollary 4.5 strongly resembles that of Zeuner's existence theorem for Sturm-Liouville hypergroups (Theorem 4.1), but its assumptions do not include condition SL1. The corollary therefore shows that it is natural to modify the definition of Sturm-Liouville hypergroup (Definition 4.1) by replacing the space

$C^\infty_{\mathrm{c,even}}$ by $\mathcal{D}^{(2,0)}$ and replacing ∂_y by $\partial_y^{[1]}$ in the initial condition, because in this way we are able to extend the class of Sturm-Liouville hypergroups to all functions A satisfying conditions SL0 and SL2.

We emphasize that condition SL1 imposes a great restriction on the behaviour of the Sturm-Liouville operator $\ell(u) = -u'' - \frac{A'}{A}u'$ near zero: in the singular case $A(0) = 0$, SL1 requires that $\frac{A'(x)}{A(x)} \sim \frac{\alpha_0}{x}$. Therefore, as shown in the next example, Corollary 4.5 leads, in particular, to a considerable extension of the class of singular operators for which an associated hypergroup exists:

Example 4.1 If A satisfies SL0 and the function $\frac{A'}{A}$ is nonnegative and decreasing, then SL2 is satisfied with $\eta := 0$. Therefore, Corollary 4.5 ensures that there exists a hypergroup associated with the operator $\ell(u) = -u'' - \frac{A'}{A}u'$. Notice that this existence result holds without any restriction on the growth of $\frac{A'(x)}{A(x)}$ as $x \downarrow 0$.

This class of examples includes the following special cases:

(a) $\ell = -\frac{d^2}{dx^2}$ $(0 < x < \infty)$; here $A(x) \equiv 1$.

As noted in Example 2.1, the Sturm-Liouville solutions are $w_\lambda(x) = \cos(\tau x)$ $(\lambda = \tau^2)$ and the $\mathcal{L}$-transform is the cosine Fourier transform $(\mathcal{F}h)(\tau) = \int_0^\infty h(x)\cos(\tau x)dx$. By elementary trigonometric identities, $w_\tau(x)w_\tau(y) = \frac{1}{2}[w_\tau(|x-y|) + w_\tau(x+y)]$, hence the $\mathcal{L}$-convolution is given by

$$\delta_x * \delta_y = \frac{1}{2}(\delta_{|x-y|} + \delta_{x+y}), \qquad x, y \geq 0.$$

In other words, $*$ is (up to identification) the ordinary convolution of symmetric measures.

(b) $\ell = -x^{2-2\alpha}\frac{d^2}{dx^2} - (1-\alpha)x^{1-2\alpha}$ $(0 < x < \infty,\ \alpha > 0)$.

This operator is of the form (4.1) with $p(x) = x^{1-\alpha}$ and $r(x) = x^{\alpha-1}$, and it is transformed into the operator $-\frac{d^2}{dx^2}$ via the change of variable $\xi = \gamma(x) = x^\alpha$ (cf. Remark 4.1). Accordingly, it follows from (a) that the $\mathcal{L}$-convolution is given by

$$\delta_x * \delta_y = \frac{1}{2}(\delta_{|x^\alpha - y^\alpha|^{1/\alpha}} + \delta_{(x^\alpha + y^\alpha)^{1/\alpha}}), \qquad x, y \geq 0.$$

This is the so-called $(\alpha, 1)$-*convolution* [190], which is a generalized convolution satisfying Urbanik's axioms (cf. Definition 2.3).

(c) $\ell = -\frac{d^2}{dx^2} - \frac{2\eta+1}{x}\frac{d}{dx}$ $(0 < x < \infty,\ \eta > -\frac{1}{2})$; here $A(x) \equiv x^{2\eta+1}$.

As noted in Sect. 2.4, the Sturm-Liouville solutions are $w_\lambda(x) = \boldsymbol{J}_\eta(\tau x)$ $(\lambda = \tau^2$, $\boldsymbol{J}_\eta$ the normalized Bessel function of the first kind) and the $\mathcal{L}$-transform is the Hankel transform $(\mathcal{F}h)(\tau) = \int_0^\infty h(x)\,\boldsymbol{J}_\eta(\tau x)\,x^{2\eta+1}dx$. The product formula for the Bessel function given in Theorem 2.6 shows that the

$\mathcal{L}$-convolution is

$$(\delta_x * \delta_y)(d\xi) = \frac{2^{1-2\eta}\Gamma(\eta+1)}{\sqrt{\pi}\,\Gamma(\eta+\frac{1}{2})}(xy)^{-2\eta} \times$$
$$\times \big[(\xi^2-(x-y)^2)((x+y)^2-\xi^2)\big]^{\eta-1/2}\mathbb{1}_{[|x-y|,x+y]}(\xi)\,\xi\,d\xi.$$

This is the Kingman convolution (cf. Definition 2.6). The corresponding hypergroup $(\mathbb{R}_0^+, *)$, which is known as the *Bessel-Kingman hypergroup*, plays a special role in the context of Sturm-Liouville hypergroups; in particular, it appears as the limit distribution in central limit theorems on hypergroups [19, Section 7.5].

(d) $\ell = -x^{2-2\alpha}\frac{d^2}{dx^2} - (2\alpha\eta+1)x^{1-2\alpha}\frac{d}{dx}$ $(0 < x < \infty, \eta > -\frac{1}{2}, \alpha > 0)$.

This operator is of the form (4.1) with $p(x) = x^{2\alpha\eta+1}$ and $r(x) = x^{2\alpha(\eta+1)-1}$; similar to (b) above, the change of variable $\xi = \gamma(x) = x^\alpha$ transforms ℓ into the operator $-\frac{d^2}{dx^2} - \frac{2\eta+1}{x}\frac{d}{dx}$. The $\mathcal{L}$-convolution is thus given by

$$(\delta_x * \delta_y)(d\xi)$$
$$= \frac{\alpha\,2^{1-2\eta}\Gamma(\eta+1)}{\sqrt{\pi}\,\Gamma(\eta+\frac{1}{2})}(xy)^{-2\alpha\eta}\big[(\xi^{2\alpha}-(x^\alpha-y^\alpha)^2)((x^\alpha+y^\alpha)^2-\xi^{2\alpha})\big]^{\eta-1/2} \times$$
$$\times \mathbb{1}_{[|x^\alpha-y^\alpha|^{1/\alpha},(x^\alpha+y^\alpha)^{1/\alpha}]}(\xi)\,\xi^{2\alpha-1}d\xi.$$

This is the so-called *(α, β)-convolution* [190]; as in (b), one can check that it satisfies axioms U1–U6 of Urbanik convolutions.

(e) $\ell = -\frac{d^2}{dx^2} - [(2\alpha+1)\coth x + (2\beta+1)\tanh x]\frac{d}{dx}$ $(0 < x < \infty,\ \alpha \geq \beta \geq -\frac{1}{2}$, $\alpha \neq -\frac{1}{2})$; here $A(x) = (\sinh x)^{2\alpha+1}(\cosh x)^{2\beta+1}$.

As noted in Example 2.5, the Sturm-Liouville solutions are $w_\lambda(x) = {}_2F_1\big(\frac{1}{2}(\eta-i\tau), \frac{1}{2}(\eta+i\tau); \alpha+1; -(\sinh x)^2\big)$ $(\eta = \alpha+\beta+1,\ \lambda = \tau^2+\eta^2$, ${}_2F_1$ the hypergeometric function) and the $\mathcal{L}$-transform is the (Fourier-)Jacobi transform (2.43). By a deep result of Koornwinder [61, 105], the measures of the product formula $w_\lambda(x)\,w_\lambda(y) = \int_{[a,b)} w_\lambda\, d(\delta_x * \delta_y)$ are given by

$$(\delta_x * \delta_y)(d\xi) = \frac{2^{-2\sigma}\Gamma(\alpha+1)(\cosh x\,\cosh y\,\cosh\xi)^{\alpha-\beta-1}}{\sqrt{\pi}\,\Gamma(\alpha+\frac{1}{2})(\sinh x\,\sinh y\,\sinh\xi)^{2\alpha}}$$
$$\times (1-Z^2)^{\alpha-1/2}\,{}_2F_1\Big(\alpha+\beta, \alpha-\beta; \alpha+\tfrac{1}{2}; \tfrac{1}{2}(1-Z)\Big)\mathbb{1}_{[|x-y|,x+y]}(\xi)A(\xi)d\xi,$$

where $Z := \frac{(\cosh x)^2+(\cosh y)^2+(\cosh\xi)^2-1}{2\cosh x\,\cosh y\,\cosh\xi}$; the corresponding hypergroup is the so-called *Jacobi hypergroup*.

For half-integer values of the parameters α, β, this hypergroup structure has various group theoretic interpretations; in particular, it is related with harmonic analysis on rank one Riemannian symmetric spaces [105]. Moreover, a remarkable property of the Jacobi hypergroup is that it admits a positive dual convolution structure, i.e. there exists a family $\{\theta_{\lambda_1,\lambda_2}\}$ of finite positive measures such that the dual product formula $w_{\lambda_1}(x)\, w_{\lambda_2}(x) = \int_0^\infty w_{\lambda_3}(x)\, \theta_{\lambda_1,\lambda_2}(d\lambda_3)$ holds, and this permits the construction of a generalized convolution which trivializes the inverse Jacobi transform [12].

(f) $\ell = -\frac{d^2}{dx^2} - \bigl(\frac{\alpha}{x} + 2\mu\bigr)\frac{d}{dx}$ $(0 < x < \infty,\ \alpha, \mu > 0)$; here $A(x) = x^\alpha e^{2\mu x}$.

As noted in Example 2.3, the solutions of the Sturm-Liouville initial value problem are $w_\lambda(x) = (2i\tau)^{-\frac{\alpha}{2}} e^{-\mu x} x^{-\frac{\alpha}{2}} M_{-\frac{\alpha\mu}{2i\tau}, \frac{\alpha-1}{2}}(2i\tau x)$ ($\lambda = \tau^2 + \mu^2$, $M_{\kappa,\nu}(\cdot)$ the Whittaker function of the first kind) and the $\mathcal{L}$-transform is the index transform $(\mathcal{F}h)(\tau) = (2i\tau)^{-\frac{\alpha}{2}} \int_0^\infty h(x)\, M_{-\frac{\alpha\mu}{2i\tau}, \frac{\alpha-1}{2}}(2i\tau x)\, x^{\frac{\alpha}{2}} e^{\mu x} dx$. It follows from Corollary 4.5 that *there exists a family of probability measures* $\{\boldsymbol{\nu}_{x,y}\}_{x,y\geq 0}$ *with support* $[|x-y|, x+y]$ *such that the Whittaker function of the first kind satisfies the product formula*

$$
\begin{aligned}
(2i\tau xy)^{-\frac{\alpha}{2}} e^{-\mu(x+y)} M_{-\frac{\alpha\mu}{2i\tau}, \frac{\alpha-1}{2}}(2i\tau x) M_{-\frac{\alpha\mu}{2i\tau}, \frac{\alpha-1}{2}}(2i\tau y) = \\
= \int \xi^{-\frac{\alpha}{2}} e^{-\mu\xi} M_{-\frac{\alpha\mu}{2i\tau}, \frac{\alpha-1}{2}}(2i\tau\xi)\, \boldsymbol{\nu}_{x,y}(d\xi).
\end{aligned}
\tag{4.53}
$$

Unlike in cases (a)–(e) above, here the function A does not satisfy condition SL1 of Zeuner's existence theorem for Sturm-Liouville hypergroups; the existence of the product formula (4.53) is, as far as we know, a novel result. It is natural to wonder whether one can determine the closed-form expression of each of the measures $\boldsymbol{\nu}_{x,y}$ in terms of classical special functions. We leave this as an open problem.

The convolutions discussed in cases (a)–(d) above are the only known examples of Sturm-Liouville convolutions which satisfy axioms U1– U6 of Urbanik.

4.6.2 The Degenerate Case: Degenerate Hypergroups of Full Support

Definition 4.11 Let K be a locally compact space and $*$ a bilinear operator on $\mathcal{M}_\mathbb{C}(K)$. The pair $(K, *)$ is said to be a *degenerate hypergroup of full support* if it satisfies the hypergroup axioms H1– H5, together with the following axiom:

DH. $\operatorname{supp}(\delta_x * \delta_y) = K$ for all $x, y \in K \setminus \{e\}$.

In order to determine the conditions under which the Sturm-Liouville convolution algebra $([a,b), *)$ is a degenerate hypergroup of full support, we need to know when the solution of the associated hyperbolic Cauchy problem (4.22) is strictly positive inside $(a,b)^2$. Our starting lemma provides an integral inequality which proves to be useful for studying the strict positivity of solution.

Lemma 4.10 *Write* $R(x) := \frac{A(x)}{B(x)}$, *where* $B(x) = \exp(\frac{1}{2}\int_\beta^x \eta(\xi)d\xi)$ *(with* $\beta > \gamma(a)$ *arbitrary). Take* $h \in \mathcal{D}^{(2,0)}$ *such that* $h \geq 0$. *Let* $u(x,y) := h(\gamma^{-1}(x), \gamma^{-1}(y))$, *where* $h \in \mathrm{C}^2\big((a,b)^2\big)$ *is the solution (4.23) of the Cauchy problem (cf. Lemma 4.9). Then the following inequality holds:*

$$\begin{aligned}
&R(x)R(y)u(x,y) \\
&\quad \geq \frac{1}{2}\int_{\gamma(a)}^{y} R(s)R(x-y+s)\big[\boldsymbol{\phi}_\eta(s) + \boldsymbol{\phi}_\eta(x-y+s)\big]u(x-y+s,s)\,ds \\
&\quad + \frac{1}{2}\int_{\gamma(a)}^{y} R(s)R(x+y-s)\big[\boldsymbol{\phi}_\eta(s) - \boldsymbol{\phi}_\eta(x+y-s)\big]u(x+y-s,s)\,ds \\
&\quad + \frac{1}{2}\int_{\Delta} R(\xi)R(\zeta)\big[\boldsymbol{\psi}_\eta(\zeta) - \boldsymbol{\psi}_\eta(\xi)\big]u(\xi,\zeta)\,d\xi d\zeta,
\end{aligned} \tag{4.54}$$

where $\Delta \equiv \Delta_{\gamma(a),x,y} = \{(\xi,\zeta) \in \mathbb{R}^2 \mid \zeta \geq \gamma(a),\ \xi + \zeta \leq x + y,\ \xi - \zeta \geq x - y\}$.

Proof Let $\{a_m\}_{m\in\mathbb{N}}$ be a sequence $b > a_1 > a_2 > \ldots$ with $\lim a_m = a$. For $m \in \mathbb{N}$, set $\tilde{a}_m = \gamma(a_m)$ and define $u_m(x,y) := h_m(\gamma^{-1}(x), \gamma^{-1}(y))$, where h_m is the function defined in (4.29). The function $v_m(x,y) = B(x)B(y)u_m(x,y)$ is a solution of

$$\begin{aligned}
(\wp_x v_m)(x,y) &= (\wp_y v_m)(x,y), && x,y > \tilde{a}_m, \\
v_m(x,\tilde{a}_m) &= B(x)B(\tilde{a}_m)h(\gamma^{-1}(x)), && x > \tilde{a}_m, \\
(\partial_y v_m)(x,\tilde{a}_m) &= \tfrac{1}{2}\eta(\tilde{a}_m)B(x)B(\tilde{a}_m)h(\gamma^{-1}(x)), && x > \tilde{a}_m,
\end{aligned}$$

where $\wp_x := -\frac{\partial^2}{\partial x^2} - \boldsymbol{\phi}_\eta(x)\frac{\partial}{\partial x} + \boldsymbol{\psi}_\eta(x)$. Clearly, $v_m(x,\tilde{a}_m), (\partial_y v_m)(x,\tilde{a}_m) \geq 0$. By Lemma 4.3 (with $\wp_1(v) = \wp_2(v) = -v'' - \boldsymbol{\phi}_\eta v' + \boldsymbol{\psi}_\eta v$), the integral equation (4.8) holds with $v = v_m$ and $c = a_m$. It is clear that we have $H \geq 0$, $I_0 \geq 0$ and $I_4 = 0$ in the right hand side of (4.8); moreover, it follows from Proposition 4.5 and Assumption MP that the integrands of I_1, I_2 and I_3 are nonnegative. Consequently, for $\alpha \in [\tilde{a}_m, y]$ we have

$$\begin{aligned}
&R(x)R(y)u_m(x,y) \\
&\quad \geq \frac{1}{2}\int_{\alpha}^{y} R(s)R(x-y+s)\big[\boldsymbol{\phi}_\eta(s) + \boldsymbol{\phi}_\eta(x-y+s)\big]u_m(x-y+s,s)\,ds \\
&\quad + \frac{1}{2}\int_{\alpha}^{y} R(s)R(x+y-s)\big[\boldsymbol{\phi}_\eta(s) - \boldsymbol{\phi}_\eta(x+y-s)\big]u_m(x+y-s,s)\,ds \\
&\quad + \frac{1}{2}\int_{\Delta_{\alpha,x,y}} R(\xi)R(\zeta)\big[\boldsymbol{\psi}_\eta(\zeta) - \boldsymbol{\psi}_\eta(\xi)\big]u_m(\xi,\zeta)\,d\xi d\zeta,
\end{aligned} \tag{4.55}$$

where $\Delta_{\alpha,x,y} = \{(\xi, \zeta) \in \mathbb{R}^2 \mid \zeta \geq \alpha,\ \xi + \zeta \leq x + y,\ \xi - \zeta \geq x - y\}$. Since by Proposition 4.4 $\lim_{m\to\infty} u_m(x, y) = u(x, y)$ pointwise for $x, y \in (\gamma(a), \infty)$, by taking the limit we deduce that for each fixed $\alpha \in (\gamma(a), y]$ the inequality (4.55) holds with u_m replaced by u. If we then take the limit $\alpha \downarrow \gamma(a)$, the desired integral inequality follows.

The next lemma will help us in verifying the strict positivity of the integrands in the integral inequality (4.54).

Lemma 4.11 *If $\gamma(a) = -\infty$, then at least one of the functions $\boldsymbol{\phi}_\eta$, $\boldsymbol{\psi}_\eta$ defined in Assumption MP is non-constant on every neighbourhood of $-\infty$.*

Proof Suppose by contradiction that $\gamma(a) = -\infty$ and $\boldsymbol{\phi}_\eta$, $\boldsymbol{\psi}_\eta$ are both constant on an interval $(-\infty, \kappa] \subset \mathbb{R}$. Recall from the proof of Proposition 4.3 that $\mathcal{L}$ is unitarily equivalent to a self-adjoint realization of $-\frac{d^2}{d\xi^2} + \mathfrak{q}$, where $\mathfrak{q}$ is given by (4.5). Clearly, $\mathfrak{q}(\xi) = \mathfrak{q}_\infty := \frac{1}{4}\boldsymbol{\phi}_\eta^2(\kappa) + \boldsymbol{\psi}_\eta(\kappa) < \infty$ for all $\xi \in (-\infty, \kappa)$. Using the theorem on the spectral properties of Sturm-Liouville operators stated in [201, Theorem 15.3], we deduce that the essential spectrum of any self-adjoint realization of ℓ restricted to an interval (a, c) (for $a < c < b$) contains $[\mathfrak{q}_\infty, \infty)$. However, we know from the proof of Proposition 4.3 that self-adjoint realizations of ℓ restricted to (a, c) have purely discrete spectrum. This contradiction proves the lemma.

We are now ready to prove that in the case $\gamma(a) = -\infty$ the solution of the (nontrivial) Cauchy problem (4.22) always has full support on $(a, b)^2$, even when the initial condition is compactly supported:

Theorem 4.7 (Strict Positivity of Solution for the Cauchy Problem (4.22)**)** *Suppose that $\gamma(a) = -\infty$. Take $h \in \mathcal{D}^{(2,0)}$. If $f \geq 0$ and $f(\tau_0) > 0$ for some $\tau_0 \in (a, b)$, then the function h given by* (4.23) *is such that*

$$h(x, y) > 0 \qquad \textit{for } x, y \in (a, b).$$

Proof Let $u(x, y) := h(\gamma^{-1}(x), \gamma^{-1}(y))$ and $\tilde{\tau}_0 = \gamma(\tau_0)$. Fix $x_0 \geq y_0 > -\infty$. Since $\lim_{y\to-\infty} u(\tilde{\tau}_0, y) = f(\tau_0) > 0$, there exists $\kappa \in (-\infty, \min\{y_0, \tau_0\})$ such that $u(\tilde{\tau}_0, y) > 0$ for all $y \leq \kappa$.

Suppose $\boldsymbol{\phi}_\eta$ is non-constant on every neighbourhood of $-\infty$. Choosing a smaller κ if necessary, we may assume that $\boldsymbol{\phi}_\eta(\kappa) > \boldsymbol{\phi}_\eta(\xi)$ for all $\xi > \kappa$. For each $x > \tilde{\tau}_0$ and $y \leq \kappa$ we have by Lemma 4.10

$$R(x)R(y)u(x, y) \geq \tfrac{1}{2} \int_{-\infty}^{y} R(s)R(x - y + s)\big[\boldsymbol{\phi}_\eta(s) + \boldsymbol{\phi}_\eta(x - y + s)\big]u(x - y + s, s)\, ds,$$

and the integrand in the right hand side is continuous and strictly positive at $s = y - x + \tilde{\tau}_0$, so the integral is positive and therefore $u(x, y) > 0$ for all $x \geq \tilde{\tau}_0$ and

$y \leq \kappa$. Again by Lemma 4.10,

$$\begin{aligned} &R(x_0)R(y_0)u(x_0, y_0) \\ &\quad \geq \tfrac{1}{2} \int_{-\infty}^{y_0} R(s)R(x_0 + y_0 - s)\bigl[\boldsymbol{\phi}_\eta(s) - \boldsymbol{\phi}_\eta(x_0 + y_0 - s)\bigr]u(x_0 + y_0 - s, s)\, ds, \end{aligned}$$

with the integrand being strictly positive for $s < \min\{\kappa, x_0 + y_0 - \tilde{\tau}_0\}$, thus $u(x_0, y_0) > 0$.

Suppose now that $\boldsymbol{\psi}_\eta$ is non-constant on every neighbourhood of $-\infty$ and that κ is chosen such that $\boldsymbol{\psi}_\eta(\kappa) > \boldsymbol{\psi}_\eta(\xi)$ for all $\xi > \kappa$. The integral inequality of Lemma 4.10 yields

$$R(x_0)R(y_0)u(x_0, y_0) \geq \tfrac{1}{2} \int_\Delta R(\xi)R(\zeta)\bigl[\boldsymbol{\psi}_\eta(\zeta) - \boldsymbol{\psi}_\eta(\xi)\bigr]u(\xi, \zeta)\, d\xi\, d\zeta,$$

where $\Delta = \{(\xi, \zeta) \in \mathbb{R}^2 \mid \xi + \zeta \leq x_0 + y_0,\ \xi - \zeta \geq x_0 - y_0\}$. Clearly, the integrand is continuous and > 0 on $\{(\tau_0, \zeta) \mid \zeta \leq \min(y_0 - |x_0 - \tau_0|, \kappa)\} \subset \Delta$, and it follows at once that $u(x_0, y_0) > 0$.

By Lemma 4.11 it follows that $u(x_0, y_0) > 0$. Since $x_0 \geq y_0 > -\infty$ are arbitrary we conclude that $h(x, y) > 0$ for $b > x \geq y > a$ and, by symmetry, for $x, y \in (a, b)$.

Corollary 4.6 (Existence Theorem for Degenerate Hypergroups of Full Support) *Let ℓ be a Sturm-Liouville expression of the form* (4.1) *which satisfies Assumption MP, and suppose that $\gamma(a) = -\infty$. Then $\bigl([a, b), *\bigr)$ is a degenerate hypergroup of full support.*

Proof By Proposition 4.9, the pair $\bigl([a, b), *\bigr)$ satisfies axioms H1–H5. As in the proof of Proposition 4.17, $z \in [a, b)$ belongs to $\operatorname{supp}(\delta_x * \delta_y)$ if and only if $\int_{[\sigma^2,\infty)} w_\lambda(x)\, w_\lambda(y)\, (\mathcal{F} f_\varepsilon)(\lambda)\, \boldsymbol{\rho}_{\mathcal{L}}(d\lambda) > 0$ for all $\varepsilon > 0$, where $\{f_\varepsilon\} \subset \mathcal{D}^{(2,0)}$ is a family of functions satisfying (4.51). But it follows from Theorem 4.7 that $h_{f_\varepsilon}(x, y) = \int_{[\sigma^2,\infty)} w_\lambda(x)\, w_\lambda(y)\, (\mathcal{F} f_\varepsilon)(\lambda)\, \boldsymbol{\rho}_{\mathcal{L}}(d\lambda) > 0$ for all $x, y \in (a, b)$. Hence each $z \in [a, b)$ belongs to all the sets $\operatorname{supp}(\delta_x * \delta_y)$, $x, y \in (a, b)$; therefore, $\bigl([a, b), *\bigr)$ satisfies axiom DH.

This corollary confirms that, as anticipated in Sect. 4.1, the Whittaker convolution studied in Chap. 3 is a particular case of a general family of Sturm-Liouville convolutions which do not satisfy the compactness axiom, but which also allow us to develop the basic notions and facts from probabilistic harmonic analysis.

Example 4.2 Let $\boldsymbol{\zeta} \in \mathrm{C}^1(\mathbb{R}^+)$ be a nonnegative decreasing function and let $\kappa > 0$. The differential expression

$$\ell = -x^2 \frac{d^2}{dx^2} - \bigl[\kappa + x\bigl(1 + \boldsymbol{\zeta}(x)\bigr)\bigr]\frac{d}{dx}, \qquad 0 < x < \infty$$

is a particular case of (4.1), obtained by considering $p(x) = xe^{-\kappa/x+I_\zeta(x)}$ and $r(x) = \frac{1}{x}e^{-\kappa/x+I_\zeta(x)}$, where $I_\zeta(x) = \int_1^x \zeta(y)\frac{dy}{y}$. (If $\kappa = 1$ and $\zeta(x) = 1-2\alpha > 0$, we recover the normalized Whittaker operator (3.2).) The change of variable $z = \log x$ transforms ℓ into the standard form $\widetilde{\ell} = -\frac{d^2}{dz^2} - \frac{A'(z)}{A(z)}\frac{d}{dz}$, where $\frac{A'(z)}{A(z)} = \kappa e^{-\kappa z} + \zeta(e^z)$. It is clear that $\gamma(a) = -\infty$ and that ℓ satisfies Assumption MP with $\eta = \lim_{y\to\infty}\zeta(y)$, and it is not difficult to show that the boundary condition $\int_a^c \int_y^c \frac{dx}{p(x)} r(y)dy < \infty$ holds. Consequently, the Sturm-Liouville operator ℓ gives rise to a convolution algebra $(\mathbb{R}_0^+, *)$ such that $\text{supp}(\delta_x * \delta_y) = \mathbb{R}_0^+$ for all $x, y > 0$.

4.7 Harmonic Analysis on L^p Spaces

Finally, we turn to the mapping properties of the $\mathcal{L}$-convolution of functions, defined in the natural way (compare with Sect. 3.7):

Definition 4.12 Let $f, g : [a, b) \longrightarrow \mathbb{C}$. If the integral

$$(f * g)(x) = \int_a^b (\mathcal{T}^y f)(x)\, g(y)\, r(y)dy = \int_a^b \int_{[a,b)} f(\xi)\, (\delta_x * \delta_y)(d\xi)\, g(y)\, r(y)dy$$

exists for almost every $x \in [a, b)$, then we call it the *$\mathcal{L}$-convolution* of the functions f and g.

As usual, the convolution is trivialized by the $\mathcal{L}$-transform and commutes with both the Sturm-Liouville operator ℓ and the associated translation operator:

Proposition 4.18 *Let $y \in [a, b)$ and $\lambda \geq 0$. Then:*

(a) *If $f \in L^2(r)$ and $g \in L^1(r)$, then $\big(\mathcal{F}(f * g)\big)(\lambda) = (\mathcal{F}f)(\lambda)\, (\mathcal{F}g)(\lambda)$;*
(b) *If $f \in L^2(r)$ and $g \in L^1(r)$, then $\mathcal{T}^y(f * g) = (\mathcal{T}^y f) * g$;*
(c) *If $f \in \mathcal{D}(\mathcal{L}^{(2)})$ and $g \in L^1(r)$, then $f * g \in \mathcal{D}(\mathcal{L}^{(2)})$ and $\ell(f * g) = (\ell f) * g$.*

Proof We know from Proposition 4.10 that $\mathcal{F}(\mathcal{T}^\mu f)(\lambda) = \widehat{\mu}(\lambda)\, (\mathcal{F}f)(\lambda)$, hence these properties can be proved using the same reasoning as in Proposition 3.20.

The $\mathcal{L}$-convolution also satisfies a Young inequality analogous to those of the previous chapters:

Proposition 4.19 (Young Inequality for the $\mathcal{L}$-Convolution) *Let $p_1, p_2 \in [1, \infty]$ such that $\frac{1}{p_1} + \frac{1}{p_2} \geq 1$. For $f \in L^{p_1}(r)$ and $g \in L^{p_2}(r)$, the $\mathcal{L}$-convolution $f * g$ is well-defined and, for $s \in [1, \infty]$ defined by $\frac{1}{s} = \frac{1}{p_1} + \frac{1}{p_2} - 1$, it satisfies*

$$\|f * g\|_s \leq \|f\|_{p_1} \|g\|_{p_2}$$

(in particular, $f * g \in L^s(r)$). Consequently, the $\mathcal{L}$-convolution is a continuous bilinear operator from $L^{p_1}(r) \times L^{p_2}(r)$ into $L^s(r)$.

Proof Identical to that of Proposition 3.18.

4.7.1 *A Family of L^1 Spaces*

We will study the $\mathcal{L}$-convolution as an operator acting on the family of Lebesgue spaces $\{L^1_\kappa\}_{-\infty<\kappa\leq\sigma^2}$, where $L^1_\kappa = L^1\bigl((a,b), w_\kappa(x)r(x)dx\bigr)$. Observe that this is an ordered family:

$$L^1_{\kappa_2} \subset L^1_{\kappa_1} \qquad \text{whenever } \; -\infty < \kappa_2 \leq \kappa_1 \leq \sigma^2. \tag{4.56}$$

This follows from the fact that (due to the Laplace-type representation (4.16)) we have $0 \leq w_{\kappa_1}(x) \leq w_{\kappa_2}(x)$ for all $x \in [a,b)$ whenever $-\infty < \kappa_2 \leq \kappa_1 \leq \sigma^2$. In particular, the space $L^1_0 \equiv L^1(r)$ is contained in the spaces L^1_κ with $0 \leq \kappa \leq \sigma^2$.

The basic properties of the $\mathcal{L}$-transform, translation and convolution on the spaces L^1_κ are as follows (we write $\|\cdot\|_{1,\kappa} := \|\cdot\|_{L^1_\kappa}$):

Proposition 4.20 *Let $-\infty < \kappa \leq \sigma^2$, let $f, g \in L^1_\kappa$, and fix $y \in [a,b)$. Then:*

(a) *The $\mathcal{L}$-transform $(\mathcal{F}f)(\lambda) := \int_a^b f(x)\, w_\lambda(x)\, r(x)dx$ is, for all $\lambda \geq \sigma^2$, well-defined as an absolutely convergent integral; in addition, f is uniquely determined by $(\mathcal{F}f)|_{[\sigma^2,\infty)}$.*

(b) *The $\mathcal{L}$-translation $(\mathcal{T}^y f)(x) := \int_{[a,b)} f\, d\boldsymbol{\nu}_{x,y}$ is well-defined and it satisfies $\|\mathcal{T}^y f\|_{1,\kappa} \leq w_\kappa(y) \|f\|_{1,\kappa}$ (in particular, $\mathcal{T}^y\bigl(L^1_\kappa\bigr) \subset L^1_\kappa$).*

(c) *The $\mathcal{L}$-convolution $(f * g)(x) := \int_a^b (\mathcal{T}^y f)(x)\, g(y)\, r(y)dy$ is well-defined and it satisfies $\|f * g\|_{1,\kappa} \leq \|f\|_{1,\kappa} \cdot \|g\|_{1,\kappa}$ (in particular, $L^1_\kappa * L^1_\kappa \subset L^1_\kappa$).*

Proof

(a) The absolute convergence of $\int_a^b f(x)\, w_\lambda(x)\, r(x)dx$ is immediate from (4.56). Letting $\mu(dx) = f(x)\, r(x)dx$, the same proof of Proposition 4.7(ii) shows that if $\widehat{\mu}(\lambda) \equiv (\mathcal{F}f)(\lambda) = 0$ for all $\lambda \geq \sigma^2$, then μ is the zero measure; consequently, the function f is uniquely determined by its $\mathcal{L}$-transform.

(b) Let $f \in L^1_\kappa$ and let $\ell^{\langle\kappa\rangle}$ be the operator defined in Lemma 4.2. We saw in the proof of Theorem 4.5 that $\boldsymbol{\nu}^{\langle\kappa\rangle}_{x,y}(d\xi) = \frac{w_\kappa(\xi)}{w_\kappa(x) w_\kappa(y)} \boldsymbol{\nu}_{x,y}(d\xi)$, hence

$$\begin{aligned}
(\mathcal{T}^y f)(x) &= \int_{[a,b)} f\, d\boldsymbol{\nu}_{x,y} \\
&= w_\kappa(x) w_\kappa(y) \int_{[a,b)} \frac{f}{w_\kappa}\, d\boldsymbol{\nu}^{\langle\kappa\rangle}_{x,y} \\
&= w_\kappa(x) w_\kappa(y) \Bigl(\mathcal{T}^y_{\langle\kappa\rangle} \frac{f}{w_\kappa}\Bigr)(x),
\end{aligned}$$

here $\boldsymbol{\nu}_{x,y}^{\langle\kappa\rangle}$ and $\mathcal{T}_{\langle\kappa\rangle}$ are, respectively, the measure of the product formula and the translation operator associated with $\ell^{\langle\kappa\rangle}$. Since $\|f\|_{1,\kappa} = \bigl\|\frac{f}{w_\kappa}\bigr\|_{L^1(r^{\langle\kappa\rangle})}$, it follows from Proposition 4.10(a) that $\mathcal{T}^y f$ is well-defined and

$$\|\mathcal{T}^y f\|_{1,\kappa} = w_\kappa(y)\bigl\|\mathcal{T}_{\langle\kappa\rangle}^y \tfrac{f}{w_\kappa}\bigr\|_{L^1(r^{\langle\kappa\rangle})} \leq w_\kappa(y)\bigl\|\tfrac{f}{w_\kappa}\bigr\|_{L^1(r^{\langle\kappa\rangle})} = w_\kappa(y)\|f\|_{1,\kappa}.$$

(c) Using part (b), we compute

$$\begin{aligned}\|f * g\|_{1,\kappa} &\leq \int_a^b \int_a^b |(\mathcal{T}^x f)(\xi)|\,|g(\xi)|\,r(\xi)d\xi\; w_\kappa(x)r(x)dx \\ &= \int_a^b \int_a^b |(\mathcal{T}^\xi f)(x)|\,w_\kappa(x)r(x)dx\,|g(\xi)|\,r(\xi)d\xi \\ &\leq \|f\|_{1,\kappa} \int_a^b |g(\xi)|\,w_\kappa(\xi)r(\xi)d\xi \;=\; \|f\|_{1,\kappa}\cdot\|g\|_{1,\kappa}.\end{aligned}$$

Corollary 4.7 *The Banach space L_κ^1, equipped with the convolution multiplication $f \cdot g \equiv f * g$, is a commutative Banach algebra without identity element.*

Proof Proposition 4.20(c) shows that the $\mathcal{L}$-convolution defines a binary operation on L_κ^1 for which the norm is submultiplicative. Since the trivialization property $\mathcal{F}(f * g) = (\mathcal{F}f)\cdot(\mathcal{F}g)$ (Proposition 4.18(a)) extends (by continuity) to all $f, g \in L_\kappa^1$, the commutativity and associativity of the $\mathcal{L}$-convolution in the space L_κ^1 is a consequence of the uniqueness of the $\mathcal{L}$-transform (Proposition 4.20(a)). An argument similar to that of the proof of Corollary 3.4 shows that the Banach algebra $(L_\kappa^1, *)$ has no identity element.

Next we state a Wiener-Lévy type theorem for the $\mathcal{L}$-convolution which includes, as a particular case, the corresponding theorem for the Whittaker convolution established in Theorem 3.5.

Theorem 4.8 (Wiener-Lévy Type Theorem) *For $-\infty < \kappa \leq \sigma^2$, write $\Pi_\kappa := \bigl\{\lambda \in \mathbb{C} \bigm| |\mathrm{Im}(\sqrt{\lambda - \sigma^2})| \leq \mathrm{Im}(\sqrt{\kappa - \sigma^2})\bigr\}$. Let $f \in L_\kappa^1$ and $\varrho \in \mathbb{C}$. The following assertions are equivalent:*

(i) *$\varrho + (\mathcal{F}f)(\lambda) \neq 0$ for all $\lambda \in \Pi_\kappa$ (including $\lambda = \infty$);*
(ii) *There exists a unique function $g \in L_\kappa^1$ and complex constant $\widetilde{\varrho}$ such that*

$$\frac{1}{\varrho + (\mathcal{F}f)(\lambda)} = \widetilde{\varrho} + (\mathcal{F}g)(\lambda) \qquad (\lambda \in \Pi_\kappa). \tag{4.57}$$

If $\lim_{\lambda\to\infty} w_\lambda(x) = 0$ for all $a < x < b$, then the unique complex constant in (4.57) *is $\widetilde{\varrho} = \varrho^{-1}$.*

We note that the hypothesis $\lim_{\lambda\to\infty} w_\lambda(x) = 0$ holds, in particular, for Sturm-Liouville operators whose Laplace representation (4.16) is such that the measures

$\pi_x \in \mathcal{M}_+(\mathbb{R})$ $(a < x < b)$ are absolutely continuous with respect to the Lebesgue measure. (This follows from the Riemann-Lebesgue lemma, as in the proof of Lemma 3.14.) In addition, it has been proved (see [30, Proposition 1]) that the property $\lim_{\lambda\to\infty} w_\lambda(x) = 0$ holds for all Sturm-Liouville operators of the form $\ell(u) = -u'' - \frac{A'}{A}u'$ on $\mathbb{R}^+$ whose coefficient A satisfies assumptions SL1.1 and SL2 with $\eta = 0$. The latter class of Sturm-Liouville operators includes those associated with the Hankel and the Fourier-Jacobi transforms (Examples 4.1(c),(e)).

The proof of Theorem 4.8 relies on the following generalization of Lemma 3.5, which characterizes the set of (suitably bounded) solutions of the functional equation determined by the Sturm-Liouville product formula:

Lemma 4.12 *Let $-\infty < \kappa \leq \sigma^2$. Assume that $\boldsymbol{\theta} : [a, b) \longrightarrow \mathbb{C}$ is a Borel measurable function such that there exists $C > 0$ for which*

$$\big|\boldsymbol{\theta}(x)\big| \leq C\, w_\kappa(x) \qquad \text{for almost every } x \in [a, b), \tag{4.58}$$

and that $\boldsymbol{\theta}(x)$ is a nontrivial solution of the functional equation

$$\boldsymbol{\theta}(x)\,\boldsymbol{\theta}(y) = \int_{[a,b)} \boldsymbol{\theta}(\xi)\, \boldsymbol{\nu}_{x,y}(d\xi) \qquad \text{for almost every } x, y \in [a, b). \tag{4.59}$$

Then $\boldsymbol{\theta}(x) = w_\lambda(x)$ for some $\lambda \in \Pi_\kappa$.

Proof For $t > 0$ and $x \in (a, b)$, let $\{T_t\}$ be the Feller semigroup generated by the Neumann realization of ℓ and write $(T_t h)(x) = \int_{[a,b)} h(\xi) p_{t,x}(d\xi)$, where $\{p_{t,x}\}_{t>0, x\in(a,b)}$ is the family of transition kernels.

Assume for the moment that $0 < \kappa \leq \sigma^2$, so that $\boldsymbol{\theta} \in L_\infty(r)$. We know from Corollary 4.4 that the measure $\boldsymbol{\nu}_{t,x,y}(d\xi) = q_t(x, y, \xi)\, r(\xi) d\xi$ of the time-shifted product formula is given by $\boldsymbol{\nu}_{t,x,y} = p_{t,x} * \delta_y$; therefore, using the functional equation (4.59) we may compute

$$\begin{aligned} \big(T_t(\mathcal{T}^y \boldsymbol{\theta})\big)(x) &= \int_{[a,b)} \boldsymbol{\theta}(\xi)\, \boldsymbol{\nu}_{t,x,y}(d\xi) \\ &= \int_{[a,b)} (\mathcal{T}^y \boldsymbol{\theta})(\xi)\, p_{t,x}(d\xi) \\ &= \boldsymbol{\theta}(y) \int_{[a,b)} \boldsymbol{\theta}(\xi)\, p_{t,x}(d\xi) \;=\; \boldsymbol{\theta}(y)\, (T_t \boldsymbol{\theta})(x), \end{aligned} \tag{4.60}$$

where we also used the fact that the transition kernels $\{p_{t,x}\}_{t>0, x\in(a,b)}$ are absolutely continuous (Proposition 2.14). On the other hand, we know from a general result on one-dimensional diffusion semigroups (see [137, Corollary 4.4]) that the Feller semigroup $\{T_t\}$ is such that

$$\frac{\partial}{\partial t}(T_t h)(x) = -\ell_x (T_t h)(x) \qquad \text{for all } h \in L_\infty(r) \qquad \big(t > 0,\; x \in (a, b)\big),$$

and we thus have (noting that $\mathcal{T}^y\boldsymbol{\theta} \in L_\infty(r)$, cf. Proposition 4.10(iv))

$$\ell_x\big(T_t(\mathcal{T}^y\boldsymbol{\theta})\big)(x) = -\frac{\partial}{\partial t}\big(T_t(\mathcal{T}^y\boldsymbol{\theta})\big)(x) = \ell_y\big(T_t(\mathcal{T}^x\boldsymbol{\theta})\big)(y), \tag{4.61}$$

where the second equality holds because $\big(T_t(\mathcal{T}^y\boldsymbol{\theta})\big)(x) = \big(T_t(\mathcal{T}^x\boldsymbol{\theta})\big)(y)$. Combining (4.60) and (4.61), we deduce that

$$\ell_y[\boldsymbol{\theta}(y)\,(T_t\boldsymbol{\theta})(x)] = -\frac{\partial}{\partial t}[\boldsymbol{\theta}(y)\,(T_t\boldsymbol{\theta})(x)] = -\frac{\partial}{\partial t}[\boldsymbol{\theta}(x)\,(T_t\boldsymbol{\theta})(y)] = \ell_x[\boldsymbol{\theta}(x)\,(T_t\boldsymbol{\theta})(y)]$$

for all $t > 0$ and almost every $x, y \in (a, b)$, and therefore

$$\frac{-\frac{\partial}{\partial t}(T_t\boldsymbol{\theta})(x)}{(T_t\boldsymbol{\theta})(x)} = \frac{\ell_x\boldsymbol{\theta}(x)}{\boldsymbol{\theta}(x)} = \frac{\ell_y\boldsymbol{\theta}(y)}{\boldsymbol{\theta}(y)} = \lambda \tag{4.62}$$

for some constant $\lambda \in \mathbb{C}$. From the last equality (which holds for almost every y) it follows that $\boldsymbol{\theta}(y) = c_1 w_\lambda(y) + c_2 \boldsymbol{u}_\lambda(y)$, where $\boldsymbol{u}_\lambda$ is a solution of $\ell(u) = \lambda u$ linearly independent of w_λ and $c_1, c_2 \in \mathbb{C}$ are constants. In particular, $\boldsymbol{\theta}$ is continuous (in fact, by (4.58) we have $\boldsymbol{\theta} \in \mathrm{C}_0[a, b)$), and by continuity the functional equation (4.59) holds for all $x, y \in [a, b)$; moreover, if we let y_0 be such that $\boldsymbol{\theta}(y_0) \neq 0$, we see that

$$\boldsymbol{\theta}(x) = \frac{1}{\boldsymbol{\theta}(y_0)} \int_{[a,b)} \boldsymbol{\theta}(\xi)\, \boldsymbol{\nu}_{x,y_0}(d\xi) \longrightarrow \frac{1}{\boldsymbol{\theta}(y_0)} \int_{[a,b)} \boldsymbol{\theta}(\xi)\, \delta_{y_0}(d\xi) = 1 \qquad \text{as } x \downarrow a.$$

In order to show that $\boldsymbol{\theta}(x) = w_\lambda(x)$, by Lemma 2.1 it only remains to prove that $\lim_{x\downarrow a}(p\boldsymbol{\theta}')(x) = 0$. We know that $\lim_{t\downarrow 0}(T_t\boldsymbol{\theta})(x) = \boldsymbol{\theta}(x)$ and, by (4.62), $\frac{\partial}{\partial t}(T_t\boldsymbol{\theta})(x) = -\lambda(T_t\boldsymbol{\theta})(x)$, hence

$$(T_t\boldsymbol{\theta})(x) = e^{-\lambda t}\boldsymbol{\theta}(x) \qquad (t \geq 0,\ x \in (a, b)),$$

and therefore

$$(\mathcal{R}_\eta\boldsymbol{\theta})(x) = \frac{\boldsymbol{\theta}(x)}{\lambda + \eta} \qquad (\eta > 0,\ x \in (a, b)),$$

where, as in (2.3), $\mathcal{R}_\eta f = \int_0^\infty e^{-\eta t}\, T_t f\, dt$ denotes the resolvent of the Feller semigroup $\{T_t\}_{t\geq 0}$. Since $\boldsymbol{\theta} \in \mathrm{C}_0[a, b)$, it follows that $(\mathcal{R}_\eta\boldsymbol{\theta})(x)$ belongs to $\mathcal{D}(\mathcal{L}^{(0)})$, and therefore we have $\lim_{x\downarrow a}(p\boldsymbol{\theta}')(x) = \lim_{x\downarrow a}(\lambda + \eta)p(x)(\mathcal{R}_\eta\boldsymbol{\theta})'(x) \longrightarrow 0$ as $x \downarrow a$, as desired.

Finally, suppose that $\kappa \leq 0$ and choose $\kappa_0 < \kappa$. Recalling Lemma 4.2, it is easily seen that the function $\boldsymbol{\theta}^{\langle\kappa_0\rangle}(x) := \frac{\boldsymbol{\theta}(x)}{w_{\kappa_0}(x)}$ is bounded almost everywhere by

$w^{\langle\kappa_0\rangle}_{\kappa-\kappa_0}(x) = \frac{w_\kappa(x)}{w_{\kappa_0}(x)} \in \mathrm{C}_0[a,b)$ and satisfies the functional equation

$$\boldsymbol{\theta}^{\langle\kappa_0\rangle}(x)\,\boldsymbol{\theta}^{\langle\kappa_0\rangle}(y) = \int_{[a,b)} \boldsymbol{\theta}^{\langle\kappa_0\rangle}(\xi)\,\boldsymbol{\nu}^{\langle\kappa_0\rangle}_{x,y}(d\xi) \qquad \text{for almost every } x, y \in [a,b),$$

where, as seen above, $\boldsymbol{\nu}^{\langle\kappa_0\rangle}_{x,y}(d\xi) = \frac{w_{\kappa_0}(\xi)}{w_{\kappa_0}(x)w_{\kappa_0}(y)}\,\boldsymbol{\nu}_{x,y}(d\xi)$. Using the proof given for the case $0 < \kappa \leq \sigma^2$ (where we replace the associated Sturm-Liouville operator by $\ell^{\langle\kappa_0\rangle}$), we deduce that $\boldsymbol{\theta}^{\langle\kappa_0\rangle}(x) = w^{\langle\kappa_0\rangle}_{\lambda_0}(x)$ for some $\lambda_0 \in \mathbb{C}$. Consequently, $\boldsymbol{\theta}(x) = w_\lambda(x)$ for some $\lambda \in \mathbb{C}$.

It only remains to show that $\lambda \in \Pi_\kappa$. Indeed, taking into account the Laplace-type representation $w_\lambda(x) = \int_\mathbb{R} \cos\bigl(s\sqrt{\lambda - \sigma^2}\bigr)\pi_x(ds)$, the inequality (4.58) holds if and only if

$$\left|\int_\mathbb{R} \cos\bigl(s\sqrt{\lambda - \sigma^2}\bigr)\pi_x(ds)\right| \leq C \int_\mathbb{R} \cos\bigl(s\sqrt{\kappa - \sigma^2}\bigr)\pi_x(ds) \qquad \text{for a.e. } x \in [a,b),$$

and clearly this takes place if and only if $\lambda \in \Pi_\kappa$.

Proof of Theorem 4.8 Following the arguments from the proof of Lemma 3.15, one can prove the following counterpart of Lemma 3.15: *if* $J : L^1_\kappa \longrightarrow \mathbb{C}$ $(-\infty < \kappa \leq \sigma^2)$ *is a linear functional satisfying*

$$J(h * g) = J(h) \cdot J(g) \qquad \text{for all } h, g \in L^1_\kappa,$$

then $J(h) = \int_{[a,b)} h(\xi) w_\lambda(\xi) r(\xi) d\xi$ *for some* $\lambda \in \Pi_\kappa$. One can then establish the Wiener-Lévy type theorem by a straightforward adaptation of the proof of Theorem 3.5 (replacing $L_{\alpha,\nu}$ and $\mathcal{W}_\alpha$ by, respectively, L^1_κ and $\mathcal{F}$, etc.).

4.7.2 Application to Convolution-Type Integral Equations

As in Sect. 3.8, the Wiener-Lévy type theorem established above is applicable to the study of a class of integral equations of convolution type.

Definition 4.13 The integral equation

$$f(x) + \int_0^\infty J(x,y) f(y)\,dy = h(x), \tag{4.63}$$

where h and J are known functions and f is to be determined, is said to be an $\mathcal{L}$*-convolution equation* if $J(x,y) = (\mathcal{T}^x\varphi)(y) r(y)$ for some function $\varphi \in L^1_{\sigma^2}$. In other words, (4.63) is an $\mathcal{L}$-convolution equation if it can be represented as

$$f(x) + (f * \varphi)(x) = h(x)$$

with $h \in L^1_{\sigma^2}$.

An argument similar to the one given in Sect. 3.8 yields the following existence and uniqueness theorem for $\mathcal{L}$-convolution equations:

Theorem 4.9 *Assume that $J(x, y) = (\mathcal{T}^x\varphi)(y)r(y)$ for some $\varphi \in L^1_\kappa$.*

If $1 + (\mathcal{F}\varphi)(\lambda) \neq 0$ for all $\lambda \in \Pi_\kappa$ (including $\lambda = \infty$), then, for each given $h \in L^1_\kappa$, the integral equation (4.63) admits a unique solution $f \in L^1_\kappa$; moreover, this solution can be written in the form

$$f(x) = \varrho h(x) + (h * g)(x) = \varrho h(x) + \int_a^b h(y)\,(\mathcal{T}^x g)(y)\, r(y) dy \tag{4.64}$$

for some function $g \in L^1_\kappa$ and some constant $\varrho \in \mathbb{C}$.

Conversely, if $1 + (\mathcal{F}\varphi)(\lambda_0) = 0$ for some $\lambda_0 \in \Pi_\kappa$, then there exists no $f \in L^1_\kappa$ satisfying the integral equation (4.63).

As an interesting particular case, we deduce the existence and uniqueness of solution for integral equations involving the density $q_t(x, y, \xi)$ of the time-shifted product formula (cf. Proposition 4.6):

Corollary 4.8 *For each fixed $t > 0$, $a < x < b$ and $\psi \in L^1(r)$, the integral equation*

$$h(y) + \int_a^b h(\xi)\, q_t(x, y, \xi) r(\xi) d\xi = \psi(y)$$

has a unique solution $h \in L^1(r)$ which can be written in the form (4.64) for some function $g \in L^1(r)$.

Proof Let us justify that this result is obtained by setting $f = f_{t,x} := p(t, x, \cdot)$ in the statement of Theorem 4.9. Notice first that by Corollary 2.2 we have $f_{t,x} \in L^1_0 \equiv L^1(r)$. Moreover, we have $(\mathcal{F} f_{t,x})(\lambda) = e^{-t\lambda} w_\lambda(x)$ (cf. (2.33)), thus $1 + (\mathcal{F} f_{t,x})(0) = 2$ and

$$|1 + (\mathcal{F} f_{t,x})(\lambda)| \geq 1 - e^{-t \operatorname{Re} \lambda} |w_\lambda(x)| > 0, \qquad \lambda \in \Pi_0 \setminus \{0\}$$

(this is easily seen to hold by setting $\lambda = \tau^2 + \sigma^2$ and recalling the estimate (4.17)). Recalling that

$$\big(\mathcal{F}(\mathcal{T}^y f_{t,x})\big)(\lambda) = (\mathcal{F} f_{t,x})(\lambda)\, w_\lambda(y) = e^{-t\lambda} w_\lambda(x) w_\lambda(y) = \big(\mathcal{F} q_t(x, y, \cdot)\big)(\lambda),$$

where we used Proposition 4.10(ii), we see that $(\mathcal{T}^y f_{t,x})(\xi) = q_t(x, y, \xi)$, so that the corollary is a particular case of the theorem.

Chapter 5
Convolution-Like Structures on Multidimensional Spaces

After having constructed convolution-like operators associated with the Shiryaev process and with a family of one-dimensional diffusion processes generated by Sturm–Liouville operators, we devote this final chapter to a more general discussion of the problem formulated in the Introduction: the construction of a generalized convolution associated with a given Feller process on a generally multidimensional state space.

We start this chapter by describing the desired properties of such a generalized convolution structure; these properties will be seen to lead to strong restrictions in the behaviour of the eigenfunctions of the Feller generator. In Sects. 5.2–5.3, after reviewing some basic notions and facts from spectral theory and differential operators, we show that such restrictions fail to hold for reflected Brownian motions on bounded smooth domains of $\mathbb{R}^d$ $(d \geq 2)$ and also for certain one-dimensional diffusions, leading to negative results on the existence of associated convolution-like structures. Finally, in Sect. 5.4 we propose the notion of a family of convolutions associated with a given Feller semigroup as a natural way of overcoming the difficulties in constructing convolutions on multidimensional spaces; such families of convolutions will be shown to exist for a general class of two-dimensional manifolds endowed with cone-like metrics.

5.1 Convolutions Associated with Conservative Strong Feller Semigroups

Our work in Chaps. 3 and 4 indicates that none of the standard axiomatic definitions in the literature on generalized harmonic analysis—hypergroups, hypercomplex systems, Urbanik generalized convolutions, stochastic convolutions, etc. (see Sect. 2.3)—is fully satisfactory in identifying the essential requirements for

R. Sousa et al., *Convolution-like Structures, Differential Operators and Diffusion Processes*, Lecture Notes in Mathematics 2315,
https://doi.org/10.1007/978-3-031-05296-5_5

constructing a convolution-like operator associated with a given transition semigroup. In light of the results of the previous chapters, we will instead consider the following notion of convolution-like structure:

Definition 5.1 Let E be a locally compact separable metric space, and let $\{T_t\}_{t\geq 0}$ be a conservative strong Feller semigroup on E. We say that a bilinear operator $\diamond$ on $\mathcal{M}_{\mathbb{C}}(E)$ is a *Feller-Lévy trivializable convolution* (FLTC) for $\{T_t\}$ if the following conditions hold:

I. $(\mathcal{M}_{\mathbb{C}}(E), \diamond)$ is a commutative Banach algebra over $\mathbb{C}$ (with respect to the total variation norm) with identity element δ_a $(a \in E)$, and $(\mu, \nu) \mapsto \mu \diamond \nu$ is continuous in the weak topology of measures;

II. $\mathcal{P}(E) \diamond \mathcal{P}(E) \subset \mathcal{P}(E)$;

III. There exists a family $\Theta \subset \mathrm{B}_\mathrm{b}(E) \setminus \{0\}$ such that for $\mu, \mu_1, \mu_2 \in \mathcal{P}(E)$ we have

$$\mu = \mu_1 \diamond \mu_2 \qquad \text{if and only if} \qquad \mu(\vartheta) = \mu_1(\vartheta) \cdot \mu_2(\vartheta) \text{ for all } \vartheta \in \Theta,$$

where $\mu(\vartheta) := \int_E \vartheta(\xi) \mu(d\xi)$;

IV. The transition kernel $\{p_{t,x}\}_{t\geq 0, x\in E}$ of the semigroup $\{T_t\}$ is of the form

$$p_{t,x} = \mu_t \diamond \delta_x,$$

where $\{\mu_t\}_{t\geq 0} \subset \mathcal{P}(E)$ is a family of measures such that $\mu_{t+s} = \mu_t \diamond \mu_s$ for all $t, s \geq 0$.

Conditions I and II in the above definition can be interpreted as basic axioms that allow us to interpret $(\mathcal{M}_{\mathbb{C}}(E), \diamond)$ as a probability-preserving convolution-like structure. Condition III requires the existence of an integral transform with bounded kernels which determines uniquely a given measure $\mu \in \mathcal{M}_{\mathbb{C}}(E)$ (in the sense that if $\mu(\vartheta) = \nu(\vartheta)$ for all $\vartheta \in \Theta$, then $\mu = \nu$) and trivializes the convolution in the same way as the Fourier transform trivializes the ordinary convolution. As noted in [197], it is possible, in principle, to study infinite divisibility of probability measures on measure algebras not satisfying Condition III; however, it is natural to require Condition III to hold, not only because, to the best of our knowledge, all known examples of convolution-like structures are constructed from a product formula of the form $\vartheta(x)\vartheta(y) = (\delta_x \diamond \delta_y)(\vartheta)$ (and therefore possess such a family of trivializing functions) but also because this trivialization property leads to a richer theory. Lastly, condition IV expresses the motivating goal discussed in the Introduction: the Feller semigroup $\{T_t\}$ should have the convolution semigroup property with respect to the operator $\diamond$ or, in other words, the Feller process $\{X_t\}_{t\geq 0}$ determined by $\{T_t\}$ is a Lévy process with respect to $\diamond$ in the sense that we have $P\big[X_t \in \cdot | X_s = x\big] = \mu_{t-s} \diamond \delta_x$ for every $0 \leq s \leq t$ and $x \in E$.

The problem of existence of an associated FLTC is meaningful for any given strong Feller semigroup on a locally compact separable metric space. Before we present some notable examples of this class of semigroups, let us recall some prerequisite notions from the theory of Dirichlet forms. Let μ be a σ-finite measure on E. We say that $(\mathcal{E}, \mathcal{D}(\mathcal{E}))$ is a *Dirichlet form* on $L^2(E, \mu)$ if $\mathcal{D}(\mathcal{E})$ is a dense subspace of $L^2(E, \mu)$ and $\mathcal{E}$ is a nonnegative, closed, Markovian symmetric sesquilinear form defined on $\mathcal{D}(\mathcal{E}) \times \mathcal{D}(\mathcal{E})$. The associated non-positive self-adjoint operator $(\mathcal{G}^{(2)}, \mathcal{D}(\mathcal{G}^{(2)}))$ is defined as

$$u \in \mathcal{D}(\mathcal{G}^{(2)}) \quad \text{if and only if}$$
$$\exists \phi \in L^2(E, \mu) \text{ such that } \mathcal{E}(u, v) = -\langle \phi, v \rangle_{L^2(E,\mu)} \text{ for all } v \in \mathcal{D}(\mathcal{E}), \tag{5.1}$$

and $\mathcal{G}^{(2)} u := \phi$ for $u \in \mathcal{D}(\mathcal{G}^{(2)})$. The semigroup determined by $\mathcal{E}$ is defined by $T_t^{(2)} := e^{t\mathcal{G}^{(2)}}$ (where the latter is obtained by spectral calculus); one can show [33, Theorem 1.1.3] that $\{T_t^{(2)}\}$ is a strongly continuous, sub-Markovian contraction semigroup on $L^2(E, \mu)$. The Dirichlet form $(\mathcal{E}, \mathcal{D}(\mathcal{E}))$ is said to be *strongly local* if $\mathcal{E}(u, v) = 0$ whenever $u \in \mathcal{D}(\mathcal{E})$ has compact support and $v \in \mathcal{D}(\mathcal{E})$ is constant on a neighbourhood of $\operatorname{supp}(u)$. It is said to be *regular* if $\mathcal{D}(\mathcal{E}) \cap \mathrm{C_c}(E)$ is dense both in $\mathcal{D}(\mathcal{E})$ with respect to the norm $\|u\|_{\mathcal{D}(\mathcal{E})} = \sqrt{\mathcal{E}(u, u) + \|u\|_{L^2(E,\mu)}}$ and in $\mathrm{C_c}(E)$ with respect to the sup norm. A well-known result [66, Theorem 7.2.1] states that if $(\mathcal{E}, \mathcal{D}(\mathcal{E}))$ is a regular Dirichlet form on $L^2(E, \mu)$ with semigroup $\{T_t^{(2)}\}_{t \geq 0}$, then there exists a Hunt process with state space E whose transition semigroup $\{P_t\}_{t \geq 0}$ is such that $P_t u$ is, for all $u \in \mathrm{C_c}(E)$, a quasi-continuous version of $T_t^{(2)} u$. (A *Hunt process* is essentially a strong Markov process whose paths are right-continuous and quasi-left-continuous; see [66, Appendix A.2] for details.) We refer to the textbooks of Fukushima, Oshima and Takeda [66] and of Chen and Fukushima [33] for further background on Dirichlet forms and related objects.

Example 5.1

(a) Let (E, g) be a complete Riemannian manifold, let $\mathbf{m}$ be the Riemannian volume on E and let ∇ denote the Riemannian gradient on (M, g). The sesquilinear form

$$\mathcal{E}(u, v) = \frac{1}{2} \int_E \langle \nabla u, \nabla v \rangle_g \, d\mathbf{m}, \qquad u, v \in \mathcal{D}(\mathcal{E})$$

with domain

$$\begin{aligned} \mathcal{D}(\mathcal{E}) &= \text{closure of } \mathrm{C_c^\infty}(E) \text{ in the Sobolev space } H^1(E) \\ &\equiv \{u \in L^2(E) \mid |\nabla u| \in L^2(E)\} \end{aligned}$$

is a strongly local regular Dirichlet form on $L^2(E) \equiv L^2(E, \mathbf{m})$. The Hunt diffusion process $\{X_t\}_{t \geq 0}$ with state space E associated with this Dirichlet form is the *Brownian motion on* (E, g). One can show that the strongly continuous contraction semigroup $\{T_t\}$ determined by $\mathcal{E}$ is such that $T_t\big(\mathrm{B_b}(E)\big) \subset \mathrm{C_b}(E)$, so that the Brownian motion $\{X_t\}$ is a strong Feller process [182, Section 6]. Moreover, it is shown in [66, Example 5.7.2] that the Feller semigroup $\{T_t\}$ is conservative provided that the Riemannian volume $\mathbf{m}$ is such that

$$\liminf_{r\to\infty} \frac{1}{r^2} \log \mathbf{m}(\mathbb{B}(x_0; r)) < \infty \text{ for some fixed } x_0 \in E.$$

Let $\mathcal{G}^{(0)} : \mathcal{D}(\mathcal{G}^{(0)}) \subset \mathrm{C}_0(E) \longrightarrow \mathrm{C}_0(E)$ be the infinitesimal generator of the Brownian motion $\{X_t\}$. Then $\mathcal{G}^{(0)}u = \frac{1}{2}\Delta u$ for $u \in \mathrm{C}_\mathrm{c}^\infty(E) \subset \mathcal{D}(\mathcal{G}^{(0)})$, where Δ is the Laplace–Beltrami operator on the Riemannian manifold (E, g).

(b) Let $E = \mathbb{R}^d$, m a positive function such that $m, \frac{1}{m} \in \mathrm{C_b}(\mathbb{R}^d)$ and $A = (a_{jk})$ a symmetric $d \times d$ matrix-valued function such that $a_{jk} \in \mathrm{C}(\mathbb{R}^d)$ (for each $j, k \in \{1, \ldots, d\}$) and

$$c^{-1}|\xi|^2 \leq \sum_{j,k=1}^{d} a_{jk}(x)\xi_j\xi_k \leq c|\xi|^2, \qquad (x, \xi) \in \mathbb{R}^d \times \mathbb{R}^d \tag{5.2}$$

for some constant $c > 0$. The sesquilinear form

$$\mathcal{E}(u, v) = \frac{1}{2} \sum_{j,k=1}^{d} \int_{\mathbb{R}^d} a_{jk}(x) \frac{\partial u}{\partial x_j} \frac{\partial \overline{v}}{\partial x_k} m(x)dx, \qquad u, v \in \mathcal{D}(\mathcal{E})$$

with domain

$$\mathcal{D}(\mathcal{E}) = \text{closure of } \mathrm{C}_\mathrm{c}^\infty(\mathbb{R}^d) \text{ under the inner product } \mathcal{E}(\cdot, \cdot) + \langle \cdot, \cdot \rangle_{L^2(\mathbf{m})}$$

is a strongly local regular Dirichlet form on the space $L^2(\mathbf{m}) \equiv L^2(\mathbb{R}^d, m(x)dx)$ [66, Section 3.1]. The Hunt diffusion $\{X_t\}_{t\geq 0}$ associated with the Dirichlet form $\mathcal{E}$ is conservative [66, Example 5.7.1]. The process $\{X_t\}$, which is called the *(A, m)-diffusion* on $\mathbb{R}^d$, is a strong Feller process, cf. [183, Example 4.C and Proposition 7.5]. If, in addition, we have $\frac{\partial(ma_{jk})}{\partial x_j} \in L^2_{\mathrm{loc}}(\mathbb{R}^d)$ for each $j, k \in \{1, \ldots, d\}$, then the infinitesimal generator $\mathcal{G}^{(0)}$ of the Feller semigroup is the elliptic operator $(\mathcal{G}^{(0)}u)(x) = \frac{1}{2m(x)} \sum_{j,k=1}^{d} \frac{\partial}{\partial x_j}\bigl(m(x)a_{jk}(x)\frac{\partial u}{\partial x_k}\bigr)$ (for $u \in \mathrm{C}_\mathrm{c}^2(\mathbb{R}^d) \subset \mathcal{D}(\mathcal{G}^{(0)})$).

(c) Let E be the closure of a bounded Lipschitz domain $\mathring{E} \subset \mathbb{R}^d$ and, as usual, let $H^k(E)$ be the Sobolev space defined as $H^k(E) := \{u \in L^2(E, dx) \mid \partial^\alpha u \in L^2(E, dx) \text{ for all } \alpha = (\alpha_1, \ldots, \alpha_d) \text{ with } |\alpha| \leq k\}$. Let $m \in H^1(E)$ be a positive function such that $m, \frac{1}{m} \in \mathrm{C}(E)$ and let $A = (a_{jk})$ be a symmetric bounded $d \times d$ matrix-valued function such that $a_{jk} \in H^1(E)$ for $j, k \in \{1, \ldots, d\}$ and the uniform ellipticity condition (5.2) holds for

$(x, \xi) \in E \times \mathbb{R}^d$. The sesquilinear form

$$\mathcal{E}(u, v) = \frac{1}{2} \sum_{j,k=1}^{d} \int_E a_{jk}(x) \frac{\partial u}{\partial x_j} \frac{\partial \overline{v}}{\partial x_k} m(x) dx, \qquad u, v \in \mathcal{D}(\mathcal{E}) = H^1(E)$$

is a strongly local regular Dirichlet form on $L^2(E, \mathbf{m}) \equiv L^2(E, m(x)dx)$ whose associated Hunt diffusion process is a conservative Feller process, cf. [32, 34]. The process $\{X_t\}$ is called the (A, m)*-reflected diffusion* on E. The infinitesimal generator $\mathcal{G}^{(0)}$ of the Feller process $\{X_t\}$ is such that $\mathrm{C}_\mathrm{c}^2(\mathring{E}) \subset \mathcal{D}(\mathcal{G}^{(0)})$ and $(\mathcal{G}^{(0)} u)(x) = \frac{1}{2m(x)} \sum_{j,k=1}^{d} \frac{\partial}{\partial x_j}\bigl(m(x) a_{jk}(x) \frac{\partial u}{\partial x_k}\bigr)$ for $u \in \mathrm{C}_\mathrm{c}^2(\mathring{E})$. In the special case $a_{ij} = \delta_{ij}$ and $m = \mathbb{1}$, the (A, m)-reflected diffusion is known as the *reflected Brownian motion* on E, whose infinitesimal generator $\mathcal{G}^{(0)} u = \frac{1}{2}\Delta u$ is the so-called *Neumann Laplacian* on E.

(d) Let E be a locally compact separable metric space with distance $\mathbf{d}$ and let $\mathbf{m}$ be a locally finite Borel measure on E with $\mathbf{m}(U) > 0$ for all nonempty open sets $U \subset E$. Suppose that the triplet $(E, \mathbf{d}, \mathbf{m})$ satisfies the *measure contraction property* introduced in [183, Definition 4.1]; roughly speaking, this means that there exists a family of quasi-geodesic maps connecting almost every pair of points $x, y \in E$ and which satisfy a contraction property which controls the distortions of the measure $\mathbf{m}$ along each quasi-geodesic. It was proved in [183] that the family of Dirichlet forms defined as

$$\mathcal{E}^r(u, u) = \frac{1}{2} \int_E \int_{\mathbb{B}(x;r)\setminus\{x\}} \left| \frac{u(z) - u(x)}{\mathbf{d}(z, x)} \right|^2 \frac{\mathbf{m}(dz)}{\sqrt{\mathbf{m}(\mathbb{B}(z; r))}} \frac{\mathbf{m}(dx)}{\sqrt{\mathbf{m}(\mathbb{B}(x; r))}}, \qquad r > 0$$

(and $\mathcal{E}^r(u, v)$ defined via the polarization identity) converges as $r \downarrow 0$ (in a suitable variational sense, see [183]) to a strongly local regular Dirichlet form on $L^2(E, \mathbf{m})$. The associated Hunt diffusion $\{X_t\}_{t \geq 0}$ is a strong Feller process with state space E. If the growth of the volumes $\mathbf{m}(\mathbb{B}(x; r))$ satisfies the condition stated in [66, Theorem 5.7.3], then the Feller process $\{X_t\}$ is conservative. This class of strong Feller processes includes, as particular cases, the diffusions of Examples (a) and (b) above, diffusions on manifolds with boundaries or corners, spaces obtained by gluing together manifolds, among others.

As discussed above, our motivating problem is that of determining, for a given Feller semigroup $\{T_t\}$, necessary or sufficient conditions for the existence of an associated FLTC satisfying the requirements of Definition 5.1. The following proposition shows that the converse problem of constructing a (strong) Feller semigroup associated with a given convolution-like semigroup of measures on the space E has a straightforward solution:

Proposition 5.1 *Let E be a locally compact separable metric space with the Heine-Borel property (i.e. where each closed bounded subset is compact), and let $\diamond$ be a bilinear operator on $\mathcal{M}_\mathbb{C}(E)$ satisfying conditions I and II of Definition 5.1. Let*

$\{\mu_t\} \subset \mathcal{P}(E)$ be a family of measures such that

$$\mu_{t+s} = \mu_t \diamond \mu_s \text{ for all } t, s \geq 0, \qquad \mu_t \overset{w}{\longrightarrow} \mu_0 \text{ as } t \downarrow 0, \qquad \mu_0 = \delta_a.$$

For $\nu \in \mathcal{M}^+(E)$, define $(\mathcal{T}^\nu f)(x) := \int_E f\, d(\nu \diamond \delta_x)$, and assume that $\mathcal{T}^{\delta_y} f \in \mathrm{C}_0(E)$ for all $f \in \mathrm{C}_0(E)$ and $y \in E$. Then the operators

$$T_t : \mathrm{C}_0(E) \longrightarrow \mathrm{C}_0(E), \qquad T_t f := \mathcal{T}^{\mu_t} f$$

constitute a conservative Feller semigroup. If, in addition, we have $(\mu_t \diamond \delta_x)(dy) = p_t(x, y)\mathbf{m}(dy)$ $(t > 0,\ x \in E)$ for some measure $\mathbf{m} \in \mathcal{M}_+(E)$ and some density function $p_t(x, y)$ which is locally bounded on $E \times E$ for each $t > 0$, then $\{T_t\}_{t\geq 0}$ is a strong Feller process.

Proof It is trivial that each operator T_t is positivity-preserving and conservative. The semigroup property $T_{t+s} = T_t T_s$ and the strong continuity $\lim_{t\downarrow 0} \|T_t f - f\|_\infty = 0$ follow as in the proof of Proposition 3.8. Next, let $\nu \in \mathcal{M}_+(E)$ and $\{x_n\} \subset E$ a sequence such that $d(x_n, a) \to \infty$ as $n \to \infty$. By assumption $\mathcal{T}^{\delta_y} f \in \mathrm{C}_0(E)$, thus by dominated convergence we have

$$|(\mathcal{T}^\nu f)(x_n)| \leq \int_E |(\mathcal{T}^{\delta_y} f)(x_n)|\, \nu(dy) \longrightarrow 0 \qquad \text{as } n \to \infty,$$

showing that for each $\varepsilon > 0$ the space $\{x : |(\mathcal{T}^\nu f)(x)| \geq \varepsilon\}$ is bounded and, therefore, compact (by the Heine-Borel property). This shows that $\mathcal{T}^\nu\bigl(\mathrm{C}_0(E)\bigr) \subset \mathrm{C}_0(E)$ for all bounded measures ν; in particular, $T_t\bigl(\mathrm{C}_0(E)\bigr) \subset \mathrm{C}_0(E)$. The fact that $\{T_t\}$ is strong Feller under the stated absolute continuity condition follows from Theorem 2.6.

We now prove an important fact concerning the family Θ of trivializing functions in the definition of an FLTC, namely that each $\vartheta \in \Theta$ is an eigenfunction of the $\mathrm{C_b}$-generator of the associated Feller semigroup. (As in (2.4), the notion of $\mathrm{C_b}$-*generator* of a semigroup $\{T_t\}_{t\geq 0}$ refers to the operator $\mathcal{G}^{(b)}$ with domain $\mathcal{D}(\mathcal{G}^{(b)}) = \mathcal{R}_\eta\bigl(\mathrm{C_b}(E)\bigr)$ $(\eta > 0)$ and given by $(\mathcal{G}^{(b)} u)(x) = \eta u(x) - g(x)$ for $u = \mathcal{R}_\eta g$, $g \in \mathrm{C_b}(E)$; recall that $\mathcal{R}_\eta f = \int_0^\infty e^{-\eta t} T_t f\, dt$ denotes the resolvent of the semigroup $\{T_t\}$.)

Proposition 5.2 *Let $\{T_t\}$ be a conservative strong Feller semigroup on a locally compact separable metric space E, and let $\diamond$ be a bilinear operator on $\mathcal{M}_\mathbb{C}(E)$ satisfying conditions I, II and IV of Definition 5.1. Suppose that $\vartheta \in \mathrm{B_b}(E)$, $\vartheta \not\equiv 0$ is a function such that*

$$(\delta_x \diamond \delta_y)(\vartheta) = \vartheta(x) \cdot \vartheta(y) \qquad \text{for all } x, y \in E. \tag{5.3}$$

Then $\vartheta(a) = \|\vartheta\|_\infty = 1$. *Moreover,* ϑ *is an eigenfunction of the* C_b*-generator* $(\mathcal{G}^{(b)}, \mathcal{D}(\mathcal{G}^{(b)}))$ *associated with an eigenvalue of nonpositive real part, in the sense that we have* $\vartheta \in \mathcal{D}(\mathcal{G}^{(b)})$ *and* $\mathcal{G}^{(b)}\vartheta = -\lambda\vartheta$ *for some* $\lambda \in \mathbb{C}$ *with* $\operatorname{Re}\lambda \geq 0$.

Proof Clearly, $\vartheta(x) = \delta_x(\vartheta) = (\delta_x \diamond \delta_a)(\vartheta) = \vartheta(x)\vartheta(a)$ for all $x \in E$. Since $\vartheta \not\equiv 0$, this implies that $\vartheta(a) = 1$.

Next, pick $\varepsilon > 0$ and choose $x_0 \in E$ such that $|\vartheta(x_0)| > \|\vartheta\|_\infty - \varepsilon$. Then $(\|\vartheta\|_\infty - \varepsilon)^2 < |\vartheta(x_0)|^2 = |(\delta_{x_0} \diamond \delta_{x_0})(\vartheta)| \leq \|\vartheta\|_\infty$ (by condition II, $\delta_{x_0} \diamond \delta_{x_0} \in \mathcal{P}(E)$, which justifies the last step). Since ε is arbitrary, $\|\vartheta\|_\infty^2 \leq \|\vartheta\|_\infty$, hence $\|\vartheta\|_\infty \leq 1$.

Since $\diamond$ is bilinear and weakly continuous, a straightforward argument yields that $(\mu \diamond \nu)(d\xi) = \int_E \int_E (\delta_x \diamond \delta_y)(d\xi)\mu(dx)\nu(dy)$ for $\mu, \nu \in \mathcal{M}_\mathbb{C}(E)$. Consequently, (5.3) implies that $(\mu \diamond \nu)(\vartheta) = \mu(\vartheta) \cdot \nu(\vartheta)$ for all $\mu, \nu \in \mathcal{M}_\mathbb{C}(E)$. In particular,

$$(T_t\vartheta)(x) \equiv p_{t,x}(\vartheta) = (\mu_t \diamond \delta_x)(\vartheta) = \mu_t(\vartheta) \cdot \vartheta(x)$$

due to condition IV. Given that $\{T_t\}$ is strong Feller, we have $T_t\vartheta \in C_b(E)$ and therefore $\vartheta = \frac{T_t\vartheta}{\mu_t(\vartheta)} \in C_b(E)$. Moreover, the fact that $\{T_t\}$ is a conservative Feller semigroup ensures that $\mu_t(\vartheta) = (T_t\vartheta)(a)$ is a continuous function of t which, by condition IV, satisfies the functional equation $\mu_{t+s}(\vartheta) = \mu_t(\vartheta)\mu_s(\vartheta)$. Therefore $\mu_t(\vartheta) = e^{-\lambda t}$ for some $\lambda \in \mathbb{C}$, and the fact that $|\mu_t(\vartheta)| \leq \|\vartheta\|_\infty = 1$ implies that $\operatorname{Re}\lambda \geq 0$. We thus have $T_t\vartheta = e^{-\lambda t}\vartheta$ and $\mathcal{R}_\eta\vartheta = \frac{\vartheta}{\lambda+\eta}$ for $\eta > 0$, so we conclude that $\vartheta \in \mathcal{D}(\mathcal{G}^{(b)})$ and $\mathcal{G}^{(b)}\vartheta = -\lambda\vartheta$.

Corollary 5.1 *Let* $\{T_t\}$ *be a conservative strong Feller semigroup on a locally compact separable metric space* E. *Let* $\diamond$ *be an FLTC for* $\{T_t\}$ *and* Θ *the corresponding family of trivializing functions. Then*

$$\Theta \subset \left\{ \omega \in \mathcal{D}(\mathcal{G}^{(b)}) \;\middle|\; \begin{array}{c} \omega(a) = \|\omega\|_\infty = 1, \\ \mathcal{G}^{(b)}\omega = -\lambda\omega \text{ for some } \lambda \in \mathbb{C} \text{ with } \operatorname{Re}\lambda \geq 0 \end{array} \right\}.$$

In particular, each $\mu \in \mathcal{M}_\mathbb{C}(E)$ *is uniquely determined by the integrals* $\mu(\omega) \equiv \int_E \omega(x)\mu(dx)$, *where* ω *belongs to the set of solutions of* $\mathcal{G}^{(b)}u = -\lambda u$ $(\operatorname{Re}\lambda \geq 0)$ *satisfying* $\omega(a) = \|\omega\|_\infty = 1$.

It is worth noting that if the strong Feller semigroup $\{T_t\}$ is symmetric with respect to a finite measure $\mathbf{m} \in \mathcal{M}_+(E)$ (i.e. if $\int_E (T_t f)(x)\, g(x)\mathbf{m}(dx) = \int_E f(x)\,(T_t g)(x)\mathbf{m}(dx)$ for $f, g \in C_c(E)$), then the space $C_b(E)$ is contained in $L^2(E, \mathbf{m})$; accordingly, the Feller semigroup $\{T_t\}_{t\geq 0}$ and the C_b-generator $\mathcal{G}^{(b)}$ extend, respectively, to a strongly continuous semigroup $\{T_t^{(2)}\}$ of symmetric operators on $L^2(E, \mathbf{m})$ and to the corresponding infinitesimal generator $\mathcal{G}^{(2)}$. In this setting, the trivializing functions $\vartheta \in \Theta$ are eigenfunctions of the L^2-generator $\mathcal{G}^{(2)}$. Applying spectral-theoretic results for self-adjoint operators on Hilbert spaces, we can deduce further properties for the trivializing functions:

Proposition 5.3 *Let* $\{T_t\}$ *be a conservative Feller semigroup on a locally compact separable metric space* E. *Suppose that the corresponding transition kernels*

$\{p_{t,x}(\cdot)\}_{t>0,x\in E}$ are of the form $p_{t,x}(dy) = p_t(x,y)\mathbf{m}(dy)$ for some finite measure $\mathbf{m} \in \mathcal{M}_+(E)$ and some density function $p_t(x,y)$ which is bounded and symmetric on $E \times E$ for each $t > 0$. Then $\{T_t\}$ is strong Feller, is symmetric with respect to $\mathbf{m}$, and admits an extension $\{T_t^{(2)}\}$ which is a strongly continuous semigroup on the space $L^2(E,\mathbf{m})$. There exists a sequence $0 \leq \lambda_1 \leq \lambda_2 \leq \lambda_3 < \ldots$ with $\lambda_j \to \infty$ and an orthonormal basis $\{\omega_j\}_{j\in\mathbb{N}}$ of $L^2(E,\mathbf{m})$ such that

$$T_t^{(2)}\omega_j = e^{-\lambda_j t}\omega_j \quad (t \geq 0), \qquad \mathcal{G}^{(2)}\omega_j = -\lambda_j\omega_j,$$

where $\mathcal{G}^{(2)}$ stands for the generator of the L^2-semigroup $\{T_t^{(2)}\}$. The sequence of eigenvalues is such that $\sum_{j=1}^\infty e^{-\lambda_j t} < \infty$ for each $t > 0$ (so that, in particular, $\lim_{j\to\infty}\lambda_j = \infty$).

Assume also that $\diamond$ is an FLTC for $\{T_t\}$ and that Θ is the family of trivializing functions for $\diamond$. Write $S_k = \{j \in \mathbb{N} \mid \lambda_j = \lambda_k\}$ and $m_k = |S_k|$ $(k \in \mathbb{N})$. Then each function $\vartheta \in \Theta$ is a solution of $\mathcal{G}^{(2)}\vartheta = -\lambda_j\vartheta$ for some $j \in \mathbb{N}$. Furthermore, there exist functions $\{\vartheta_j\}_{j\in\mathbb{N}} \subset \Theta$ such that

$$\mathrm{span}(\{\omega_j\}_{j\in S_k}) = \mathrm{span}(\{\vartheta_j\}_{j\in S_k})$$

and, consequently, $\overline{\mathrm{span}}(\Theta) = L^2(E,\mathbf{m})$.

Proof The strong Feller property follows from Theorem 2.6. The symmetry with respect to $\mathbf{m}$ is obvious, and it is straightforward to show that for $f \in \mathrm{C_c}(E)$ we have $\|T_t f\|_{L^2(E,\mathbf{m})} \leq \|f\|_{L^2(E,\mathbf{m})}$ and $\|T_t f - f\|_{L^2(E,\mathbf{m})} \to 0$ as $t \downarrow 0$, so that the extension $\{T_t^{(2)}\}$ is a strongly continuous semigroup on $L^2(E,\mathbf{m})$.

Let $\langle\cdot,\cdot\rangle$ be the inner product on $L^2(E,\mathbf{m})$. By the spectral theorem for compact self-adjoint operators (cf. e.g. [185, Theorem 6.7]), the operator $T_1^{(2)}$ can be written as $T_1^{(2)} = \sum_{j=1}^\infty \mu_j\langle\omega_j,\cdot\rangle\,\omega_j$, where $\mu_1 \geq \mu_2 \geq \ldots$ are the eigenvalues of $T_1^{(2)}$ and $\{\omega_j\}_{j\in\mathbb{N}}$ is an orthonormal basis of $L^2(E,\mathbf{m})$ such that each ω_j is an eigenfunction of $T_1^{(2)}$ associated with the eigenvalue μ_j; in addition, we have $\mu_1 \leq \|T_1^{(2)}\|$ and $\mu_j \downarrow 0$ as $j \to \infty$. If we define $\lambda_j = -\log\mu_j$, then it follows that $T_t^{(2)} = \sum_{j=1}^\infty e^{-\lambda_j t}\langle\omega_j,\cdot\rangle\,\omega_j$. (This can be justified as follows, cf. [72, pp. 463–464] for further details: we know that $(T_1^{(2)} - \mu_j)\omega_j = (T_{1/2}^{(2)} + \mu_j^{1/2})(T_{1/2}^{(2)} - \mu_j^{1/2})\omega_j = 0$, and all the eigenvalues of $(T_{1/2}^{(2)} + \mu_j^{1/2})$ are positive, hence $T_{1/2}^{(2)}\omega_j = \mu_j^{1/2}\omega_j$; similarly $T_t^{(2)}\omega_j = e^{-\lambda_j t}\omega_j$ for all $t = m/2^k$ and thus, by strong continuity, for all $t > 0$.) Consequently, $\mathcal{G}^{(2)}\omega_j = \lim_{t\downarrow 0}\frac{1}{t}(T_t^{(2)}\omega_j - \omega_j) = -\lambda_j\omega_j$. Since $\mathbf{m}$ is a finite measure and the densities $p_t(\cdot,\cdot)$ are bounded, the operator $T_t^{(2)}$ is, for each $t > 0$, a Hilbert-Schmidt operator, and therefore we have $\sum_{j=1}^\infty e^{-\lambda_j t} < \infty$ for all $t > 0$.

By Corollary 5.1 each $\vartheta \in \Theta$ is such that $\mathcal{G}^{(2)}\vartheta = -\lambda\vartheta$ for some $\lambda \in \mathbb{C}$. Given that $\Theta \subset L^2(E,\mathbf{m})$ and eigenfunctions associated with different eigenvalues are orthogonal, we have $\lambda = \lambda_j$ because otherwise we get a contradiction with the basis property of $\{\omega_{jk}\}$.

For the last part, fix $t > 0$, $k \in \mathbb{N}$ and let $\Theta_k := \{\vartheta \in \Theta \mid T_t^{(2)}\vartheta = e^{-\lambda_k t}\vartheta\} \subset L^2(E,\mathbf{m})$. Given that $\{\omega_j\}_{j\in S_k}$ is a basis of the eigenspace associated with λ_k, we have $\mathrm{span}(\Theta_k) \subset \mathrm{span}(\{\omega_j\}_{j\in S_k})$. To prove the reverse inclusion, let $h \in \mathrm{span}(\{\omega_j\}_{j\in S_k}) \cap \mathrm{span}(\Theta_k)^\perp$, write $\nu_h(dx) := h(x)\mathbf{m}(dx)$ and observe that (since $\mathbf{m}$ is a finite measure) $\nu_h \in \mathcal{M}_\mathbb{C}(E)$. Then the integral

$$\nu_h(\vartheta) = \int_E \vartheta(x)h(x)\mathbf{m}(dx)$$

is equal to zero for $\vartheta \in \Theta_k$ because $h \in \mathrm{span}(\Theta_k)^\perp$, and is also equal to zero for $\vartheta \in \Theta \setminus \Theta_k$ because then h and ϑ are eigenfunctions of $T_t^{(2)}$ associated with different eigenvalues. Since measures $\nu \in \mathcal{M}_\mathbb{C}(E)$ are uniquely determined by the integrals $\{\nu(\vartheta)\}_{\vartheta\in\Theta}$, it follows that $\nu_h = 0$ and therefore $h = 0$; this shows that $\mathrm{span}(\Theta_k) = \mathrm{span}(\{\omega_j\}_{j\in S_k})$. It follows at once that there exist linearly independent functions $\{\vartheta_j\}_{j\in\mathbb{N}} \subset \Theta$ such that $\mathrm{span}(\{\omega_j\}_{j\in S_k}) = \mathrm{span}(\{\vartheta_j\}_{j\in S_k})$.

The conclusions of Proposition 5.3 are valid, in particular, for the Feller semigroups associated with the Brownian motion on a compact Riemannian manifold or with an (A, m)-reflected diffusion on a bounded Lipschitz domain, cf. Examples 5.1(a) and (c) respectively. (Indeed, it follows from e.g. [183, Theorem 7.4] that in both cases we have $p_{t,x}(dy) = p_t(x, y)\mathbf{m}(dy)$ with $p_t(x, y)$ bounded and symmetric; recall also that compact Riemannian manifolds have finite volume, cf. e.g. [75, Theorem 3.11].)

It is worth emphasizing that, by Corollary 5.1 and Proposition 5.3, the existence of an FLTC for a Feller semigroup $\{T_t\}$ satisfying the assumptions above implies that the following *common maximizer property* holds:

CM. There exists a set $\{\vartheta_j\}_{j\in\mathbb{N}}$ of eigenfunctions of $\mathcal{G}^{(2)}$ which is dense in $L^2(E,\mathbf{m})$ and such that $\vartheta_j(a) = \|\vartheta_j\|_\infty = 1$ for some point $a \in E$.

(The functions ϑ_j associated with a common eigenvalue need not be orthogonal in $L^2(E,\mathbf{m})$.) This necessary condition will play a fundamental role in the proof of the inexistence results established in Sect. 5.2.

The next proposition and corollary show that the existence of an FLTC is closely related with the positivity of a regularized kernel which resembles the density (4.39) of the time-shifted product formula for Sturm–Liouville convolutions.

Proposition 5.4 *In the conditions of the first paragraph of Proposition 5.3, assume that the metric space E is compact. Let $0 \le \lambda_1 \le \lambda_2 \le \lambda_3 < \ldots$ be the eigenvalues of $-\mathcal{G}^{(2)}$ and let $\{\varphi_j\}_{j\in\mathbb{N}} \subset L^2(E,\mathbf{m})$ be an orthogonal set of functions such that*

$$T_t^{(2)}\varphi_j = e^{-\lambda_j t}\varphi_j, \qquad \varphi_1 = \mathbb{1}, \qquad \sup_j \|\varphi_j\|_2 < \infty,$$

where $\|\cdot\|_2$ denotes the norm of the space $L^2(E,\mathbf{m})$. Then $\varphi_j \in \mathrm{C}(E)$ for all $j \in \mathbb{N}$, and the series $\sum_{j=1}^\infty \frac{1}{\|\varphi_j\|_2^2} e^{-\lambda_j t}\varphi_j(x)\,\varphi_j(y)\,\varphi_j(\xi)$ is absolutely convergent *for all $t > 0$ and $x, y, \xi \in E$. Moreover, the following assertions are equivalent:*

(i) *We have*

$$q_t(x, y, \xi) := \sum_{j=1}^{\infty} \frac{1}{\|\varphi_j\|_2^2} e^{-\lambda_j t} \varphi_j(x)\, \varphi_j(y)\, \varphi_j(\xi) \geq 0 \tag{5.4}$$

for all $t > 0$ and $x, y, \xi \in E$.

(ii) *For each $x, y \in E$ there exists a measure $\boldsymbol{\nu}_{x,y} \in \mathcal{P}(E)$ such that the product $\varphi_j(x)\, \varphi_j(y)$ admits the integral representation*

$$\varphi_j(x)\, \varphi_j(y) = \int_E \varphi_j(\xi)\, \boldsymbol{\nu}_{x,y}(d\xi), \qquad x, y \in E,\ j \in \mathbb{N}. \tag{5.5}$$

If these equivalent conditions hold and, in addition, there exists a $a \in E$ such that $\varphi_j(a) = 1$ for all $j \in \mathbb{N}$, then the bilinear operator $\diamond$ on $\mathcal{M}_{\mathbb{C}}(E)$ defined as

$$(\mu \diamond \nu)(d\xi) = \int_E \int_E \boldsymbol{\nu}_{x,y}(d\xi)\, \mu(dx)\, \nu(dy) \tag{5.6}$$

is an FLTC for $\{T_t\}$ with trivializing family $\Theta = \{\varphi_j\}_{j \in \mathbb{N}}$.

Proof Denote the inner product of the space $L^2(E, \mathbf{m})$ by $\langle \cdot, \cdot \rangle$. For each $\varepsilon > 0$ we have

$$|\varphi_j(x)| = e^{\lambda_j \varepsilon} |(T_\varepsilon^{(2)} \varphi_j)(x)| = e^{\lambda_j \varepsilon} \langle \varphi_j, p_\varepsilon(x, \cdot) \rangle \leq c_\varepsilon \sqrt{\mathbf{m}(E)}\, e^{\lambda_j \varepsilon} \|\varphi_j\|_2 < \infty \tag{5.7}$$

for $\mathbf{m}$-almost every $x \in E$, where $c_\varepsilon = \sup_{(x,y) \in E \times E} p_\varepsilon(x, y)$. This shows that the function φ_j belongs to the space $\mathrm{B_b}(E)$ (possibly after redefining φ_j on a $\mathbf{m}$-null set). Since $\{T_t\}$ is strong Feller (Proposition 5.3), it follows that $\varphi_j = e^{\lambda_j \varepsilon} T_\varepsilon \varphi_j \in \mathrm{C}(E)$. The assumption that $\sup_j \|\varphi_j\|_2 < \infty$, together with the estimate (5.7), ensures that the series $\sum_{j=1}^{\infty} \frac{1}{\|\varphi_j\|_2^2} e^{-\lambda_j t} \varphi_j(x)\, \varphi_j(y)\, \varphi_j(\xi)$ is absolutely convergent.

Suppose that (5.4) holds and fix $x, y \in E$. For $t > 0$, let $\boldsymbol{\nu}_{t,x,y} \in \mathcal{M}_+(E)$ be the measure defined by $\boldsymbol{\nu}_{t,x,y}(d\xi) = q_t(x, y, \xi)\mathbf{m}(d\xi)$. We have

$$\begin{aligned}
\int_E \varphi_j(\xi)\, \boldsymbol{\nu}_{t,x,y}(d\xi) &= \int_E \varphi_j(\xi) \sum_{k=1}^{\infty} \frac{1}{\|\varphi_j\|_2^2} e^{-\lambda_k t} \varphi_k(x)\, \varphi_k(y)\, \varphi_k(\xi)\, \mathbf{m}(d\xi) \\
&= \sum_{k=1}^{\infty} \frac{1}{\|\varphi_j\|_2^2} e^{-\lambda_k t} \varphi_k(x)\, \varphi_k(y)\, \langle \varphi_j, \varphi_k \rangle \\
&= e^{-\lambda_j t} \varphi_j(x)\, \varphi_j(y).
\end{aligned} \tag{5.8}$$

It then follows from (5.8) (with $j = 1$) that $\boldsymbol{\nu}_{t,x,y}(E) = 1$, so that

$$\boldsymbol{\nu}_{t,x,y} \in \mathcal{P}(E) \qquad \text{for all } t > 0,\; x, y \in E.$$

Now, let $\{t_n\}_{n\in\mathbb{N}}$ be an arbitrary decreasing sequence with $t_n \downarrow 0$. Since any uniformly bounded sequence of finite positive measures contains a vaguely convergent subsequence, there exists a subsequence $\{t_{n_k}\}$ and a measure $\boldsymbol{\nu}_{x,y} \in \mathcal{M}_+(E)$ such that $\boldsymbol{\nu}_{t_{n_k},x,y} \overset{v}{\longrightarrow} \boldsymbol{\nu}_{x,y}$ as $k \to \infty$. Let us show that all such subsequences $\{\boldsymbol{\nu}_{t_{n_k},x,y}\}$ have the same vague limit. Suppose that t_k^1, t_k^2 are two different sequences with $t_k^s \downarrow 0$ and that $\boldsymbol{\nu}_{t_k^s,x,y} \overset{v}{\longrightarrow} \boldsymbol{\nu}_{x,y}^s$ as $k \to \infty$ $(s = 1, 2)$. Recalling that E is compact, it follows that for all $h \in \mathrm{C}(E)$ and $\varepsilon > 0$ we have

$$\begin{aligned}
\int_E (T_\varepsilon h)(\xi)\, \boldsymbol{\nu}_{x,y}^s(d\xi) &= \lim_{k\to\infty} \int_E (T_\varepsilon h)(\xi)\, \boldsymbol{\nu}_{t_k^s,x,y}(d\xi) \\
&= \lim_{k\to\infty} \sum_{j=1}^\infty \frac{1}{\|\varphi_j\|_2^2} e^{-\lambda_j(t_k^s+\varepsilon)} \varphi_j(x)\, \varphi_j(y)\, \langle h, \varphi_j\rangle \\
&= \sum_{j=1}^\infty \frac{1}{\|\varphi_j\|_2^2} e^{\;\lambda_j\varepsilon} \varphi_j(x)\, \varphi_j(y)\, \langle h, \varphi_j\rangle,
\end{aligned}$$

where the second equality follows from the identities $\langle q_t(x, y, \cdot), \varphi_j\rangle = e^{-\lambda_j t}\varphi_j(x)\,\varphi_j(y)$ and $\langle T_\varepsilon h, \varphi_j\rangle = \langle h, T_\varepsilon\varphi_j\rangle = e^{-\lambda_j\varepsilon}\langle h, \varphi_j\rangle$. Consequently, we have

$$\int_E (T_\varepsilon h)(\xi)\, \boldsymbol{\nu}_{x,y}^1(d\xi) = \int_E (T_\varepsilon h)(\xi)\, \boldsymbol{\nu}_{x,y}^2(d\xi) \qquad \text{for all } \varepsilon > 0. \tag{5.9}$$

Since $h \in \mathrm{C}(E)$, by strong continuity of the Feller semigroup $\{T_t\}$ we have $\lim_{\varepsilon\downarrow 0} \|T_\varepsilon h - h\|_\infty = 0$, so by taking the limit $\varepsilon \downarrow 0$ in both sides of (5.9) we deduce that $\boldsymbol{\nu}_{x,y}^1(h) = \boldsymbol{\nu}_{x,y}^2(h)$, where $h \in \mathrm{C}(E)$ is arbitrary; therefore, $\boldsymbol{\nu}_{x,y}^1 = \boldsymbol{\nu}_{x,y}^2$. Thus all subsequences have the same vague limit, and from this we conclude that $\boldsymbol{\nu}_{t,x,y} \overset{v}{\longrightarrow} \boldsymbol{\nu}_{x,y}$ as $t \downarrow 0$. The product formula (5.5) is then obtained by taking the limit $t \downarrow 0$ in the leftmost and rightmost sides of (5.8).

Conversely, suppose that (5.5) holds for some measure $\boldsymbol{\nu}_{x,y} \in \mathcal{M}_+(E)$. Noting that for $h \in \mathrm{C}(E)$ we have

$$\begin{aligned}
\big\langle h, p_t(x, \cdot)\big\rangle = (T_t h)(x) &= \sum_{j=1}^\infty \frac{1}{\|\varphi_j\|_2^2} \langle T_t h, \varphi_j\rangle\, \varphi_j(x) \\
&= \sum_{j=1}^\infty \frac{1}{\|\varphi_j\|_2^2} e^{-\lambda_j t} \langle h, \varphi_j\rangle\, \varphi_j(x) \\
&= \Big\langle h, \sum_{j=1}^\infty \frac{1}{\|\varphi_j\|_2^2} e^{-\lambda_j t} \varphi_j(x)\varphi_j(\cdot)\Big\rangle,
\end{aligned}$$

we see that $p_t(x,y) = \sum_{j=1}^{\infty} \frac{1}{\|\varphi_j\|_2^2} e^{-\lambda_j t} \varphi_j(x)\, \varphi_j(y)$. Consequently, for $t > 0$ and $x, y \in E$ we have

$$q_t(x,y,\xi) = \sum_{j=1}^{\infty} \frac{1}{\|\varphi_j\|_2^2} e^{-\lambda_j t} \varphi_j(x) \int_E \varphi_j(z)\, \boldsymbol{\nu}_{y,\xi}(dz) = \int_E p_t(x,z)\, \boldsymbol{\nu}_{y,\xi}(dz) \geq 0,$$

because both the density $p_t(x,\boldsymbol{\cdot})$ and the measures $\boldsymbol{\nu}_{y,\xi}$ are nonnegative.

Finally, assume that $\varphi_j(a) = 1$ for all j and that (ii) holds. Let $\diamond$ be the operator defined by (5.6). To prove that $\Theta = \{\varphi_j\}_{j\in\mathbb{N}}$ satisfies condition III in Definition 5.1, it only remains to show that each $\mu \in \mathcal{M}_\mathbb{C}(E)$ is uniquely characterized by $\{\mu(\varphi_j)\}_{j\in\mathbb{N}}$. Indeed, if we take $\mu \in \mathcal{M}_\mathbb{C}(E)$ such that $\mu(\varphi_j) = 0$ for all j, then for $h \in \mathrm{C}(E)$ and $t > 0$ we have

$$\begin{aligned} \int_E (T_t h)(x)\, \mu(dx) &= \int_E \sum_{j=1}^{\infty} \frac{1}{\|\varphi_j\|_2^2} e^{-\lambda_j t} \langle h, \varphi_j\rangle\, \varphi_j(x)\, \mu(dx) \\ &= \sum_{j=1}^{\infty} e^{-\lambda_j t} \langle h, \varphi_j\rangle\, \mu(\varphi_j) \\ &= 0, \end{aligned}$$

and this implies that $\mu(h) = 0$ for all $h \in \mathrm{C}(E)$, so that $\mu \equiv 0$. Using the fact that Θ satisfies condition III, we can easily check that $\diamond$ is commutative, associative, bilinear and has identity element δ_a. It is also straightforward that $\|\mu \diamond \nu\| \leq \|\mu\| \cdot \|\nu\|$ and that $\mathcal{P}(E) \diamond \mathcal{P}(E) \subset \mathcal{P}(E)$. If $x_n \to x$ and $y_n \to y$, then

$$(\delta_{x_n} \diamond \delta_{y_n})(\varphi_j) = \varphi_j(x_n)\varphi_j(y_n) \longrightarrow \varphi_j(x)\varphi_j(y) = (\delta_x \diamond \delta_y)(\varphi_j) \qquad (j \in \mathbb{N})$$

and therefore (by compactness of E and a vague convergence argument similar to that of Remark 4.3.II) $\delta_{x_n} \diamond \delta_{y_n} \overset{w}{\longrightarrow} \delta_x \diamond \delta_y$; arguing as in the proof of Proposition 4.9 it follows that $(\mu, \nu) \mapsto \mu \diamond \nu$ is continuous in the weak topology. Noting that $p_{t,x}(\varphi_j) = e^{-\lambda_j t}\varphi_j(x) = p_{t,a}(\varphi_j)\delta_x(\varphi_j)$, we conclude that $\diamond$ is an FLTC for $\{T_t\}$.

Corollary 5.2 *In the conditions of Proposition 5.4, assume that the operator* $T_1^{(2)}$ *has simple spectrum (i.e. all the eigenvalues* $e^{-\lambda_j}$ *have multiplicity* 1*). Let* $\{(\lambda_j, \omega_j)\}_{j\in\mathbb{N}}$ *be the eigenvalue-eigenfunction pairs defined in Proposition 5.3. Then the following are equivalent:*

- ***(i)*** *There exists an FLTC for* $\{T_t\}_{t\geq 0}$*;*
- ***(ii)*** *There exists* $a \in E$ *such that* $|\omega_j(a)| = \|\omega_j\|_\infty$ *for all* $j \in \mathbb{N}$*, and the positivity condition* (5.4) *holds for the nonnormalized eigenfunctions* $\varphi_j(x) := \frac{\omega_j(x)}{\omega_j(a)}$.

Proof The implication (ii) $\Longrightarrow$ (i) follows from the final statement in Proposition 5.4.

Conversely, if (i) holds then the common maximizer property discussed above implies that $\Theta = \{\varphi_j\}_{j\in\mathbb{N}}$ where $\varphi_j(x) := \frac{\omega_j(x)}{\omega_j(a)}$; from this it follows (by condition III of Definition 5.1) that the product formula (5.5) holds with $\boldsymbol{\nu}_{x,y} = \delta_x \diamond \delta_y$ and therefore (by Proposition 5.4) the φ_j satisfy the positivity condition (5.4).

We note here that the assumption that $T_t^{(2)}$ (or, equivalently, the generator $\mathcal{G}^{(2)}$) has no eigenvalues with multiplicity greater than 1 is known to hold for many strong Feller semigroups of interest. In fact, it is proved in [81, Example 6.4] that the property that *all the eigenvalues of the Neumann Laplacian are simple* is a generic property in the set of all bounded connected C^2 domains $E \subset \mathbb{R}^d$. (The meaning of this is the following: given a bounded connected C^2 domain E, consider the collection of domains $\mathfrak{M}_3(E) = \{h(E) \mid h : E \longrightarrow \mathbb{R}^d$ is a C^3-diffeomorphism$\}$, which is a separable Banach space, see [81] for details concerning the appropriate topology. Let $\mathfrak{M}_{\text{simp}} \subset \mathfrak{M}_3(E)$ be the subspace of all $\widetilde{E} \in \mathfrak{M}_3(E)$ such that all the eigenvalues of the Neumann Laplacian on $\widetilde{E}$ are simple. Then $\mathfrak{M}_{\text{simp}}$ can be written as a countable intersection of open dense subsets of $\mathfrak{M}_3(E)$.) Similar results hold for the Laplace–Beltrami operator on a compact Riemannian manifold: it was proved in [189] that, given a compact manifold M, the set of Riemannian metrics g for which all the eigenvalues of the Laplace–Beltrami operator on (M, g) are simple is a generic subset of the space of Riemannian metrics on M.

However, one should not expect the property of simplicity of spectrum to hold for Euclidean domains or Riemannian manifolds with symmetries. For instance, if a bounded domain $E \subset \mathbb{R}^2$ is invariant under the natural action of the dihedral group $\boldsymbol{D}_n$, then one can show (see [79]) that the Dirichlet or Neumann Laplacian on E has infinitely many eigenvalues with multiplicity ≥ 2.

Remark 5.1 (Connection with Ultrahyperbolic Equations) Assume that there exists a dense orthogonal set of eigenfunctions $\{\varphi_j\}_{j\in\mathbb{N}}$ such that $\varphi_1 = \mathbb{1}$ and $\varphi_j(a) = \|\varphi_j\|_\infty = 1$ for all $j \in \mathbb{N}$. By the above, in order to prove the existence of an FLTC for $\{T_t\}$ we need to ensure that

$$Q_{t,h}(x,y) := \sum_{j=1}^{\infty} \frac{1}{\|\varphi_j\|_2^2} e^{-t\lambda_j} \varphi_j(x)\, \varphi_j(y)\, \langle h, \varphi_j\rangle \geq 0 \qquad \text{for all } h \in \mathrm{C}(E),\ t > 0.$$

The function $Q_{t,h}(x, y)$ is a solution of $\mathcal{G}_x^{(0)} u = \mathcal{G}_y^{(0)} u$ (where $\mathcal{G}_x^{(0)}$ is the Feller generator $\mathcal{G}^{(0)}$ acting on the variable x) satisfying the boundary condition $Q_{t,h}(x, a) = (T_t h)(x)$. Since the point a is a maximizer of all the eigenfunctions φ_j, the function $Q_{t,h}(x, y)$ also satisfies (at least formally) the boundary condition $(\nabla_y Q_{t,h})(x, a) = 0$. (This could be justified e.g. by proving that the series can be differentiated term by term. In the next section we will see that this argument can be applied to the Neumann Laplacian on suitable bounded domains of $\mathbb{R}^d$.) This indicates that, as in Sect. 4.3.1, the (positivity) properties of the boundary value problem

$$\mathcal{G}_x^{(0)} u = \mathcal{G}_y^{(0)} u, \qquad u(x,a) = u_0(x), \qquad (\nabla_y u)(x,a) = 0 \tag{5.10}$$

are related with the problem of constructing a convolution associated with the given strong Feller semigroup.

Consider Examples 5.1(b)–(c) or, more generally, any example of a strong Feller semigroup generated by a uniformly elliptic differential operator on $E \subset \mathbb{R}^d$ ($d > 2$). In this context, the principal part of the differential operator $\mathcal{G}_x^{(0)} - \mathcal{G}_y^{(0)}$ has d terms $\frac{\partial^2}{\partial x_j^2}$ with positive coefficient and d terms $\frac{\partial^2}{\partial y_j^2}$ with negative coefficient. Such partial differential operators are often said to be of *ultrahyperbolic type* (cf. e.g. [150, §I.5] and [161, Definition 2.6]). According to the results of [41] and [40, §VI.17], the solution for the boundary value problem (5.10) is, in general, not unique. The existing theory on well-posedness of ultrahyperbolic boundary value problems is, in many other respects, rather incomplete; in particular, as far as we know, no maximum principles have been determined for such problems. Adapting the integral equation technique which was used in Chap. 4 to problems determined by Feller semigroups on multidimensional spaces is, therefore, a highly nontrivial problem.

5.2 Nonexistence of Convolutions: Diffusion Processes on Bounded Domains

The results of the previous section show that the existence of an FLTC for a given Feller process depends on two conditions—the common maximizer property and the positivity of an ultrahyperbolic boundary value problem—for which there are no reasons to hope that they can be established other than in special cases. In particular, the common maximizer property becomes less natural when we move from one-dimensional diffusions to multidimensional diffusions: while in the first case it is natural that the properties of a differential operator enforce one of the endpoints of the interval to be a common maximizer, this is no longer the case on a bounded domain of $\mathbb{R}^d$ with differentiable boundary because one no longer expects that one of the points of the boundary will play a special role.

In fact, as we will see in Sect. 5.2.3, under certain conditions one can prove that (reflected) Brownian motions on bounded domains of $\mathbb{R}^d$ or on compact Riemannian manifolds (cf. Examples 5.1(a) and (c) respectively) do not satisfy the common maximizer property and, therefore, it is not possible to construct an associated FLTC. In Sect. 5.2.1 we start by presenting some examples which allow us to understand the geometrical properties of the eigenfunctions that are usually encountered in the multidimensional case. Section 5.2.2 contains some useful auxiliary results.

5.2.1 Special Cases and Numerical Examples

The first example illustrates the fact that the construction of the convolution becomes trivial if the generator of the multidimensional diffusion decomposes trivially (via separation of variables) into one-dimensional Sturm–Liouville operators for which an associated FLTC exists.

Example 5.2 (Neumann Eigenfunctions of a d-Dimensional Rectangle) Consider the d-dimensional rectangle $E = [0, \beta_1] \times \ldots \times [0, \beta_d] \subset \mathbb{R}^d$. The (nonnormalized) eigenfunctions of the Neumann Laplacian on E and the associated eigenvalues are given by

$$\varphi_{j_1,\ldots,j_d}(x_1,\ldots,x_d) = \prod_{\ell=1}^{d} \cos\Bigl(\pi j_\ell \frac{x_\ell}{\beta_\ell}\Bigr), \qquad \lambda_{j_1,\ldots,j_d} = \pi^2 \sum_{\ell=1}^{d} \frac{j_\ell^2}{\beta_\ell^2}$$

($j_1, \ldots, j_d \in \mathbb{N}_0$). These eigenfunctions constitute an orthogonal basis of $L^2(E, dx)$. The point $(0, \ldots, 0)$ is, obviously, a maximizer of all the functions $\varphi_{j_1,\ldots,j_d}$, thus the common maximizer property holds. Moreover, we can trivially construct an FLTC for the (reflected Brownian) semigroup generated by the Neumann Laplacian: indeed, it is easy to check that the product of the hypergroups $([0, \beta_1], \underset{\beta_1}{\circledcirc}), \ldots, ([0, \beta_d], \underset{\beta_d}{\circledcirc})$ satisfies all the requirements of Definition 5.1. (The product of the hypergroups is taken as in Sect. 2.3; recall also that $([0, \beta_\ell], \underset{\beta_\ell}{\circledcirc})$ is the Sturm–Liouville hypergroup of compact type associated with the operator $\frac{d^2}{dx_\ell^2}$ on $[0, \beta_\ell]$, cf. Remark 4.2.)

Example 5.3 (Neumann Eigenfunctions of Disks and Balls) Let $E \subset \mathbb{R}^2$ be the closed disk of radius R. It is well-known that the eigenfunctions of the Neumann Laplacian on E are given, in polar coordinates, by

$$\varphi_{0,k}(r,\theta) = J_0(j'_{0,k}\tfrac{r}{R}),$$
$$\varphi_{m,k,1}(r,\theta) = J_m(j'_{m,k}\tfrac{r}{R})\cos(m\theta),$$
$$\varphi_{m,k,2}(r,\theta) = J_m(j'_{m,k}\tfrac{r}{R})\sin(m\theta),$$

where $m, k \in \mathbb{N}$ and $j'_{m,k}$ stands for the k-th (simple) zero of the derivative of the Bessel function of the first kind $J_m(\cdot)$ (see [104, Section 7.2] and [80, Proposition 2.3]). The corresponding eigenvalues are $\lambda_{0,k} = (j'_{0,k}/R)^2$ (with multiplicity 1) and $\lambda_{m,k} = (j'_{m,k}/R)^2$ (with multiplicity 2). A classical result on the Bessel function [200, pp. 485, 488] ensures that for $m \geq 1$ we have $|J_m(j'_{m,k})| > |J_m(x)|$ for all $x > j'_{m,k}$, hence the eigenfunctions $\varphi_{m,k,1}$ and $\varphi_{m,k,2}$ ($m \geq 1$) attain their global maximum on the circle $\{r = \frac{j'_{m,1}}{j'_{m,k}} R\}$. This shows, in particular, that no orthogonal basis of $L^2(E, dx)$ composed of Neumann eigenfunctions can satisfy the common maximizer property.

More generally, if $E \subset \mathbb{R}^d$ is a closed d-ball with radius R, then the eigenfunctions of the Neumann Laplacian on E are

$$\varphi_{m,k}(r, \theta) = r^{1-\frac{d}{2}} J_{m-1+\frac{d}{2}}(c_{m,k} \tfrac{r}{R})\, H_m(\theta),$$

where (r, θ) are hyperspherical coordinates, $m \in \mathbb{N}_0$, $k \in \mathbb{N}$, H_m is a spherical harmonic of order m (see [104]) and $c_{m,k}$ is the k-th zero of the function $\xi \mapsto (1-\frac{d}{2})J_{m-1+\frac{d}{2}}(\xi) + \xi J'_{m-1+\frac{d}{2}}(\xi)$. The corresponding eigenvalues are $\lambda_{m,k} = c_{m,k}^2$, whose multiplicity is equal to the dimension of the space of spherical harmonics of order m. By similar arguments we conclude that the common maximizer property does not hold.

Example 5.4 (Neumann Eigenfunctions of a Circular Sector) Let $E \subset \mathbb{R}^2$ be the sector of angle $\frac{\pi}{q}$, $E = \{(r \cos\theta, r \sin\theta) \mid 0 \le r \le 1,\ 0 \le \theta \le \frac{\pi}{q}\}$, where $q \in \mathbb{N}$. The eigenfunctions of the Neumann Laplacian on E and the associated eigenvalues (which have multiplicity 1, cf. [20]) are given by

$$\varphi_{m,k}(r, \theta) = \cos(qm\theta)\, J_{qm}(j'_{qm,k} \tfrac{r}{R}), \qquad \lambda_{m,k} = (j'_{qm,k}/R)^2.$$

As in the previous example it follows that the global maximizer of $\varphi_{m,k}$ lies in the arc $\bigl\{r = \frac{j'_{qm,1}}{j'_{qm,k}} R\bigr\}$, so that the common maximizer property does not hold.

Example 5.5 (Neumann Eigenfunctions of a Circular Annulus) If $E \subset \mathbb{R}^2$ is the annulus $\{(r, \theta) \mid r_0 \le r \le R,\ 0 \le \theta < 2\pi\}$, where $0 < r_0 < R < \infty$, then the Neumann eigenfunctions on E are

$$\begin{aligned}
\varphi_{0,k}(r, \theta) &= J_0(c_{0,k} \tfrac{r}{R}) Y_0'(c_{0,k}) - J_0'(c_{0,k}) Y_0(c_{0,k} \tfrac{r}{R}),\\
\varphi_{m,k,1}(r, \theta) &= \bigl(J_m(c_{m,k} \tfrac{r}{R}) Y_m'(c_{m,k}) - J_m'(c_{m,k}) Y_m(c_{m,k} \tfrac{r}{R})\bigr) \cos(m\theta), \qquad (5.11)\\
\varphi_{m,k,2}(r, \theta) &= \bigl(J_m(c_{m,k} \tfrac{r}{R}) Y_m'(c_{m,k}) - J_m'(c_{m,k}) Y_m(c_{m,k} \tfrac{r}{R})\bigr) \sin(m\theta),
\end{aligned}$$

where $m, k = 1, 2, \ldots$, $Y_m(\cdot)$ is the Bessel function of the second kind [145, §10.2] and $c_{m,k}$ is the k-th zero of the function $\xi \mapsto J_m'(\frac{r_0}{R}\xi) Y_m'(\xi) - J_m'(\xi) Y_m'(\frac{r_0}{R}\xi)$. The associated eigenvalues are $\lambda_{m,k} = (c_{m,k}^2/R^2)$. Figure 5.1 presents the contour plots of some of the Neumann eigenfunctions, obtained in two different ways: in panel (a) using the explicit representations (5.11), where the constants $c_{m,k}$ are computed numerically with the help of the `NSolve` function of *Wolfram Mathematica*; and in panel (b) using a numerical approximation of the eigenvalues and eigenfunctions which was computed via the `NDEigensystem` routine of *Wolfram Mathematica*. Since the eigenvalues $\lambda_{m,k}$ with $m \ge 1$ have multiplicity 2, the plots obtained by these two approaches differ by a rotation. The results indicate that some of the eigenfunctions (those associated with the first zero $c_{m,1}$) attain their maximum at the outer circle $\{r = R\}$, while other eigenfunctions (those associated with the higher zeros $c_{m,k}$, $k \geq 2$) attain their maximum either at the inner circle $\{r = r_0\}$ or at the interior of the annulus. It is therefore clear that the Neumann eigenfunctions do not satisfy the common maximizer property.

There are few other examples of domains of $\mathbb{R}^d$ for which the Neumann eigenfunctions can be computed in closed form. However, in the general case of an arbitrary domain $E \subset \mathbb{R}^2$ it is still possible to assess whether the common maximizer property holds by analysing the contour plots of the eigenfunctions; these can be computed, for a given bounded domain of $\mathbb{R}^2$, by the same procedure which was used to produce the plots in panel (b) of Fig. 5.1.

This is illustrated in Figs. 5.2 and 5.3, which present the contour plots of the first eigenfunctions of two non-symmetric bounded regions of $\mathbb{R}^2$ with smooth boundary. As we can see, the eigenfunctions attain their maximum values at different points which lie either on the boundary or at the interior of the domain. Note also that the

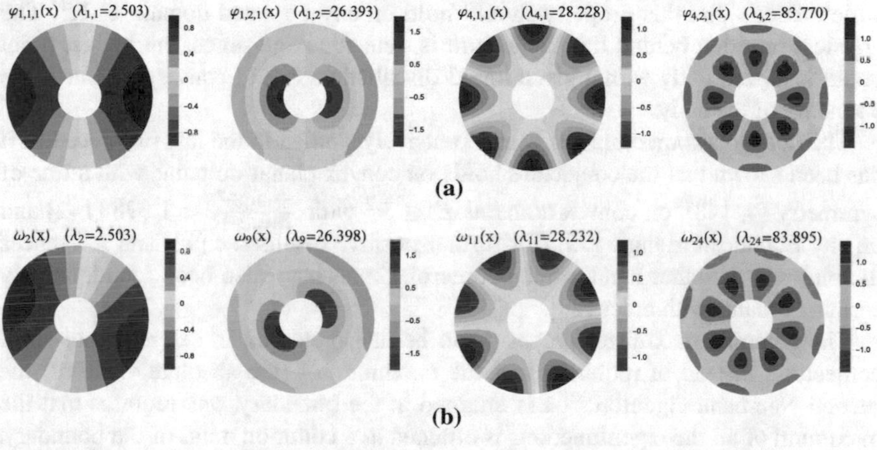

Fig. 5.1 Contour plots of the Neumann eigenfunctions of a circular annulus with inner radius $r_0 = 0.3$ and outer radius $R = 1$. (**a**) Closed form expressions. (**b**) Numerical approximation. In panel (**b**), the notation ω_k refers to the orthogonal eigenfunction associated with the k-th largest eigenvalue λ_k. In both panels the eigenfunctions were normalized so that their L^2 norm equals 1. Similar results were obtained for other values of $\frac{r_0}{R}$

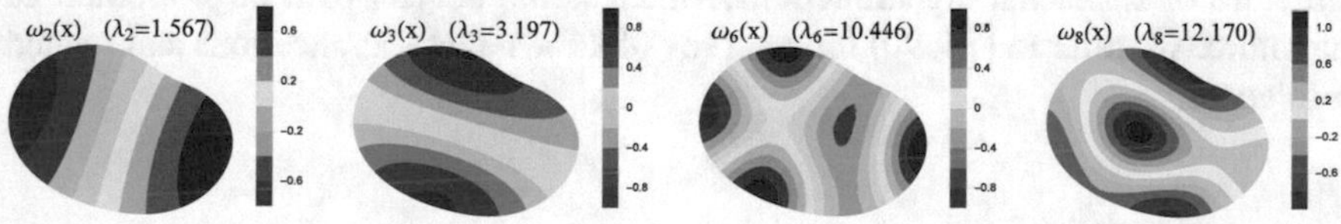

Fig. 5.2 Contour plots of some eigenfunctions of a region obtained by a non-symmetric deformation of an ellipse. (As above, we denote by ω_k the Neumann eigenfunction associated with the k-th largest eigenvalue λ_k, and the plots were produced using the `NDEigensystem` function of *Wolfram Mathematica*)

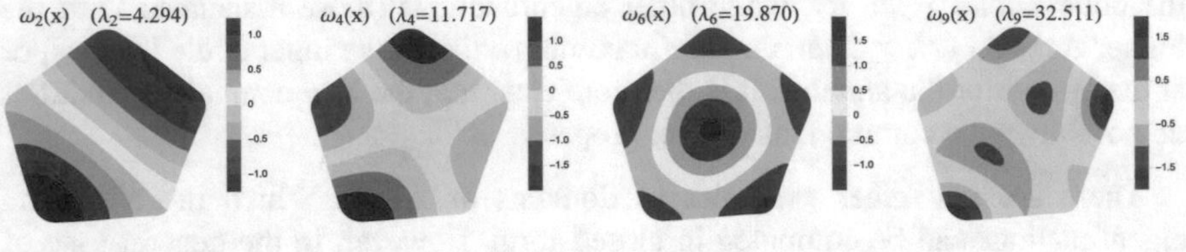

Fig. 5.3 Contour plots of the Neumann eigenfunctions of a region obtained by a non-symmetric deformation of a pentagon with smoothed corners

associated eigenvalues are simple, which is unsurprising since the domain has no symmetries (cf. comment after Corollary 5.2).

Remark 5.2 (Connection with the hot spots conjecture) All the examples presented above have the property that *if φ_2 is an eigenfunction associated with the smallest nonzero Neumann eigenvalue λ_2, then the maximum and minimum of φ_2 are attained at the boundary ∂E.* This is the so-called *hot spots conjecture* of J. Rauch, which asserts that this property should hold on any bounded domain of $\mathbb{R}^d$. The physical intuition behind this conjecture is that, for large times, the hottest point on an insulated body with a given initial distribution should converge towards the boundary of the body.

The hot spots conjecture has been extensively studied in the last two decades: it has been shown that the conjecture holds on convex planar domains with a line of symmetry [4, 148], on convex domains $E \subset \mathbb{R}^2$ with $\frac{\operatorname{diam}(E)^2}{\operatorname{Area}(E)} < 1.378$ [142] and on any Euclidean triangle [95] (for further positive results see [95] and references therein). On the other hand, some counterexamples have also been found, namely certain domains with holes [28].

The common maximizer property can be interpreted as an extended hot spots conjecture: instead of requiring that the maximum of (the absolute value of) the second Neumann eigenfunction is attained at the boundary, one requires that the maximum of all the eigenfunctions is attained at a common point of the boundary. The negative result of Corollary 5.3 below shows that the location of the hottest point in the limiting distribution (as time goes to infinity) of the temperature of an insulated body depends on the initial temperature distribution.

The common maximizer property and the hot spots conjecture are subtopics of the more general problem of understanding the topological and geometrical structure of Laplacian eigenfunctions, which is the subject of a huge amount of literature. We refer to [74, 90] for a survey of known facts, applications and related references.

5.2.2 Some Auxiliary Results

Before moving on to a discussion of the common maximizer property on more general spaces, we collect some results from functional analysis and PDE theory which will be needed in the ensuing analysis. We begin by stating a version of the Sobolev embedding theorem and of the trace (and inverse trace) theorem.

Theorem 5.1

(a) *[203, Corollary 6.1] Let $E \subset \mathbb{R}^d$ be the closure of a bounded convex domain, and let $j, k \in \mathbb{N}_0$ with $k - j > \frac{d}{2}$. Then any function in $H^k(E)$ is j times continuously differentiable, and the embedding $H^k(E) \hookrightarrow \mathrm{C}^j(E)$ is continuous.*

(b) *[75, Theorem 7.1] Let (E, g) be a compact Riemannian manifold of dimension d, and let $j, k \in \mathbb{N}$ with $k > \frac{j}{2} + \frac{d}{4}$. Then any function $f \in H^{2k}(E) := \{u \mid u, \Delta u, \ldots, \Delta^k u \in L^2(E, \mathbf{m})\}$ belongs to $\mathrm{C}^j(E)$, and the embedding $H^{2k}(E) \hookrightarrow \mathrm{C}^j(E)$ is continuous.*

Theorem 5.2 ([203, Theorems 8.7 and 8.8]) *Let $k \in \mathbb{N}$ and let $E \subset \mathbb{R}^d$ be the closure of a bounded C^{k+1} domain. There exists a linear continuous* trace operator $\mathrm{Tr}_{\partial E} : H^k(E) \longrightarrow \prod_{j=0}^{k-1} H^{k-j-1/2}(\partial E)$ *such that*

$$\mathrm{Tr}_{\partial E} f = \left(f\,|_{\partial E}, \frac{\partial f}{\partial \boldsymbol{n}}\bigg|_{\partial E}, \ldots, \frac{\partial^{k-1} f}{\partial \boldsymbol{n}^{k-1}}\bigg|_{\partial E} \right) \qquad \text{for } f \in \mathrm{C}^{2k-1}(E),$$

where $\boldsymbol{n}$ denotes the unit outer normal vector orthogonal to ∂E, and $H^{k-j-1/2}(\partial E)$ is the fractional Sobolev space defined as the set of functions $\varphi \in H^{k-j-1}(\partial E)$ such that for each $\alpha = (\alpha_1, \ldots, \alpha_d)$ with $|\alpha| = k - j - 1$ we have

$$\int_{\mathbb{R}^d \times \mathbb{R}^d} \frac{|\partial^\alpha \varphi(x) - \partial^\alpha \varphi(y)|^2}{|x - y|^{d+1}} dx\, dy < \infty.$$

In addition, there exists a linear continuous extension operator $\mathcal{Z} : \prod_{j=0}^{k-1} H^{k-j-1/2}(\partial E) \longrightarrow H^k(E)$ such that

$$\mathrm{Tr}_{\partial E}(\mathcal{Z}\varphi) = \varphi \qquad \text{for all } \varphi \in \prod_{j=0}^{k-1} H^{k-j-1/2}(\partial E).$$

We will also make use of the following upper bounds on the heat kernel (i.e. the transition density of the semigroup $\{T_t^{(2)}\}$ generated by the Neumann Laplacian or the Laplace–Beltrami operator) and on its gradient:

Proposition 5.5 *Assume that $E \subset \mathbb{R}^d$ is the closure of a bounded Lipschitz domain, and let $p_t(x, y)$ be the heat kernel determined by the Neumann Laplacian on E.*

(a1) *[8, Theorem 3.1] There exist constants* $c_1, c_2 > 0$ *such that*

$$p_t(x,y) \leq c_1\, t^{-d/2} \exp\!\left(-\frac{|x-y|^2}{c_2\, t}\right) \qquad \textit{for all } t > 0.$$

(a2) *[199, Lemma 3.1] There exist constants* $c_3, c_4 > 0$ *such that*

$$|\nabla_y p_t(x,y)| \leq c_3\, t^{-(d+1)/2} \exp\!\left(-\frac{|x-y|^2}{c_4\, t}\right) \qquad \textit{for all } t > 0.$$

Assume now that (E, g) *is a compact Riemannian manifold of dimension* d*, let* $p_t(x, y)$ *be the heat kernel determined by the Laplace–Beltrami operator on* E*, and let* $\mathbf{d}$ *be the Riemannian distance function.*

(b1) *[75, Corollary 15.17] There exist constants* $c_5, c_6 > 0$ *such that*

$$p_t(x,y) \leq c_5\, t^{-d/2} \exp\!\left(-\frac{\mathbf{d}(x,y)^2}{c_6\, t}\right) \qquad \textit{for all } t > 0.$$

(b2) *[87] There exist constants* $c_7, c_8 > 0$ *such that*

$$|\nabla_y p_t(x,y)| \leq c_7\, t^{-(d+1)/2} \exp\!\left(-\frac{\mathbf{d}(x,y)^2}{c_8\, t}\right) \qquad \textit{for all } t > 0.$$

Lastly, we present a version of Mercer's theorem, which provides an uniformly convergent eigenfunction expansion representation for positive-definite integral kernels:

Theorem 5.3 ([170, Theorem 3.11.9]) *Let* E *be a compact metric space endowed with a positive Borel measure* μ*. Assume* $L^2(E, \mu)$ *is infinite-dimensional. Let* $K \in \mathrm{C}(E \times E)$*, let* $A_K : L^2(E, \mu) \longrightarrow L^2(E, \mu)$ *be the operator defined as*

$$(A_K f)(x) = \int_E K(x,y)\, f(y)\, \mu(dy),$$

and suppose that A_K *is a positive operator. Then the eigenvalues* $\{\lambda_j\}_{j\in\mathbb{N}}$ *of* A_K *are associated with continuous eigenfunctions* $\{\varphi_j\}_{j\in\mathbb{N}} \subset \mathrm{C}(E) \cap L^2(E, \mu)$*, and we have*

$$K(x,y) = \sum_{j=1}^{\infty} \lambda_j\, \varphi_j(x) \overline{\varphi_j(y)},$$

where the series converges absolutely and uniformly on $E \times E$*.*

5.2.3 Eigenfunction Expansions, Critical Points and Nonexistence Theorems

We are now ready to discuss the (failure of the) common maximizer property for reflected Brownian motions on general bounded domains of $\mathbb{R}^d$ $(d \geq 2)$. We begin with two important observations, the first of which is quite obvious: if a is a common maximizer for the eigenfunctions $\{\varphi_j\}_{j\in\mathbb{N}}$, then it is a common critical point, i.e. we have $(\nabla\varphi_j)(a) = 0$ for all j. The second observation is that the usual eigenfunction expansion

$$f = \sum_{j=1}^{\infty} \frac{1}{\|\varphi_j\|_2^2} \langle f, \varphi_j \rangle \, \varphi_j \tag{5.12}$$

suggests that the point a will also be a critical point of any function f which is sufficiently regular so that the expansion (5.12) is convergent in the pointwise sense and can be differentiated term by term. Thus if we prove that such pointwise convergence and differentiation is admissible for a class of functions whose derivatives are not restricted to vanish at any given point, then the common maximizer property cannot hold. The next proposition and corollary make this rigorous.

Proposition 5.6 *Let $E \subset \mathbb{R}^d$ be the closure of a bounded convex domain. Let $\{X_t\}$ be the reflected Brownian motion on E, let $\{\omega_j\}_{j\in\mathbb{N}}$ be an orthonormal basis of $L^2(E) \equiv L^2(E, dx)$ consisting of eigenfunctions of the Neumann Laplacian $-\mathcal{G}^{(2)} \equiv -\Delta_N : \mathcal{D}(\Delta_N) \longrightarrow L^2(E)$ and let $0 \leq \lambda_1 \leq \lambda_2 \leq \lambda_3 \leq \ldots$ be the associated eigenvalues. Let $m \in \mathbb{N}$, $m > \frac{d}{2} + 1$ and let $h \subset H^m(E)$ be a function such that $\Delta^k h \in \mathcal{D}(\Delta_N)$ for $k = 0, 1, \ldots, m-1$. Then*

$$h(x) = \sum_{j=0}^{\infty} \langle h, \omega_j \rangle \, \omega_j(x) \quad \textit{and} \quad (\nabla h)(x) = \sum_{j=0}^{\infty} \langle h, \omega_j \rangle (\nabla\omega_j)(x) \quad \textit{for all } x \in E, \tag{5.13}$$

where both series converge absolutely and uniformly on E.

Proof Let $\{T_t^{(2)}\}_{t\geq 0}$ and $\{\mathcal{R}_\eta^{(2)}\}_{\eta>0}$ be, respectively, the strongly continuous semigroup and resolvent on $L^2(E)$ generated by the Neumann Laplacian and let $p_t(x, y)$ be the Neumann heat kernel, i.e. the transition density of the semigroup $\{T_t^{(2)}\}$. Using the Sobolev embedding theorem (Theorem 5.1(a)), one can prove (cf. [43, proof of Theorem 5.2.1]) that the heat kernel is C^∞ jointly in the variables $(t, x, y) \in (0, \infty) \times E \times E$. Denote by $\partial_{\vec{v},x}$ the directional derivative with respect to the variable $x \in \mathbb{R}^d$ in a given direction $\vec{v} \in \mathbb{R}^d \setminus \{0\}$. By Proposition 5.5, there exist constants $c_1, c_2, c_3, c_4 > 0$ such that the following estimates hold:

$$p_t(x,y) \le c_1\, t^{-d/2} \exp\biggl(-\frac{|x-y|^2}{c_2\, t}\biggr), \tag{5.14}$$

$$|\partial_{\vec{v},y}\, p_t(x,y)| \le c_3\, t^{-(d+1)/2} \exp\biggl(-\frac{|x-y|^2}{c_4\, t}\biggr). \tag{5.15}$$

Using the basic semigroup identity for the heat kernel, we obtain

$$\begin{aligned} |\partial_{\vec{v},x}\partial_{\vec{v},y}\, p_t(x,y)| &\le \int_E |\partial_{\vec{v},x}\, p_{t/2}(x,\xi)\, \partial_{\vec{v},y}\, p_{t/2}(\xi,y)|\, d\xi \\ &\le c_5\, t^{-(d+1)} \exp\biggl(-\frac{|x-y|^2}{2c_4\, t}\biggr). \end{aligned} \tag{5.16}$$

Next we recall that [42, Problem 2.9]

$$\begin{aligned} (\mathcal{R}_\alpha^{(2)})^k h &\equiv (\alpha - \Delta_N)^{-k} h & & (h \in L^2(E),\ \alpha > 0,\ k = 1, 2, \ldots), \\ &= \frac{1}{(k-1)!} \int_0^\infty e^{-\alpha t}\, t^{k-1}\, T_t^{(2)} h\, dt & & \end{aligned}$$

and therefore the k-th power $(\mathcal{R}_\alpha^{(2)})^k$ of the resolvent is an integral operator with kernel

$$G_{\alpha,k}(x,y) = \frac{1}{(k-1)!} \int_0^\infty e^{-\alpha t}\, t^{k-1}\, p_t(x,y)\, dt. \tag{5.17}$$

If $k = 2m > d+1$, then using the estimate (5.14) we see that $G_{\alpha,2m}(x,x) < \infty$ and, furthermore, $G_{\alpha,2m}$ is a continuous function of $(x,y) \in E \times E$. Since E is compact and $(\mathcal{R}_\alpha^{(2)})^{2m} : L^2(E) \longrightarrow L^2(E)$ is nonnegative and has a continuous kernel, an application of Mercer's theorem (Theorem 5.3) yields that the kernel $G_{\alpha,2m}$ can be represented by the absolutely and uniformly convergent spectral expansion

$$G_{\alpha,2m}(x,y) = \sum_{j=1}^\infty \frac{\omega_j(x)\omega_j(y)}{(\alpha+\lambda_j)^{2m}}, \tag{5.18}$$

where the series converges absolutely and uniformly in $(x,y) \in E \times E$. (Note that $\bigl((\alpha+\lambda_j)^{-2m}, \omega_j\bigr)$ are the eigenvalue-eigenfunction pairs for $\mathcal{R}_\alpha^{(2)}$.) In addition, it follows from (5.17) and the estimates (5.15)–(5.16) that

$$\partial_{\vec{v},x}\partial_{\vec{v},y}\, G_{\alpha,2m}(x,y) = \frac{1}{(2m-1)!} \int_0^\infty e^{-\alpha t}\, t^{2m-1}\, \partial_{\vec{v},x}\partial_{\vec{v},y}\, p_t(x,y)\, dt,$$

where the integral converges absolutely and uniformly and defines a continuous function of $(x, y) \in E \times E$. (The function $\partial_{\vec{v},y} G_{\alpha,2m}$ is also continuous on $E \times E$.) Using standard arguments (cf. [144, §21.2, proof of Corollary 3]), one can then deduce from (5.18) that

$$\partial_{\vec{v},x} \partial_{\vec{v},y} G_\alpha^{(2m)}(x, y) = \sum_{j=1}^{\infty} \frac{(\partial_{\vec{v}}\omega_j)(x)(\partial_{\vec{v}}\omega_j)(y)}{(\alpha + \lambda_j)^{2m}}, \tag{5.19}$$

again with absolute and uniform convergence in $(x, y) \in E \times E$.

Let $h \in H^m(E)$ be such that $\Delta^k h \in \mathcal{D}(\Delta_N)$ for $k = 0, 1, \ldots, m-1$, and write $h = \mathcal{R}_\alpha^m g$ where $g := (\alpha - \Delta_N)^m h \in L^2(E)$. Since $m > \frac{d}{2} + 1$, we have $h \in C^1(E)$ by the Sobolev embedding theorem. We thus have

$$\begin{aligned}
\sum_{j=0}^{\infty} \big|\langle h, \omega_j\rangle\, \omega_j(x)\big| &= \sum_{j=0}^{\infty} \frac{|\langle g, \omega_j\rangle|}{(\alpha + \lambda_j)^m} |\omega_j(x)| \\
&\le \Big(\sum_{j=0}^{\infty} |\langle g, \omega_j\rangle|^2\Big)^{\frac{1}{2}} \cdot \Big(\sum_{j=0}^{\infty} \frac{|\omega_j(x)|^2}{(\alpha + \lambda_j)^{2m}}\Big)^{\frac{1}{2}} \\
&= \|g\| \cdot \big(G_\alpha^{(2m)}(x, x)\big)^{1/2} < \infty,
\end{aligned}$$

and similarly

$$\sum_{j=0}^{\infty} \big|\langle h, \omega_j\rangle (\partial_{\vec{v}}\omega_j)(x)\big| \le \|g\| \cdot \big|\partial_{\vec{v},x} \partial_{\vec{v},y} G_\alpha^{(2m)}(x, x)\big|^{1/2} < \infty.$$

This shows that the series in the right-hand sides of (5.13) converge absolutely and uniformly in $x \in E$, and the result immediately follows.

Corollary 5.3 (Nonexistence of Common Critical Points) *Let $m \in \mathbb{N}$, $m > \frac{d}{2} + 1$ and let $E \subset \mathbb{R}^d$ $(d \ge 2)$ be the closure of a bounded convex domain with C^{2m+2} boundary ∂E. Let $\{\omega_j\}_{j \in \mathbb{N}}$ be an orthonormal basis of $L^2(E)$ consisting of eigenfunctions of $-\Delta_N$. Then for each $x_0 \in E$ there exists $j \in \mathbb{N}$ such that $(\nabla\omega_j)(x_0) \ne 0$.*

Proof If $x_0 \in \mathring{E}$, it is clearly possible to choose $h \in C_c^\infty(E) \subset \{u \in H^m(E) \mid u, \Delta u, \ldots, \Delta^{m-1} u \in \mathcal{D}(\Delta_N)\}$ such that $(\nabla h)(x_0) \ne 0$. If x_0 belongs to ∂E, let $\vec{v} \in T_{\partial E}(x_0) \setminus \{0\}$ and choose $\varphi \in C^\infty(\partial E)$ such that $d\varphi_{x_0}(\vec{v}) \ne 0$. Combining the trace theorem (Theorem 5.2) with the Sobolev embedding theorem, we find that that there exists $h \in H^{2m}(E) \subset C^1(E)$ such that

$$h|_{\partial E} = \varphi \qquad \text{and} \qquad \mathrm{Tr}_{\partial E}\Big(\frac{\partial^j h}{\partial \boldsymbol{n}^j}\Big) = 0, \quad j = 1, 2, \ldots, 2m-1.$$

Consequently, h is such that $(\nabla h)(x_0) \neq 0$ and $h, \Delta h, \ldots, \Delta^{m-1} h \in \mathcal{D}(\Delta_N) = \{u \in H^2(E) \mid \mathrm{Tr}_{\partial E}(\frac{\partial h}{\partial n}) = 0\}$. (This characterization of $\mathcal{D}(\Delta_N)$ is well-known, see [167, Section 10.6.2].) Therefore, given any $x_0 \in E$ we can apply Proposition 5.6 to the function h defined above to conclude that

$$\sum_{j=0}^{\infty} \langle h, \omega_j \rangle (\nabla \omega_j)(x_0) = (\nabla h)(x_0) \neq 0,$$

which implies that $(\nabla \omega_j)(x_0) \neq 0$ for at least one j.

The conclusions of Proposition 5.6 and Corollary 5.3 are also valid for the eigenfunctions of the Laplace–Beltrami operator on a compact Riemannian manifold (Example 5.1(a)):

Proposition 5.7 (Nonexistence of Common Critical Points on Compact Riemannian Manifolds) *Let (E, g) be a compact Riemannian manifold (without boundary) of dimension d and $\{\omega_j\}_{j \in \mathbb{N}}$ an orthonormal basis of $L^2(E, \mathbf{m})$ consisting of eigenfunctions of the Laplace–Beltrami operator on (E, g). Then* (5.13) *holds for all functions $h \in H^{2m}(E)$ ($m \in \mathbb{N}$, $m > \frac{d}{4} + \frac{1}{2}$), with the series converging absolutely and uniformly. Furthermore, for each $x_0 \in E$ there exists $j \in \mathbb{N}$ such that $(\nabla \omega_j)(x_0) \neq 0$.*

Proof It is well-known that the heat kernel $p_t(x, y)$ for the Laplace–Beltrami operator is C^∞ jointly in the variables $(t, x, y) \in (0, \infty) \times E \times E$ [43, Theorem 5.2.1]. By Proposition 5.5, the heat kernel $p_t(x, y)$ and its gradient satisfy, for $0 < t \leq 1$ and $x, y \in E$, the upper bounds

$$p_t(x, y) \leq c_1 t^{-d/2} \exp\biggl(-\frac{\mathbf{d}(x, y)^2}{c_2 t}\biggr), \qquad |\nabla_y p_t(x, y)| \leq c_3 t^{-(d+1)/2} \exp\biggl(-\frac{\mathbf{d}(x, y)^2}{c_4 t}\biggr).$$

Let $U \subset E$ be a coordinate neighbourhood. Arguing as in the proof of Proposition 5.6, we find that for $x, y \in U$ the kernel of the $2m$-th power of the resolvent admits the spectral representation (5.18) and can be differentiated term by term as in (5.19). (The directional derivatives are defined in local coordinates.) By the Sobolev embedding theorem (Theorem 5.1(b)), $H^{2m}(E) \subset \mathrm{C}^1(E)$; therefore, the estimation carried out above yields that the expansions (5.13) hold. We have $\partial E = \emptyset$, thus for each $x_0 \in E$ we can choose $h \in \mathrm{C}^\infty(E)$ such that $(\nabla h)(x_0) \neq 0$. As in the proof of Corollary 5.3 it follows that $(\nabla \omega_j)(x_0) \neq 0$ for at least one j.

As noted above, the existence of a common critical point is a necessary condition for the common maximizer property to hold; in turn, this is (under the assumption that the spectrum is simple, cf. Corollary 5.2) a necessary condition for the existence of an FLTC. Therefore, the following nonexistence theorem is a direct consequence of the preceding results.

Theorem 5.4 *Let $\{T_t\}_{t \geq 0}$ be either the Feller semigroup on a bounded domain $E \subset \mathbb{R}^d$ with C^{2m+2} boundary ($m > \frac{d}{2}+1$) associated with the reflected Brownian motion* *on E or the Feller semigroup associated with the Brownian motion on a compact Riemannian manifold. Assume that the operator* $T_1^{(2)}$ *has simple spectrum. Then there exists no FLTC for the semigroup* $\{T_t\}$.

This theorem is not applicable to regular polygons and other domains which are invariant under reflection or rotation (i.e. under the natural action of a dihedral group), as this invariance enforces the presence of eigenvalues with multiplicity greater than 1. On the other hand, we know that the eigenspaces on such symmetric domains can be associated to the different symmetry subspaces of the irreducible representations of the dihedral group [79]. In most cases, the multiplicity of all the eigenspaces corresponding to the one-dimensional irreducible representations is equal to 1 [134]; therefore, an adaptation of the proofs presented above should allow us to establish the nonexistence of common critical points among the eigenfunctions associated to the one-dimensional eigenspaces.

The nonexistence theorem established above strongly depends on the discreteness of the spectrum of the generator of the Feller process. Extending Theorem 5.4 to Brownian motions on unbounded domains on $\mathbb{R}^d$ or on noncompact Riemannian manifolds is a challenging problem, as these diffusions generally have a nonempty continuous spectrum. We leave this topic for future research.

5.3 Nonexistence of Convolutions: One-Dimensional Diffusions

As we saw in the previous sections, the construction of an FLTC is a difficult problem for which there is little hope of finding a solution unless the generator can be decomposed into a product of one-dimensional operators. Motivated by this, we now return to the one-dimensional setting in order to demonstrate that the necessary conditions for the existence of an FLTC determined in Sect. 5.1 also give rise to nonexistence theorems for a class of one-dimensional diffusion processes.

The following result shows that a necessary and sufficient condition for existence of an FLTC similar to that of Corollary 5.2 holds for Sturm–Liouville operators whose spectrum is not necessarily discrete:

Proposition 5.8 *Consider a Sturm–Liouville operator* ℓ *of the form* (4.1) *whose coefficients are such that* $p(x), r(x) > 0$ *for all* $x \in (a,b)$, $p, p', r, r' \in \mathrm{AC_{loc}}(a,b)$ *and* $\int_a^c \int_y^c \frac{dx}{p(x)} r(y)dy < \infty$. *Let* $w_\lambda(\cdot)$ $(\lambda \in \mathbb{C})$ *be the unique solution of* (2.12), $\boldsymbol{\rho}_{\mathcal{L}}$ *the spectral measure of Theorem 2.5,* $\boldsymbol{\Lambda} = \mathrm{supp}(\boldsymbol{\rho}_{\mathcal{L}})$, *and* $\{T_t\}_{t\geq 0}$ *the Feller semigroup generated by the realization of* ℓ *defined in* (2.34). *Assume that the endpoint b is not exit and that* $e^{-t\cdot} \in L^2(\boldsymbol{\Lambda}; \boldsymbol{\rho}_{\mathcal{L}})$ *for all* $t > 0$. *Set* $I = [a,b)$ *if b is natural and* $I = [a,b]$ *if b is regular or entrance. Then the following are equivalent:*

(i) There exists an FLTC for $\{T_t\}_{t\geq 0}$ *with trivializing family* $\Theta = \{w_\lambda\}_{\lambda\in\boldsymbol{\Lambda}}$.

(ii) *We have $w_\lambda \in \mathrm{C_b}(I)$ for all $\lambda \in \mathbf{\Lambda}$, and the function*

$$q_t(x, y, \xi) := \int_{\mathbf{\Lambda}} e^{-t\lambda}\, w_\lambda(x)\, w_\lambda(y)\, w_\lambda(\xi)\, \boldsymbol{\rho}_{\mathcal{L}}(d\lambda) \qquad \bigl(t > 0,\ x, y, \xi \in (a, b)\bigr) \tag{5.20}$$

is well-defined as an absolutely convergent integral; moreover, the measures defined as $\boldsymbol{\nu}_{t,x,y}(d\xi) = q_t(x, y, \xi)\, r(\xi) d\xi$ are such that $\{\boldsymbol{\nu}_{t,x,y}\}_{0<t\leq 1,\, x,y\in(a,\beta]}$ is, for each $\beta < b$, a tight family of probability measures on I.

Proof (i)$\Longrightarrow$(ii): Let $\diamond$ be an FLTC for $\{T_t\}_{t\geq 0}$ and $p_{t,x} = \mu_t \diamond \delta_x$ the transition kernel of $\{T_t\}_{t\geq 0}$. Since the w_λ are multiplicative linear functionals on the Banach algebra $(\mathcal{M}_\mathbb{C}(I), \diamond)$, we have $\|w_\lambda\|_\infty = 1$ for all $\lambda \in \mathbf{\Lambda}$, hence the right-hand side of (5.20) is absolutely convergent. Moreover,

$$(\mu_t \diamond \delta_x \diamond \delta_y)(w_\lambda) = e^{-t\lambda}\, w_\lambda(x)\, w_\lambda(y) = \mathcal{F}[q_t(x, y, \cdot)](\lambda) \qquad \bigl(t > 0,\ x, y \in (a, b)\bigr)$$

and it follows that for $g \in \mathrm{C_c}(I)$

$$\begin{aligned}
\int_I g(\xi)\, (\mu_t \diamond \delta_x \diamond \delta_y)(d\xi) &= \lim_{s\downarrow 0} \int_I (T_s g)(\xi)\, (\mu_t \diamond \delta_x \diamond \delta_y)(d\xi) \\
&= \lim_{s\downarrow 0} \int_{\mathbf{\Lambda}} (\mathcal{F}g)(\lambda)\, e^{-(t+s)\lambda}\, w_\lambda(x)\, w_\lambda(y)\, \boldsymbol{\rho}(d\lambda) \\
&= \int_I g(\xi)\, q_t(x, y, \xi)\, r(\xi) d\xi,
\end{aligned}$$

where we used Fubini's theorem, Proposition 2.14 and the isometric property of $\mathcal{F}$. Since g is arbitrary, this shows that the measures $(\mu_t \diamond \delta_x \diamond \delta_y)(d\xi)$ and $\boldsymbol{\nu}_{t,x,y}(d\xi) := q_t(x, y, \xi)\, r(\xi) d\xi$ coincide. Consequently, $\boldsymbol{\nu}_{t,x,y} \in \mathcal{P}(I)$ for all $t > 0$ and $x, y \in (a, b)$. Since $\{T_t\}_{t\geq 0}$ is a Feller process, the mapping $(t, x) \mapsto p_{t,x} = \mu_t \diamond \delta_x$ is continuous on $\mathbb{R}_0^+ \times I$ with respect to the weak topology of measures, and therefore the family $\{\boldsymbol{\nu}_{t,x,y}\}_{0<t\leq 1,\, x,y\in(a,\beta]}$ is relatively compact, hence tight.

(ii)$\Longrightarrow$(i): In the case where b is regular or entrance, this implication follows from Proposition 5.4. Assume that (ii) holds and that b is natural. By Theorem 2.5

$$e^{-t\lambda}\, w_\lambda(x)\, w_\lambda(y) = \int_I w_\lambda(\xi)\, \boldsymbol{\nu}_{t,x,y}(d\xi) \qquad \bigl(t > 0,\ x, y \in (a, b),\ \lambda \in \mathbf{\Lambda}\bigr), \tag{5.21}$$

where the integral converges absolutely. Since $\{\boldsymbol{\nu}_{t,x,y}\}_{0<t\leq 1,\, x,y\in(a,\beta]}$ is tight, given $x, y \in I$ there exists a sequence $\{(t_n, x_n, y_n)\}_{n\in\mathbb{N}} \subset \mathbb{R}^+ \times (a, b) \times (a, b)$ such that $(t_n, x_n, y_n) \to (0, x, y)$ and the measures $\boldsymbol{\nu}_{t_n,x_n,y_n}$ converge weakly to a measure $\boldsymbol{\nu}_{x,y} \in \mathcal{P}(I)$ as $n \to \infty$. Moreover, if $(x, y) \neq (a, a)$ and $\boldsymbol{\nu}^1_{x,y}$, $\boldsymbol{\nu}^2_{x,y}$

denote two such limits, then arguing as in the proof of Theorem 4.5 we find that the equalities

$$\int_I g(\xi)\, \boldsymbol{\nu}^1_{x,y}(d\xi) = \int_{\boldsymbol{\Lambda}} w_\lambda(x)\, w_\lambda(y)\, (\mathcal{F}g)(\lambda)\, \boldsymbol{\rho}_{\mathcal{L}}(d\lambda) = \int_I g(\xi)\, \boldsymbol{\nu}^2_{x,y}(d\xi)$$

hold for all $g \in \mathcal{D}^{(2,0)}$, so that $\boldsymbol{\nu}^1_{x,y} = \boldsymbol{\nu}^2_{x,y}$; hence the measures $\boldsymbol{\nu}_{t,\widetilde{x},\widetilde{y}}$ converge weakly as $(t, \widetilde{x}, \widetilde{y}) \to (0, x, y)$ to a unique limit $\boldsymbol{\nu}_{x,y}$ which is characterized by the identity $\int_I g(\xi)\, \boldsymbol{\nu}_{x,y}(d\xi) = \int_{\boldsymbol{\Lambda}} w_\lambda(x)\, w_\lambda(y)\, (\mathcal{F}g)(\lambda)\, \boldsymbol{\rho}_{\mathcal{L}}(d\lambda)$ ($g \in \mathcal{D}^{(2,0)}$, $x, y \in I$, $(x, y) \neq (a, a)$). Using this fact and the reasoning in the proof of Proposition 4.7(ii), we can verify that each measure $\mu \in \mathcal{M}_{\mathbb{C}}(I)$ is uniquely determined by the family of integrals $\{\mu(w_\lambda)\}_{\lambda \in \boldsymbol{\Lambda}}$. From this it follows, by taking limits in both sides of (5.21), that the measure $\boldsymbol{\nu}_{a,a} = \delta_a$ is the unique weak limit of $\boldsymbol{\nu}_{t_n,x_n,y_n}$ as $(t_n, x_n, y_n) \to (0, a, a)$.

For $\mu, \nu \in \mathcal{M}_{\mathbb{C}}(I)$, define $(\mu \diamond \nu)(d\xi) := \int_I \int_I \boldsymbol{\nu}_{x,y}(d\xi)\, \mu(dx)\, \nu(dy)$, where $\boldsymbol{\nu}_{x,y} \in \mathcal{P}(I)$ is the unique weak limit described above. Then (5.21) yields that

$$w_\lambda(x)\, w_\lambda(y) = \int_I w_\lambda(\xi)\, (\delta_x \diamond \delta_y)(d\xi) \qquad \bigl(x, y \in I,\ \lambda \in \boldsymbol{\Lambda}\bigr) \tag{5.22}$$

and, consequently, condition III of Definition 5.1 holds with $\Theta = \{w_\lambda\}_{\lambda \in \boldsymbol{\Lambda}}$. It is also clear that $(\mathcal{M}_{\mathbb{C}}(I), \diamond)$ is a commutative Banach algebra over $\mathbb{C}$ with identity element δ_a and that $\mathcal{P}(I) \diamond \mathcal{P}(I) \subset \mathcal{P}(I)$. From the tightness of $\{\boldsymbol{\nu}_{t,x,y}\}_{0<t\leq 1,\, x,y\in(a,\beta]}$ ($\beta < b$) it follows that the family of limits $\{\boldsymbol{\nu}_{x,y}\}_{x,y\in(a,\beta]}$ is also tight; it then follows from (5.22) that the map $(x, y) \mapsto \boldsymbol{\nu}_{x,y}$ is continuous with respect to the weak topology and, consequently, $(\mu, \nu) \mapsto \mu \diamond \nu$ is also continuous. Finally, the same reasoning of the proof of Proposition 3.11 yields that the transition kernel $\{p_{t,x}\}_{t>0,x\in I}$ satisfies condition IV of Definition 5.1.

We note that the assumption that $e^{-t\cdot} \in L^2(\boldsymbol{\Lambda}; \boldsymbol{\rho}_{\mathcal{L}})$ holds for all Sturm–Liouville operators whose left endpoint a is regular, and it also holds for a fairly large class of operators for which the endpoint a is entrance [107, 108].

Using the proposition above, one can show that the positivity of the limit $\lim_{\xi\to\infty} \frac{A'(\xi)}{2A(\xi)}$ (cf. Lemma 4.1) is a crucial condition for the construction of Sturm–Liouville convolutions presented in Chap. 4:

Proposition 5.9 *Consider a Sturm–Liouville operator ℓ of the form (4.1) whose coefficients are such that $p(x), r(x) > 0$ for all $x \in (a, b)$, $p, p', r, r' \in$ $\mathrm{AC}_{\mathrm{loc}}(a, b)$ and $\int_a^c \int_y^c \frac{dx}{p(x)} r(y) dy < \infty$. Suppose that $\gamma(b) = \int_c^b \sqrt{\frac{r(y)}{p(y)}} dy = \infty$ and the function A defined in (4.2) is such that $\frac{A'}{A}$ is of bounded variation on $[\tilde{c}, \infty)$ for some $\tilde{c} > \gamma(a)$ and $\lim_{\xi\to\infty} \frac{A'(\xi)}{2A(\xi)} = \sigma \in (-\infty, 0)$. Then the unique solution $w_\lambda(\cdot)$ of (2.12) is such that*

$$\sup_{x\in[a,b)} |w_\lambda(x)| = \infty \qquad \text{for all } \lambda > \sigma^2.$$

Consequently, there exists no FLTC for the Feller semigroup $\{T_t\}$ generated by the realization of ℓ defined in (2.34).

Proof The Sturm–Liouville equation $-\frac{1}{A}(Au')' = \lambda u$ is of the form $(\boldsymbol{p}u')' - \boldsymbol{q}u = 0$, where the coefficients $\boldsymbol{p} = A$ and $\boldsymbol{q} = -\lambda A$ are such that

$$\frac{(\boldsymbol{pq})'}{\boldsymbol{pq}} = \kappa i\sqrt{\lambda}(1+\boldsymbol{\phi}) \qquad \text{with } \kappa := -\frac{4\sigma i}{\sqrt{\lambda}}, \quad \boldsymbol{\phi} := \frac{A'}{2\sigma A} - 1,$$

so that $\boldsymbol{\phi}(\xi) = o(1)$ as $\xi \to \infty$ and $\boldsymbol{\phi}' = (\frac{A'}{2\sigma A})'$ is integrable near $+\infty$ (this follows from the assumption that $\frac{A'}{A}$ is of bounded variation, cf. [62, Proposition 3.30]). After applying a result on the asymptotic behaviour of solutions of second-order differential equations stated in [52, Theorem 2.6.1], we conclude that for $\lambda \neq \sigma^2$ the equation $-\frac{1}{A}(Au')' = \lambda u$ has two linearly independent solutions u_+ and u_- such that

$$u_\pm(\xi) \sim [-\lambda A(\xi)^2]^{-\frac{1}{4}\pm\frac{1}{4}\sqrt{1-\lambda/\sigma^2}} \exp\biggl(\pm\frac{i\lambda}{\sqrt{\lambda-\sigma^2}}\int_{\tilde{c}}^{\xi}\frac{A'(z)}{2\sigma A(z)}\biggl(\frac{A'(z)}{2\sigma A(z)}-1\biggr)dz\biggr).$$

The function $A(\xi)^{-1/2} = A(c)^{-1/2}\exp\bigl(-\frac{1}{2}\int_{\tilde{c}}^{\xi}\frac{A'(z)}{A(z)}dz\bigr)$ is clearly unbounded as $\xi \to \infty$; therefore, u_+ and u_- are both unbounded for $\lambda > \sigma^2$. Since $w_\lambda(x)$ is a real-valued linear combination of $u_-(\gamma(x))$ and $u_+(\gamma(x))$, it follows that $\sup_{x\in[a,b)}|w_\lambda(x)| = \infty$ for $\lambda > \sigma^2$.

Combining the above with Proposition 5.8, we find that $\{w_\lambda\}_{\lambda>\sigma^2}$ cannot belong to any trivializing family for an FLTC, hence by Corollary 5.1 the trivializing family must be contained in $\{w_\lambda\}_{\lambda\in[0,\sigma^2]} \cap \mathrm{C_b}[a,b)$. On the other hand, it follows from Theorem 2.5 that there exist nonzero measures $\mu \in \mathcal{M}_\mathbb{C}[a,b)$ such that $\mu(w_\lambda) = 0$ whenever $\lambda \in [0,\sigma^2]$ and w_λ is bounded. (Indeed, we have $(\sigma^2,\infty) \subset \boldsymbol{\Lambda}$ by the same argument in the proof of Proposition 4.3; if we let $\varphi \in L^2(\boldsymbol{\Lambda};\boldsymbol{\rho}_{\mathcal{L}}) \setminus \{0\}$ with $\mathrm{supp}(\varphi) \subset (\sigma^2,\infty)$, then $\mu(dx) = (\mathcal{F}^{-1}\varphi)(x)r(x)dx$ defines a measure $\mu \in \mathcal{M}_\mathbb{C}[a,b)$, and we have $\mu(w_\lambda) = 0$ for $\lambda \notin (\sigma^2,\infty)$.) Consequently, no family $\Theta \subset \{w_\lambda\}_{\lambda\in[0,\sigma^2]} \cap \mathrm{C_b}[a,b)$ can satisfy condition III of Definition 5.1. This contradiction shows that there exists no FLTC for $\{T_t\}$.

Example 5.6 Proposition 5.9 shows, in particular, that the following operators do not admit an associated (positivity-preserving) Sturm–Liouville convolution structure:

(a) $\ell = -\frac{d^2}{dx^2} - \bigl(\frac{\alpha}{x} + 2\mu\bigr)\frac{d}{dx}$, with $\alpha > 0$ and $\mu < 0$.

This is the generator of a mean-reverting Bessel process with negative drift (Example 2.3).

(b) $\ell = -\frac{d}{dx^2} - [(2\alpha+1)\coth x + (2\beta+1)\tanh x]\frac{d}{dx}$, with $\alpha > -1$ and $\alpha + \beta + 1 < 0$.

This is the Jacobi operator, which is the generator of the hypergeometric diffusion (Example 2.5).

(c) $\ell = -x^2\frac{d^2}{dx^2} - (c + 2(1-\alpha)x)\frac{d}{dx}$, with $c > 0$ and $\alpha > \frac{1}{2}$.

This is the Whittaker operator, which is the generator of a nonstandardized Shiryaev process (Example 2.4 and Remark 3.11).

Sturm–Liouville operators with two natural endpoints were excluded from the discussion in Chap. 4 because, given the absence of a natural candidate for the identity element, the construction of generalized convolutions would require a different approach. The ordinary convolution is a Sturm–Liouville convolution for the operator $\frac{d^2}{dx^2}$ on $\mathbb{R}$; as far as we know, this is the only known example of a convolution associated with a Sturm–Liouville operator with two natural boundaries. Let us note some examples in which the nonexistence of an associated FLTC follows from the results above:

Example 5.7 (Sturm–Liouville Operators with Two Natural Endpoints)

(a) Let $\theta > 0$ and $c \in \mathbb{R}$. The differential operator

$$\ell = -\frac{d^2}{dx^2} - (c - \theta x)\frac{d}{dx}, \qquad -\infty < x < \infty$$

is the generator of the Ornstein–Uhlenbeck process [3, 122]. Both endpoints $x = -\infty$ and $x = +\infty$ are natural. It is well-known that the self-adjoint realization of ℓ has a purely discrete spectrum, with eigenvalues $\lambda_n = n\theta$ and orthogonal eigenfunctions $\varphi_n(x) = H_n^{(\theta,c)}(x)$ $(n \in \mathbb{N}_0)$, where $H_n^{(\theta,c)}$ are the Hermite polynomials defined as

$$H_n^{(\theta,c)}(x) := e^{-cx+\frac{\theta}{2}x^2}\frac{d^n}{dx^n}\bigl(\theta^{-n}e^{cx-\frac{\theta}{2}x^2}\bigr).$$

Since each $H_n^{(\theta,c)}$ is a polynomial of degree n, it is clear that the L^2-extension of the Feller semigroup generated by ℓ (and therefore the Feller semigroup itself) has no bounded eigenfunctions other than $\varphi_0 \equiv 1$. Consequently, by Corollary 5.1, one cannot construct a Sturm–Liouville convolution for the transition semigroup of the Ornstein–Uhlenbeck process.

(b) Let $\kappa > 0$ and $\alpha \in \mathbb{R}$. The differential operator

$$\ell = -(1+x^2)\frac{d^2}{dx^2} - \kappa(\alpha - x)\frac{d}{dx}, \qquad -\infty < x < \infty$$

is the generator of the *Student diffusion process* [3, 122]. Both endpoints $x = \pm\infty$ are natural, and the self-adjoint realization of ℓ has a purely absolutely continuous spectrum on the interval $\bigl((\kappa+1)^2/2, \infty\bigr)$, together with a finite set of eigenvalues below $(\kappa+1)^2/2$. The operator ℓ can be transformed, via the change of variables $\xi = \operatorname{arcsinh}(x)$, into the standard form $-\frac{d^2}{d\xi^2} - \frac{A'(\xi)}{A(\xi)}\frac{d}{d\xi}$, where $\frac{A'(\xi)}{A(\xi)} = \frac{\kappa\alpha}{\cosh(\xi)} - (\kappa+1)\tanh(x)$. Since $\lim_{\xi\to\infty}\frac{A'(\xi)}{A(\xi)} = -(\kappa+1) < 0$

and $\lim_{\xi\to-\infty}\frac{A'(\xi)}{A(\xi)} = \kappa + 1 > 0$, it follows from the proof of Proposition 5.9 that if $\lambda > (\kappa+1)^2/2$ then the equation $\ell(u) = \lambda u$ has no nonzero bounded solutions. Arguing in the same way we conclude that the transition semigroup of the Student diffusion does not admit an associated Sturm–Liouville convolution.

The next examples, related with the Laguerre and Jacobi polynomials, show that Sturm–Liouville operators which do not admit an associated FLTC may, nevertheless, admit a convolution structure satisfying a weaker set of axioms:

Example 5.8 Let $\alpha, \kappa > 0$. The Laguerre differential operator

$$\ell = -x\frac{d^2}{dx^2} - \kappa(\alpha - x)\frac{d}{dx}, \qquad 0 < x < \infty$$

is the generator of the Cox–Ingersoll–Ross (CIR) process [3, 122]. The endpoint $x = 0$ is classified as regular if $\kappa\alpha < 1$ and entrance if $\kappa\alpha \geq 1$, while the endpoint $x = +\infty$ is natural. The spectrum of the Neumann realization of ℓ is purely discrete, with the orthogonal eigenfunctions being given by $\varphi_n(x) = L_n^{(\beta)}(\kappa x)$ $(n \in \mathbb{N}_0)$, where $\beta = \kappa\alpha - 1$ and $L_n^{(\beta)}$ are the Laguerre polynomials defined as

$$L_n^{(\beta)}(x) := \frac{1}{(\beta+1)_n} x^{-\beta} e^x \frac{d^n}{dx^n}\bigl(x^{\beta+n}e^{-x}\bigr).$$

As in Example 5.7(a), the fact that each $L_n^{(\beta)}$ is a polynomial of degree n means that the nonconstant eigenfunctions of the Feller semigroup generated by ℓ (i.e. of the transition semigroup of the CIR diffusion) are unbounded, and from this it follows that there exists no FLTC for this Feller semigroup. For $\beta \geq -\frac{1}{2}$, one can also reach the same conclusion by recalling the well-known product formula for the Laguerre polynomials, which is given by [73, p. 149]

$$\begin{aligned} &L_n^{(\beta)}(x) L_n^{(\beta)}(y) \\ &= \begin{cases} \int_0^\infty L_n^{(\beta)}(\xi)\, K_\beta(x,y,\xi)\, \xi^\beta e^{-\xi} d\xi, & \beta > -\frac{1}{2}, \\ \frac{1}{2}\Bigl[e^{\sqrt{xy}} L_n^{(-1/2)}\bigl((\sqrt{x}-\sqrt{y})^2\bigr) + e^{-\sqrt{xy}} L_n^{(-1/2)}\bigl((\sqrt{x}+\sqrt{y})^2\bigr)\Bigr] & \beta = -\frac{1}{2}, \\ \qquad + \int_0^\infty L_n^{(-1/2)}(\xi)\, K_{-1/2}(x,y,\xi)\, \xi^{-1/2} e^{-\xi} d\xi, & \end{cases} \end{aligned}$$

where

$$\begin{aligned} &K_\beta(x,y,\xi) \\ &= \begin{cases} \frac{2^{\beta-1}\Gamma(\beta+1)}{\sqrt{2\pi}(xy\xi)^\beta} \exp(\frac{x}{2}+\frac{y}{2}+\frac{\xi}{2})\, J_{\beta-1/2}(Z)\, Z^{\beta-1/2}\, \mathbb{1}_{[(\sqrt{x}-\sqrt{y})^2,(\sqrt{x}+\sqrt{y})^2]}(\xi), & \beta > -\frac{1}{2}, \\ -\frac{1}{4}(xy\xi)^{1/2} \exp(\frac{x}{2}+\frac{y}{2}+\frac{\xi}{2})\, J_1(Z)\, Z^{-1} \mathbb{1}_{[(\sqrt{x}-\sqrt{y})^2,(\sqrt{x}+\sqrt{y})^2]}(\xi), & \beta = -\frac{1}{2}. \end{cases} \end{aligned}$$

The kernel of this product formula fails to be nonnegative, so it does not lead to a positivity-preserving convolution structure.

The normalized Laguerre functions defined as $\boldsymbol{L}_n^{(\beta)}(\kappa x) := e^{-\kappa x/2} L_n^{(\beta)}(\kappa x)$ are the eigenfunctions of the modified Laguerre operator

$$\boldsymbol{\ell} = -x\frac{d^2}{dx^2} - \kappa\alpha\frac{d}{dx} + \frac{\kappa^2 x}{4}.$$

It is clear that $\boldsymbol{L}_n^{(\beta)}(0) = 1$, and by [73, Remark 1.10.88] we have $|\boldsymbol{L}_n^{(\beta)}(x)| \leq 1$ for all $\beta \geq 0$ and $n \in \mathbb{N}_0$. Moreover, the normalized product formula

$$\boldsymbol{L}_n^{(\beta)}(x)\,\boldsymbol{L}_n^{(\beta)}(y) = \int_0^\infty \boldsymbol{L}_n^{(\beta)}(\xi)\,\boldsymbol{k}_\beta(x, y, \xi)\,\xi^\beta d\xi,$$

where

$$\boldsymbol{k}_\beta(x, y, \xi) := \exp\Bigl(-\frac{x}{2} - \frac{y}{2} - \frac{\xi}{2}\Bigr)\, k_\beta(x, y, \xi)$$

is such that $\int_0^\infty |\boldsymbol{k}_\beta(x, y, \xi)|\,\xi^\beta d\xi \leq 1$ for all $\beta \geq 0$ and $x, y \in \mathbb{R}^+$, cf. [73, Lemma 1.10.25]. This property allows us to define the *Laguerre convolution* of finite complex measures as $(\mu \star \nu)(d\xi) := \int_{\mathbb{R}^+}\int_{\mathbb{R}^+}(\delta_x \star \delta_y)(d\xi)\,\mu(dx)\,\nu(dy)$, where

$$(\delta_x \star \delta_y)(d\xi) := \boldsymbol{k}_\beta(x, y, \xi)\,\xi^\beta d\xi \quad \text{for } x, y > 0, \qquad \delta_x \star \delta_0 = \delta_0 \star \delta_x = \delta_x.$$

It is clear that $\delta_x \star \delta_y$ is, for $x, y > 0$, a nonpositive signed measure, and therefore this convolution does not preserve the space $\mathcal{P}(\mathbb{R}_0^+)$. But, as noted in [165, Section 2], the Laguerre convolution is a bilinear operator on $\mathcal{M}_\mathbb{C}(\mathbb{R}_0^+)$ such that:

(i) The space $(\mathcal{M}_\mathbb{C}(\mathbb{R}_0^+), \star)$, equipped with the total variation norm, is a commutative Banach algebra over $\mathbb{C}$ whose identity element is the Dirac measure δ_0;
(ii) The map $(\mu, \nu) \mapsto \mu \star \nu$ is continuous in the vague topology;
(iii) $\{\boldsymbol{L}_n^{(\beta)}\}_{n\in\mathbb{N}}$ is a trivializing family for $\star$, i.e. we have

$$\mu = \mu_1 \star \mu_2 \qquad \text{if and only if} \qquad \mu(\boldsymbol{L}_n^{(\beta)}) = \mu_1(\boldsymbol{L}_n^{(\beta)}) \cdot \mu_2(\boldsymbol{L}_n^{(\beta)}) \text{ for all } n \in \mathbb{N}.$$

In addition, one can check that $\boldsymbol{p}_{t,x} = \boldsymbol{p}_{t,0} \star \delta_x$, where $\boldsymbol{p}_{t,x}$ stands for the (nonconservative) Feller semigroup generated by $\boldsymbol{\ell}$.

Example 5.9 For $\alpha, \beta > -1$, consider the Jacobi differential operator $\ell = -(1 - x^2)\frac{d^2}{dx^2} - (\beta - \alpha - (\alpha + \beta + 1)x)\frac{d}{dx}$, cf. Example 2.2. Recall from Remark 4.2 that if $\beta \leq \alpha$ and either $\beta \geq -\frac{1}{2}$ or $\alpha + \beta \geq 0$, then the product formula (4.4) for the Jacobi polynomials $R_n^{(\alpha,\beta)}$ gives rise to a hypergroup structure and, therefore, to an FLTC for the transition semigroup generated by the Neumann realization of ℓ. The stated condition on the parameters is in fact necessary and sufficient for the existence of an FLTC: this follows from the fact that, by the results of Gasper [70], the product formula (4.4) exists for all $\alpha, \beta > -1$, with the measures $\boldsymbol{\nu}_{x,y}^{(\alpha,\beta)}$ satisfying

$$\int_{[-1,1]} d\big|\boldsymbol{\nu}_{x,y}^{(\alpha,\beta)}\big| \leq M \quad (x, y \in [-1,1]) \qquad \text{if and only if} \qquad \beta \leq \alpha,\ \alpha + \beta \geq -1 \tag{5.23}$$

and such that $\boldsymbol{\nu}_{x,y}^{(\alpha,\beta)}$ is a nonpositive signed measure when $\beta < \frac{1}{2}$ and $\alpha + \beta > 0$. If the equivalent conditions in (5.23) hold, then the *Jacobi convolution* defined by $\delta_x \underset{\alpha,\beta}{\circledast} \delta_y := \boldsymbol{\nu}_{x,y}^{(\alpha,\beta)}$ endows $\mathcal{M}_{\mathbb{C}}[-1,1]$ with a structure of commutative Banach algebra such that the convolution semigroup property $p_{t,x} = p_{t,0} \underset{\alpha,\beta}{\circledast} \delta_x$ holds for the transition probabilities $p_{t,x}$ of the diffusion generated by ℓ.

The notion of a signed hypergroup was introduced in [165] as a generalization of hypergroups where the convolution satisfies a weaker form of axioms H2–H6 of Definition 2.4, and it is no longer required that probability measures are preserved by the convolution. The Laguerre convolution is an example of a signed hypergroup, as well as the Jacobi convolution with $-1 < \beta \leq \alpha$ and $\alpha + \beta \geq -1$. Example 5.8 demonstrates that such signed hypergroups are also a tool for endowing diffusion semigroups with the convolution semigroup property, hence it would be desirable to establish sufficient conditions for the existence of a signed hypergroup associated with a given Sturm–Liouville operator. However, as far as we are aware, the known examples of non-positivity-preserving signed hypergroups of Sturm–Liouville type are limited to differential operators whose eigenfunctions can be written in terms of classical special functions. (See [131] for related work on a class of perturbed Bessel and Laguerre differential operators.)

Most of the nonexistence theorems presented above are based on the unboundedness of the solutions of the Sturm–Liouville equation $\ell(u) = \lambda u$. We will not address the more difficult problem of establishing such theorems for Sturm–Liouville operators such that $\ell(u) = \lambda u$ admits bounded solutions (such as those described in Lemma 2.3); however, we stress that this would require an investigation of the (non)positivity of the kernel (5.20) and the closely related (non)positivity of the hyperbolic Cauchy problem (4.22).

5.4 Families of Convolutions on Riemannian Structures with Cone-Like Metrics

The discussion in Sects. 5.1–5.2 indicates that the construction of convolution structures on multidimensional spaces only becomes feasible once we are able to decompose the problem into simpler one-dimensional problems. One should realize that this is, in particular, a key requirement for the associated ultrahyperbolic PDE (Remark 5.1) to become tractable. It is therefore natural to restrict the attention to Feller semigroups generated by elliptic operators on two-dimensional manifolds which admit *separation of variables* in the sense that the eigenfunctions are of the form $\omega_\lambda(\xi) = \psi_\lambda(x)\,\phi_\lambda(y)$ $(\xi = (x, y) \in M)$.

We have already noted (cf. Sect. 2.3 and Example 5.2) the trivial fact that if $M = M_1 \times M_2$ is a product of Riemannian manifolds endowed with the product metric (so that the Laplace–Beltrami operator on $M_1 \times M_2$ obviously admits separation of variables) and if there exists a convolution for the Laplace–Beltrami operator on both M_1 and M_2, then we can define a convolution associated with the Laplace–Beltrami operator on M by taking the product of the convolutions on M_1 and M_2.

In the present section we will introduce a nontrivial generalization of the notion of product convolution which is suited to manifolds of the form $M = \mathbb{R}^+ \times M_2$ endowed with the so-called *cone-like metric structures*. Such cone-like metrics are possibly singular metrics of the form $g = dx^2 + A(x)^2 g_{M_2}$; this is a natural generalization of the metric cone, cf. [31]. The Laplace–Beltrami operator of (M, g) is

$$\Delta = \partial_x^2 + \frac{A'(x)}{A(x)}\partial_x + \frac{1}{A(x)^2}\Delta_2$$

(where Δ_2 stands for the Laplace–Beltrami operator of M_2) and admits separation of variables, as its eigenfunctions can be written as $\omega_{k,\lambda}(\xi) = \psi_{k,\lambda}(x)\,\phi_{k,\lambda}(y)$, where $\phi_{k,\lambda}(y)$ are the eigenfunctions of Δ_2 and $\psi_{k,\lambda}(x)$ are eigenfunctions of the Sturm–Liouville operators $\Delta_{1,k} := \frac{d^2}{dx^2} + \frac{A'(x)}{A(x)}\frac{d}{dx} - \frac{\eta_k}{A(x)^2}$, where $\eta_k \geq 0$ are separation constants. If the eigenfunctions of Δ_2 and of each of the operators $\Delta_{1,k}$ admit a product formula, then this gives rise to a product formula of the form

$$\omega_{k,\lambda}(\xi_1)\,\omega_{k,\lambda}(\xi_2) = \int_M \omega_{k,\lambda}\, d\gamma_{k,\xi_1,\xi_2}.$$

The distinctive feature of this product formula (in comparison with those of Eqs. (3.19) and (4.21)) is that in general the measure in the right-hand side also depends on the multiplicity parameter k. This naturally leads to the notion of a *family of convolutions* associated with a given elliptic operator. One of our goals is to demonstrate that the convolution semigroup property for the reflected Brownian motion on the manifold (M, g), together with other properties of FLTCs, can be extended to the families of convolutions discussed in this section.

For simplicity, in the following presentation we take $M_2 = \mathbb{T}$ (the torus) and $M_2 = I$ (an interval). Most results can be straightforwardly generalized to the case where M_2 is a compact Lie group or, more generally, a compact Riemannian manifold endowed with a suitable convolution structure [1, 176].

5.4.1 The Eigenfunction Expansion of the Laplace–Beltrami Operator

In the sequel we consider the (possibly singular) Riemannian manifold (M, g), where $M = \mathbb{R}_0^+ \times \mathbb{T}$ (with $\mathbb{T} = \mathbb{R}/\mathbb{Z}$) and the ($\mathrm{C}^1$, possibly non-smooth) Riemannian metric is given by

$$g = dx^2 + A(x)^2 d\theta^2 \qquad (0 \leq x < \infty,\ \theta \in \mathbb{T}),$$

where the function A is such that

$$\begin{aligned} & A \in \mathrm{C}(\mathbb{R}_0^+) \cap \mathrm{C}^1(\mathbb{R}^+), && A(x) > 0 \text{ for } x > 0, \\ & \int_0^1 \frac{dx}{A(x)} < \infty, && \frac{A'}{A} \text{ is nonnegative and decreasing.} \end{aligned} \tag{5.24}$$

The Riemannian volume form on M is $d\Omega_g = \sqrt{\det g}\, dx d\theta = A(x) dx d\theta$. Thus, the Riemannian gradient of a function $u : M \longrightarrow \mathbb{C}$ is

$$\nabla u = \left(\partial_x u,\, \frac{1}{A^2} \partial_\theta u \right),$$

and the Laplace–Beltrami operator is

$$\Delta = \operatorname{div} \circ \nabla = \partial_x^2 + \frac{A'(x)}{A(x)} \partial_x + \frac{1}{A(x)^2} \partial_\theta^2. \tag{5.25}$$

The closure of Δ with reflecting boundary at $x = 0$ is introduced in the standard way. Consider the Sobolev space

$$H^1(M) \equiv H^1(M, \Omega_g) = \left\{ u \in L^2(M, \Omega_g) \mid \nabla u \in L^2(M, \Omega_g) \right\}$$

and the sesquilinear form $\mathcal{E} : H^1(M) \times H^1(M) \longrightarrow \mathbb{C}$ defined as

$$\begin{aligned} \mathcal{E}(u, v) &= \langle \nabla u, \nabla v \rangle_{L^2(M,\Omega_g)} \\ &= \int_M \left(\partial_x u \overline{\partial_x v} + \frac{1}{A(x)^2} \partial_\theta u \overline{\partial_\theta v} \right) d\Omega_g. \end{aligned} \tag{5.26}$$

It is clear from (5.26) that $\mathcal{E}$ is a symmetric, nonnegative, closed sesquilinear form. The associated self-adjoint operator $(\Delta_N, \mathcal{D}_N)$, defined (as in (5.1)) by

$$\mathcal{D}_N = \left\{ u \in H^1(M) \;\middle|\; \begin{array}{l} \exists v \in L^2(M, \Omega_g) \text{ such that} \\ \quad \langle \nabla u, \nabla z \rangle = -\langle v, z \rangle \text{ for all } z \in H^1(M) \end{array} \right\}, \qquad \Delta_N u = v$$

is called the *Neumann Laplacian* on M. It is an extension of the Laplace–Beltrami operator defined in a domain of smooth functions satisfying the reflective boundary condition at $x = 0$

$$(A\partial_x u)(0, \theta) = 0 \qquad \forall \theta \in \mathbb{T}.$$

We use the notations $L^p(M) = L^p(M, \Omega_g)$, $L^p(A) = L^p(\mathbb{R}^+, A(x)dx)$, and consider the Fourier decomposition

$$L^2(M) = \bigoplus_{k \in \mathbb{Z}} H_k, \qquad H_k = \left\{ e^{i2k\pi\theta} v(x) \mid v \in L^2(A) \right\}, \tag{5.27}$$

where H_k are regarded as Hilbert spaces with inner product $\left\langle e^{i2k\pi\theta} u, e^{i2k\pi\theta} v \right\rangle_{H_k} = \langle u, v \rangle_{L^2(A)}$. The direct sum is also regarded as a Hilbert space with inner product $\langle \{u_k\}, \{v_k\} \rangle_{\bigoplus H_k} = \sum_{k \subset \mathbb{Z}} \langle u_k, v_k \rangle_{H_k}$, such that for $u(x, \theta) = \sum_{k \in \mathbb{Z}} e^{i2k\pi\theta} u_k(x)$, $v(x, \theta) = \sum_{k \in \mathbb{Z}} e^{i2k\pi\theta} v_k(x)$, we have

$$\langle u, v \rangle_{L^2(M)} = \sum_{k \in \mathbb{Z}} \langle u_k, v_k \rangle_{L^2(A)},$$

$$\mathcal{E}(u, v) = \sum_{k \in \mathbb{Z}} \mathcal{E}_k(u_k, v_k),$$

where $\mathcal{E}_k(u_k, v_k) = \int_{\mathbb{R}^+} \left(u_k'(x) v_k'(x) + \frac{(2k\pi)^2}{A(x)^2} u_k(x) \overline{v_k(x)} \right) A(x) dx$ are sesquilinear forms with domains

$$\mathcal{D}(\mathcal{E}_k) = \left\{ u \in L^2(A) \cap \mathrm{AC}_{\mathrm{loc}}(\mathbb{R}^+) \;\middle|\; \frac{2k\pi}{A} u \in L^2(A),\ u' \in L^2(A) \right\}.$$

Thus, we obtain the decomposition, compatible with (5.27):

$$H^1(M) = \bigoplus_{k \in \mathbb{Z}} \mathcal{D}(\mathcal{E}_k).$$

It can be checked that the forms $\mathcal{E}_k$ are symmetric, nonnegative, and closed. Therefore, a similar argument allows us to construct self-adjoint realizations of the Sturm–Liouville operators $\Delta_k u(x) = u''(x) + \frac{A'(x)}{A(x)} u'(x) - \frac{(2k\pi)^2}{A(x)^2} u(x)$, whose domain is

$$\mathcal{D}(\Delta_k) = \left\{ u \in L^2(A) \mid u, u' \in \mathrm{AC}_{\mathrm{loc}}(\mathbb{R}^+),\ \Delta_k u \in L^2(A),\ (Au')(0) = 0 \right\},$$

for $k \in \mathbb{Z}$. This provides a decomposition of the Neumann Laplacian:

$$\mathcal{D}_N = \bigoplus_{k\in\mathbb{Z}} \mathcal{D}(\Delta_k), \qquad \Delta_N\Bigl(\sum_{k\in\mathbb{Z}} e^{i2k\pi\theta} u_k(x)\Bigr) = \sum_{k\in\mathbb{Z}} e^{i2k\pi\theta}\,\Delta_k u_k(x). \tag{5.28}$$

The first ingredient for the convolution operators associated with Δ is the following characterization of the solutions of the eigenfunction equation $-\Delta u = \lambda u$ with Neumann boundary condition at $x = 0$.

Lemma 5.1 *For each $(k, \lambda) \in \mathbb{Z}\times\mathbb{C}$, there exists a unique solution $V_{k,\lambda} \in H_{k,\infty} := \bigl\{e^{i2k\pi\theta} w(x) \bigm| w \in \mathrm{C}(\mathbb{R}_0^+)\bigr\}$ of the boundary value problem*

$$-\Delta v = \lambda v, \qquad v(0,\theta) = e^{i2k\pi\theta}, \qquad v^{[1]}(0,\theta) = 0, \tag{5.29}$$

where $v^{[1]}(x,\theta) = A(x)\,(\partial_x v)(x,\theta)$. Moreover, $\lambda \mapsto V_{k,\lambda}(x,\theta)$ is, for each fixed $(x,\theta) \in M$ and $k \in \mathbb{Z}$, an entire function of exponential type.

Proof Clearly, any such solution must be of the form $V_{k,\lambda}(x,\theta) = e^{i2k\pi\theta} w(x)$, where w is a solution of

$$-\Delta_k w(x) = \lambda w(x), \qquad w(0) = 1, \qquad (Aw')(0) = 0. \tag{5.30}$$

Therefore, Proposition 2.2 yields the result.

The unique solution of (5.30) will be denoted by $w_{k,\lambda}(x)$, so that $V_{k,\lambda}(x,\theta) = e^{i2k\pi\theta} w_{k,\lambda}(x)$. Throughout the paper we will make frequent use of the change of dependent variable described in the following elementary lemma (which results from Remark 2.1):

Lemma 5.2 *Define $\widetilde{w}_{k,\lambda}(x) := \frac{w_{k,\lambda}(x)}{\zeta_k(x)}$, where $\zeta_k(x) := \cosh\bigl(2k\pi \int_0^x \frac{dy}{A(y)}\bigr)$. Then $\widetilde{w}_{k,\lambda}(\cdot)$ is a solution of*

$$\ell_k(\widetilde{w}) = \lambda\widetilde{w}, \qquad \widetilde{w}(0) = 1, \qquad (B_k\widetilde{w}')(0) = 0, \tag{5.31}$$

where

$$\ell_k(g) := -\frac{1}{B_k}(B_k g')', \qquad B_k(x) := A(x)\,\zeta_k(x)^2. \tag{5.32}$$

Moreover, we have $\frac{B_k'}{B_k} = \eta + \phi$, where $\eta(x) = \frac{4k\pi}{A(x)}\tanh(2k\pi \int_0^x \frac{dy}{A(y)}) \geq 0$ and the functions $\phi = \frac{A'}{A}$ and $\psi := \frac{1}{2}\eta' - \frac{1}{4}\eta^2 + \frac{B_k'}{2B_k}\eta = \frac{(2k\pi)^2}{A^2}$ are both decreasing and nonnegative.

The final assertion of the above lemma implies, in particular, that Assumption MP of Chap. 4 holds for the coefficients of the Sturm–Liouville operator ℓ_k.

In what follows, to lighten the notation, points of M are denoted by $\boldsymbol{\xi} = (x, \theta)$, $\boldsymbol{\xi}_1 = (x_1, \theta_1)$, etc.

It is well-known that the classical Weyl–Titchmarsh–Kodaira theory of eigenfunction expansions of Sturm–Liouville operators can be generalized to elliptic partial differential operators on higher-dimensional spaces, see e.g. [69], [48, Theorem XIV.6.6]. As remarked in [48, p. 1713], the knowledge about the boundary conditions satisfied by the kernels of the eigenfunction expansion is much smaller in the (general) multidimensional case, when compared to the one-dimensional setting. However, in the special case where separation of variables can be applied to the eigenvalue problem for the elliptic operator and therefore the eigenvalue equation reduces to a system of ordinary differential equations, further information on the eigenfunction expansion can be obtained from the theory of multiparameter eigenvalue problems. This connection will be further discussed in Remark 5.4 below.

In particular, the Fourier decomposition (5.28), combined with the eigenfunction expansion of the Sturm–Liouville operator $-\Delta_k$, gives rise to an eigenfunction expansion of $(\Delta_N, \mathcal{D}_N)$ in terms of the separable solutions $V_{k,\lambda}$ defined in Lemma 5.1:

Proposition 5.10 *There exists a sequence of locally finite positive Borel measures $\boldsymbol{\rho}_{\boldsymbol{k}}$ on $\mathbb{R}_0^+$ such that the map $h \mapsto \mathcal{F}h$, where*

$$(\mathcal{F}h)_k(\lambda) := \int_M h(\boldsymbol{\xi})\, V_{-k,\lambda}(\boldsymbol{\xi})\, \Omega_g(d\boldsymbol{\xi}) \qquad \bigl(k \in \mathbb{Z},\ \lambda \geq 0\bigr), \tag{5.33}$$

is an isometric isomorphism $\mathcal{F} : L^2(M) \longrightarrow \bigoplus_{k\in\mathbb{Z}} L^2(\mathbb{R}_0^+, \boldsymbol{\rho}_{\boldsymbol{k}})$ whose inverse is given by

$$(\mathcal{F}^{-1}\{\varphi_k\})(\boldsymbol{\xi}) = \sum_{k\in\mathbb{Z}} \int_{\mathbb{R}_0^+} \varphi_k(\lambda)\, V_{k,\lambda}(\boldsymbol{\xi})\, \boldsymbol{\rho}_{\boldsymbol{k}}(d\lambda). \tag{5.34}$$

The convergence of the integral in (5.33) *is understood with respect to the norm of $L^2(\mathbb{R}_0^+, \boldsymbol{\rho}_{\boldsymbol{k}})$, and the convergence of the inner integrals and the series in* (5.34) *is understood with respect to the norm of $L^2(M)$. Moreover, the operator $\mathcal{F}$ is a spectral representation of $(\Delta_N, \mathcal{D}_N)$ in the sense that*

$$\mathcal{D}_N = \biggl\{ h \in L^2(M) \biggm| \sum_{k\in\mathbb{Z}} \int_{\mathbb{R}_0^+} \lambda^2 \bigl|(\mathcal{F}h)_k(\lambda)\bigr|^2 \boldsymbol{\rho}_{\boldsymbol{k}}(d\lambda) < \infty \biggr\}, \tag{5.35}$$

$$\bigl(\mathcal{F}(-\Delta_N h)\bigr)_k(\lambda) = \lambda \cdot (\mathcal{F}h)_k(\lambda), \qquad h \in \mathcal{D}_N,\ k \in \mathbb{Z}. \tag{5.36}$$

Proof Let $h \in L^2(M)$. By Fubini, $h(x,\cdot) \in L^2(\mathbb{T})$ for a.e. $x \in \mathbb{R}^+$. For these points x we have

$$h(x,\theta) = \sum_{k\in\mathbb{Z}} \widehat{h}_k(x)\, e^{i2k\pi\theta}, \qquad \text{where } \widehat{h}_k(x) := \int_0^1 e^{-i2k\pi\vartheta} h(x,\vartheta)d\vartheta, \tag{5.37}$$

the series converging in the norm of $L^2(\mathbb{T})$. It is straightforward to check that $\widehat{h}_k \in L^2(A)$ for all $k \in \mathbb{Z}$, and therefore the function $\widehat{h}_k$ can be represented in terms of the eigenfunction expansion of the Sturm–Liouville operator Δ_k (Theorem 2.5): denoting the spectral measure of Δ_k by $\boldsymbol{\rho_k}$, we have

$$\widehat{h}_k(x) = \int_{\mathbb{R}_0^+} \big(\mathcal{F}_{\Delta_k}\widehat{h}_k\big)(\lambda)\, w_{k,\lambda}(x)\, \boldsymbol{\rho_k}(d\lambda),$$

where

$$\big(\mathcal{F}_{\Delta_k}\widehat{h}_k\big)(\lambda) := \int_{\mathbb{R}^+} \widehat{h}_k(y)\, w_{k,\lambda}(y)\, A(y)dy,$$

the integrals converging in the norms of $L^2(A)$ and $L^2(\boldsymbol{\rho_k}) \equiv L^2(\mathbb{R}_0^+, \boldsymbol{\rho_k})$ respectively.

By definition of $\widehat{h}_k$, we have $\big(\mathcal{F}_{\Delta_k}\widehat{h}_k\big)(\lambda) = \int_M h(\boldsymbol{\xi})\, V_{-k,\lambda}(\boldsymbol{\xi})\, \Omega_g(d\boldsymbol{\xi}) \equiv (\mathcal{F}h)_k(\lambda)$, with equality in the $L^2(\boldsymbol{\rho_k})$-sense. Furthermore, by a dominated convergence argument it is clear that $\widehat{h}_k(x)e^{i2k\pi\theta} = \int_{\mathbb{R}_0^+} \big(\mathcal{F}_{\Delta_k}\widehat{h}_k\big)(\lambda)\, V_{k,\lambda}(\boldsymbol{\xi})\, \boldsymbol{\rho_k}(d\lambda)$ with equality in the $L^2(M)$-sense; therefore,

$$h(x,\theta) = \sum_{k\in\mathbb{Z}} \int_{\mathbb{R}_0^+} \big(\mathcal{F}_{\Delta_k}\widehat{h}_k\big)(\lambda)\, V_{k,\lambda}(\boldsymbol{\xi})\, \boldsymbol{\rho_k}(d\lambda),$$

proving the inversion formula (5.33)–(5.34). Finally, the fact that the integral operator $\mathcal{F}$ is isometric follows from the identities

$$\|h\|^2_{L^2(M)} = \sum_{k\in\mathbb{Z}} \|\widehat{h}_k\|^2_{L^2(A)} = \sum_{k\in\mathbb{Z}} \big\|\mathcal{F}_{\Delta_k}\widehat{h}_k\big\|^2_{L^2(\boldsymbol{\rho_k})} = \big\|\{(\mathcal{F}h)_k\}\big\|^2_{\bigoplus L^2(\boldsymbol{\rho_k})},$$

where the first and second steps follow from the isometric properties of the classical Fourier series and the eigenfunction expansion of Δ_k respectively.

It only remains to justify the identities (5.35)–(5.36). Using (5.28) we obtain that for $h \in \mathcal{D}_N$

$$\big(\mathcal{F}(-\Delta_N h)\big)_k(\lambda) = \big(\mathcal{F}_{\Delta_k}(\widehat{-\Delta_N h})_k\big)(\lambda) = \big(\mathcal{F}_{\Delta_k}(-\Delta_k \widehat{h}_k)\big)(\lambda) = \lambda \cdot \big(\mathcal{F}h\big)_k(\lambda),$$

which proves (5.36). Now, if $h \in L^2(M)$ is such that $\sum_{k\in\mathbb{Z}} \int_{\mathbb{R}_0^+} \lambda^2 \left|(\mathcal{F}h)_k(\lambda)\right|^2 \rho_k(d\lambda) < \infty$ then by (2.23) we have

$$(\widehat{h}_k)_{k\in\mathbb{Z}} \in \bigoplus_{k\in\mathbb{Z}} \mathcal{D}(\Delta_k) \equiv \bigoplus_{k\in\mathbb{Z}} \left\{ u \in L^2(A) \,\middle|\, \int_{\mathbb{R}_0^+} \lambda^2 |(\mathcal{F}_{\Delta_k} u)(\lambda)|^2 \rho_k(d\lambda) < \infty \right\}$$

and therefore $h = \sum_{k\in\mathbb{Z}} e^{i2k\pi\theta} \widehat{h}_k \in \mathcal{D}_N$. Conversely, if $h \in \mathcal{D}_N$ then by (5.36) we have $\{\lambda \cdot (\mathcal{F}h)_k(\lambda)\} \in \bigoplus_{k\in\mathbb{Z}} L^2(\rho_k)$, and we conclude that (5.35) holds.

Since Δ_N is a negative self-adjoint operator, it is the infinitesimal generator of a strongly continuous semigroup in $L^2(M)$, denoted by $\{e^{t\Delta_N}\}_{t\geq 0}$. It is not difficult to check that the sesquilinear form $\mathcal{E}$ is Markovian (in the sense that if $u \in H^1(M)$ then $v := \max(\min(u, 1), 0) \in H^1(M)$ and $\mathcal{E}(v, v) \leq \mathcal{E}(u, u)$); in other words, $(\mathcal{E}, H^1(M))$ is a Dirichlet form. Therefore, as mentioned in Sect. 5.1, $\{e^{t\Delta_N}\}_{t\geq 0}$ is a sub-Markovian contraction semigroup on $L^2(M)$. Furthermore, for every $p \in [1, +\infty]$ the subspace $L^2(M) \cap L^p(M)$ is invariant under $e^{t\Delta_N}$ for every $t \geq 0$, and the semigroup $\left\{e^{t\Delta_N}\right\}_{t\geq 0}$ can be extended into a strongly continuous contraction semigroup in $L^p(M)$ (see e.g. [43, Sections 1.3–1.4]). The analogous statement holds for the semigroup $\left\{e^{t\Delta_k}\right\}_{t\geq 0}$ in $L^p(A)$, for every $k \in \mathbb{Z}$.

Proposition 5.11 *Assume that the action of $e^{t\Delta_N}$ on $L^2(M)$ is given by a symmetric heat kernel, i.e. there exists a measurable function $p : \mathbb{R}^+ \times M \times M \longrightarrow \mathbb{R}_0^+$ such that:*

I. *For all $t, s > 0$ and $\xi_1, \xi_2 \in M$,*

$$p(t, \xi_1, \xi_2) = p(t, \xi_2, \xi_1), \tag{5.38}$$

$$p(t + s, \xi_1, \xi_2) = \int_M p(t, \xi_1, \xi_3)\, p(s, \xi_3, \xi_2)\, \Omega_g(d\xi_3); \tag{5.39}$$

II. *For $t > 0$, $h \in L^2(M)$ and Ω_g-a.e. $\xi_1 \in M$,*

$$(e^{t\Delta_N} h)(\xi_1) = \int_M h(\xi_2)\, p(t, \xi_1, \xi_2)\, \Omega_g(d\xi_2).$$

Then, for $t > 0$ and Ω_g-a.e. $\xi_1, \xi_2 \in M$, the heat kernel admits the spectral representation

$$p(t, \xi_1, \xi_2) = \sum_{k\in\mathbb{Z}} \int_{\mathbb{R}_0^+} e^{-t\lambda} V_{k,\lambda}(\xi_1)\, V_{-k,\lambda}(\xi_2)\, \rho_k(d\lambda), \tag{5.40}$$

where the integral and the sum are absolutely convergent.

Proof Fix $t > 0$. It follows from condition I that

$$\int_M p(t, \boldsymbol{\xi_1}, \boldsymbol{\xi_2})^2\, \Omega_g(d\boldsymbol{\xi_2}) = p(2t, \boldsymbol{\xi_1}, \boldsymbol{\xi_1}) < \infty \qquad (\boldsymbol{\xi_1} \in M),$$

meaning in particular that $p(t, \boldsymbol{\xi_1}, \cdot) \in L^2(M)$ for all $\boldsymbol{\xi_1} \in M$. Moreover, by the spectral representation property (5.35)–(5.36) we have $[\mathcal{F}(e^{t\Delta_N} h)]_k(\lambda) = e^{-t\lambda}(\mathcal{F}h)_k(\lambda)$ for all $h \in L^2(M)$, hence

$$\sum_{k\in\mathbb{Z}} \int_{\mathbb{R}_0^+} (\mathcal{F}h)_k(\lambda)\, [\mathcal{F}p(t, \boldsymbol{\xi_1}, \cdot)]_k(\lambda)\, \boldsymbol{\rho_k}(d\lambda) = (e^{t\Delta_N} h)(\boldsymbol{\xi_1})$$
$$= \mathcal{F}^{-1}\bigl\{e^{-t\cdot}(\mathcal{F}h)_k(\cdot)\bigr\}(\boldsymbol{\xi_1}) = \sum_{k\in\mathbb{Z}} \int_{\mathbb{R}_0^+} e^{-t\lambda}(\mathcal{F}h)_k(\lambda)\, V_{k,\lambda}(\boldsymbol{\xi_1})\, \boldsymbol{\rho_k}(d\lambda) \tag{5.41}$$

for Ω_g-a.e. $\boldsymbol{\xi_1} \in M$. Since $h \in L^2(M)$ is arbitrary, from (5.41) we deduce that $\{e^{-t\lambda} V_{k,\lambda}(\boldsymbol{\xi_1})\} = \{[\mathcal{F}p(t, \boldsymbol{\xi_1}, \cdot)]_k(\lambda)\} \in \bigoplus_{k\in\mathbb{Z}} L^2(\boldsymbol{\rho_k})$ for Ω_g-a.e. $\boldsymbol{\xi_1} \in M$. Therefore

$$\sum_{k\in\mathbb{Z}} \int_{\mathbb{R}_0^+} e^{-t\lambda} |V_{k,\lambda}(\boldsymbol{\xi_1})|^2\, \boldsymbol{\rho_k}(d\lambda) = \sum_{k\in\mathbb{Z}} \bigl\| e^{-t\lambda/2} V_{k,\lambda}(\boldsymbol{\xi_1}) \bigr\|_{L^2(\boldsymbol{\rho_k})}^2 < \infty,$$

and it follows (by the Cauchy-Schwarz inequality) that the right-hand side of (5.40) is absolutely convergent for Ω_g-a.e. $\boldsymbol{\xi_1}, \boldsymbol{\xi_2} \in M$. Moreover, the isometric property of $\mathcal{F}$ yields

$$p(t, \boldsymbol{\xi_1}, \boldsymbol{\xi_2}) = \bigl\langle p(t/2, \boldsymbol{\xi_1}, \cdot), p(t/2, \boldsymbol{\xi_2}, \cdot) \bigr\rangle_{L^2(M)}$$
$$= \sum_{k\in\mathbb{Z}} \bigl\langle e^{-t\lambda/2} V_{k,\lambda}(\boldsymbol{\xi_1}), e^{-t\lambda/2} V_{k,\lambda}(\boldsymbol{\xi_2}) \bigr\rangle_{L^2(\boldsymbol{\rho_k})}$$

and therefore the identity (5.40) holds for Ω_g-a.e. $\boldsymbol{\xi_1}, \boldsymbol{\xi_2} \in M$.

Corollary 5.4 *If the assumptions of Proposition 5.11 are satisfied, then for* $t \geq 0$, $k \in \mathbb{Z}$, $\lambda \in \operatorname{supp}(\boldsymbol{\rho_k})$ *and* Ω_g*-a.e.* $\boldsymbol{\xi_1} \in M$ *we have*

$$e^{-t\lambda} V_{k,\lambda}(\boldsymbol{\xi_1}) = \int_M V_{k,\lambda}(\boldsymbol{\xi_2})\, p(t, \boldsymbol{\xi_1}, \boldsymbol{\xi_2})\, \Omega_g(d\boldsymbol{\xi_2}).$$

Proof Fix $t \geq 0$ and $k \in \mathbb{Z}$. Notice that

$$p(t, \boldsymbol{\xi_1}, \boldsymbol{\xi_2}) = \sum_{j\in\mathbb{Z}} e^{i2j\pi(\theta_1-\theta_2)}\, p_{\Delta_j}(t, x_1, x_2),$$

where $p_{\Delta_j}(t, x_1, x_2) = \int_{\mathbb{R}_0^+} e^{-t\lambda}\, w_{j,\lambda}(x_1)\, w_{j,\lambda}(x_2)\, \boldsymbol{\rho_j}(d\lambda)$ is the heat kernel for the semigroup $\{e^{t\Delta_j}\}$ on $L^2(\mathbb{R}^+, A(x)dx)$ (Proposition 2.14), and the sum converges absolutely. Hence for $\lambda \in \operatorname{supp}(\boldsymbol{\rho_k})$ and Ω_g-a.e. $\boldsymbol{\xi_1} \in M$ we can write

$$
\begin{aligned}
&\int_M V_{k,\lambda}(\boldsymbol{\xi_2})\, p(t, \boldsymbol{\xi_1}, \boldsymbol{\xi_2})\, \Omega_g(d\boldsymbol{\xi_2}) \\
&= \int_M e^{i2k\pi\theta_2} w_{k,\lambda}(x_2) \sum_{j\in\mathbb{Z}} e^{i2j\pi(\theta_1-\theta_2)} p_{\Delta_j}(t, x_1, x_2)\, A(x_2)dx_2 d\theta_2 \\
&= \int_0^\infty w_{k,\lambda}(x_2) \sum_{j\in\mathbb{Z}} e^{i2j\pi\theta_1} \int_0^1 e^{i2(k-j)\pi\theta_2} d\theta_2\, p_{\Delta_j}(t, x_1, x_2)\, A(x_2)dx_2 \\
&= e^{i2k\pi\theta_1} \int_0^\infty w_{k,\lambda}(x_2)\, p_{\Delta_k}(t, x_1, x_2)\, A(x_2)dx_2 \\
&= e^{i2k\pi\theta_1} \zeta_k(x_1) \int_0^\infty \widetilde{w}_{k,\lambda}(x_2) \int_{\mathbb{R}_0^+} e^{-t\lambda_0}\, \widetilde{w}_{k,\lambda_0}(x_1)\, \widetilde{w}_{k,\lambda_0}(x_2)\, \boldsymbol{\rho_k}(d\lambda_0)\, B_k(x_2)dx_2 \\
&= e^{i2k\pi\theta_1} e^{-t\lambda} w_{k,\lambda}(x_1) \\
&= e^{-t\lambda} V_{k,\lambda}(\boldsymbol{\xi_1}).
\end{aligned}
$$

The second to last equality follows from the eigenfunction expansion of the Sturm–Liouville operator ℓ_k defined in Lemma 5.2, considering that the double integral can be recognized as $\mathcal{F}_{\ell_k}[\mathcal{F}_{\ell_k}^{-1} e^{-t\cdot}\, \widetilde{w}_{k,\cdot}(x_1)](\lambda)$, where $(\mathcal{F}_{\ell_k} g)(\lambda) := \int_{\mathbb{R}^+} g(y)\, \widetilde{w}_{k,\lambda}(y)\, B_k(y)dy \equiv (\mathcal{F}_{\Delta_k}(\zeta_k \cdot g))(\lambda)$. It should also be noted that

$$
e^{-t\lambda} \in L^2(\boldsymbol{\rho_k}), \qquad \mathcal{F}_{\ell_k}^{-1} e^{-t\cdot} \in L^1(\mathbb{R}^+, B_k(x)dx), \qquad \widetilde{w}_{k,\lambda} \in \mathrm{C_b}(\mathbb{R}_0^+)
$$

(cf. Proposition 2.14 and Lemma 2.3), and therefore the second to last equality, which holds initially for $\boldsymbol{\rho_k}$-a.e. $\lambda \geq 0$, can be extended by continuity to all $\lambda \in \operatorname{supp}(\boldsymbol{\rho_k})$.

The two results above depend on the assumption that the heat kernel exists. In the general framework of metric measure spaces, the existence of the heat kernel for the semigroup $\{e^{t\mathcal{G}^{(2)}}\}$ determined by a given Dirichlet form is equivalent to the ultracontractivity property $\|e^{t\mathcal{G}^{(2)}} h\|_\infty \leq \gamma(t)\|h\|_{L^1}$, where γ is a positive left-continuous function on $\mathbb{R}^+$ (the proof of this can be found in [5, Theorem 3.1]). A discussion of geometric conditions which ensure the existence of a heat kernel satisfying Gaussian estimates can be found in [76] and references therein.

In particular, it is known that the Laplace–Beltrami operator with Dirichlet or Neumann boundary conditions on a domain of a complete Riemannian manifold admits a heat kernel [36, 43]. We can thus state:

Proposition 5.12 *Let $\mathring{M} = \mathbb{R}^+ \times \mathbb{T}$. If A belongs to $\mathrm{C}^\infty(\mathbb{R}_0^+)$ and $A(0) > 0$, then there exists a heat kernel $p(t, x, y) \in \mathrm{C}^\infty(\mathbb{R}^+ \times \mathring{M} \times \mathring{M})$ satisfying the assumptions of Proposition 5.11.*

Proof Observe that under the stated condition we can regard (M, g) as a submanifold of the complete smooth Riemannian manifold $(\mathbb{R} \times \mathbb{T}, \widetilde{g})$, where $\widetilde{g} = dx^2 + \widetilde{A}(x)^2 d\theta^2$ and $\widetilde{A} \in \mathrm{C}^\infty(\mathbb{R})$ is a positive extension of the function A; therefore [36, Theorem 1.1] yields the result.

5.4.2 Product Formulas and Convolutions

Taking advantage of the separability of the eigenfunctions and the results of Chap. 4, we derive the following product formula for the eigenfunctions of the Neumann Laplacian:

Proposition 5.13 (Product Formula for $V_{k,\lambda}$) *For each $k \in \mathbb{N}_0$ and $\boldsymbol{\xi}_1, \boldsymbol{\xi}_2 \in M$ there exists a positive measure $\boldsymbol{\gamma}_{k,\boldsymbol{\xi}_1,\boldsymbol{\xi}_2}$ on M such that the product $V_{k,\lambda}(\boldsymbol{\xi}_1)\, V_{k,\lambda}(\boldsymbol{\xi}_2)$ admits the integral representation*

$$V_{k,\lambda}(\boldsymbol{\xi}_1)\, V_{k,\lambda}(\boldsymbol{\xi}_2) = \int_M V_{k,\lambda}(\boldsymbol{\xi}_3)\, \boldsymbol{\gamma}_{k,\boldsymbol{\xi}_1,\boldsymbol{\xi}_2}(d\boldsymbol{\xi}_3), \qquad \boldsymbol{\xi}_1, \boldsymbol{\xi}_2 \in M, \ \lambda \in \mathbb{C}. \tag{5.42}$$

Since $V_{-k,\lambda} = \overline{V_{k,\lambda}}$, this result trivially extends to all $k \in \mathbb{Z}$.

Proof Fix $k \in \mathbb{N}_0$. Recall that $V_{k,\lambda}(x, \theta) = e^{i2k\pi\theta} \zeta_k(x)\, \widetilde{w}_{k,\lambda}(x)$, where $\widetilde{w}_{k,\lambda}$ is a solution of (5.31). We saw in Lemma 5.2 that the operator ℓ_k satisfies Assumption MP, hence we can apply the existence theorem for Sturm–Liouville type product formulas (Theorem 4.2) and conclude that there exists a family of measures $\{\pi^{[k]}_{x_1,x_2}\}_{x_1,x_2 \in \mathbb{R}_0^+} \subset \mathcal{P}(\mathbb{R}_0^+)$ with $\mathrm{supp}(\pi^{[k]}_{x_1,x_2}) \subset [|x_1 - x_2|, x_1 + x_2]$ (cf. Proposition 4.17) and such that

$$\widetilde{w}_{k,\lambda}(x_1)\, \widetilde{w}_{k,\lambda}(x_2) = \int_{\mathbb{R}_0^+} \widetilde{w}_{k,\lambda}\, d\pi^{[k]}_{x_1,x_2} \qquad (x_1, x_2 \in \mathbb{R}_0^+, \ \lambda \in \mathbb{C}).$$

Consequently, the product formula (5.42) holds for the positive measures $\boldsymbol{\gamma}_{k,\boldsymbol{\xi}_1,\boldsymbol{\xi}_2}$ defined by

$$\boldsymbol{\gamma}_{k,\boldsymbol{\xi}_1,\boldsymbol{\xi}_2}(d\boldsymbol{\xi}_3) = \frac{\zeta_k(x_1)\zeta_k(x_2)}{\zeta_k(x_3)}\, \boldsymbol{\nu}_{k,\boldsymbol{\xi}_1,\boldsymbol{\xi}_2}(d\boldsymbol{\xi}_3), \tag{5.43}$$

where $\boldsymbol{\nu}_{k,\boldsymbol{\xi}_1,\boldsymbol{\xi}_2} := \pi^{[k]}_{x_1,x_2} \otimes \delta_{\theta_1+\theta_2}$.

It follows at once from the definition that the measures $\boldsymbol{\nu}_{k,\boldsymbol{\xi}_1,\boldsymbol{\xi}_2}$ are probability measures on M. The convolution operators determined by these measures are defined in the natural way:

Definition 5.2 Let $k \in \mathbb{Z}$ and $\mu, \nu \in \mathcal{M}_{\mathbb{C}}(M)$. The measure

$$(\mu \underset{k}{*} \nu)(\cdot) = \int_M \int_M \boldsymbol{\nu}_{k,\xi_1,\xi_2}(\cdot)\, \mu(d\xi_1)\, \nu(d\xi_2)$$

is called the Δ_k-*convolution* of the measures μ and ν.

In other words, the convolution algebra $(M, \underset{k}{*})$ is a product of hypergroups, namely the Sturm–Liouville hypergroup associated with ℓ_k (Theorem 4.6) and the hypergroup determined by the ordinary convolution on the torus.

The analogue of the trivialization property for the family of Δ_k-convolutions is described next.

Definition 5.3 Let $\mu \in \mathcal{M}_{\mathbb{C}}(M)$. The Δ-*Fourier transform* of the measure μ is the function defined by the integral

$$(\mathcal{F}\mu)(k, \lambda) = \int_M \frac{V_{-k,\lambda}(\xi)}{\zeta_k(x)}\, \mu(d\xi), \qquad k \in \mathbb{Z},\ \lambda \geq 0.$$

It follows from Lemma 2.3 that $\| \frac{V_{-k,\lambda}}{\zeta_k} \|_\infty \leq 1$, hence $(\mathcal{F}\mu)(k, \lambda)$ is well-defined for all $\mu \in \mathcal{M}_{\mathbb{C}}(M)$ and $(k, \lambda) \in \mathbb{Z} \times \mathbb{R}_0^+$.

Proposition 5.14 *Let $\mu, \nu \in \mathcal{M}_{\mathbb{C}}(M)$. We have*

$$\bigl(\mathcal{F}(\mu \underset{k}{*} \nu)\bigr)(k, \lambda) = (\mathcal{F}\mu)(k, \lambda)\cdot(\mathcal{F}\nu)(k, \lambda) \qquad \textit{for all } k \in \mathbb{Z} \textit{ and } \lambda \geq 0. \tag{5.44}$$

Moreover, for fixed $k \in \mathbb{Z}$ we have

$$(\mathcal{F}\alpha)(k, \cdot) = (\mathcal{F}\mu)(k, \cdot)\cdot(\mathcal{F}\nu)(k, \cdot) \quad \textit{if and only if} \quad \widehat{\alpha}_k = \widehat{\mu}_k \underset{k}{\diamond} \widehat{\nu}_k,$$

where $\widehat{\tau}_k$ ($\tau = \alpha, \mu, \nu$) is the complex measure on $\mathbb{R}_0^+$ defined by $\widehat{\tau}_k(J) = \int_M e^{-i2k\pi\theta}\, \mathbb{1}_J(x)\, \tau(d\xi)$, and $\underset{k}{\diamond}$ is the convolution defined as

$$\widehat{\mu}_k \underset{k}{\diamond} \widehat{\nu}_k(\cdot) := \int_{\mathbb{R}_0^+} \int_{\mathbb{R}_0^+} \pi^{[k]}_{x_1,x_2}(\cdot)\, \widehat{\mu}_k(dx_1)\, \widehat{\nu}_k(dx_2)$$

(here $\pi^{[k]}_{x_1,x_2} \in \mathcal{P}(\mathbb{R}_0^+)$ are the measures from the proof of Proposition 5.13).

Proof Applying the product formula (5.42), we obtain

$$\begin{aligned}\bigl(\mathcal{F}(\mu \underset{k}{*} \nu)\bigr)(k, \lambda) &= \int_M \frac{V_{-k,\lambda}(\xi)}{\zeta_k(x)}\, (\mu \underset{k}{*} \nu)(d\xi)\\ &= \int_M \int_M \int_M \frac{V_{-k,\lambda}(\xi_3)}{\zeta_k(x_3)}\, (\delta_{\xi_1} \underset{k}{*} \delta_{\xi_2})(d\xi_3)\, \mu(d\xi_1)\nu(d\xi_2)\end{aligned}$$

$$= \int_M \int_M \frac{V_{-k,\lambda}(\boldsymbol{\xi}_1)}{\zeta_k(x_1)} \frac{V_{-k,\lambda}(\boldsymbol{\xi}_2)}{\zeta_k(x_2)} \mu(d\boldsymbol{\xi}_1)\nu(d\boldsymbol{\xi}_2)$$

$$= (\mathcal{F}\mu)(k,\lambda) \cdot (\mathcal{F}\nu)(k,\lambda),$$

so that (5.44) holds. Since $\underset{k}{\diamond}$ is the Sturm–Liouville convolution for the operator $\ell_k = -\frac{1}{B_k}\frac{d}{dx}(B_k \frac{d}{dx})$, the second statement is a consequence of Proposition 4.8.

Proposition 5.15 *The Δ-Fourier transform $\mathcal{F}\mu$ of $\mu \in \mathcal{M}_{\mathbb{C}}(M)$ has the following properties:*

- ***(i)*** *For each $k \in \mathbb{Z}$, $(\mathcal{F}\mu)(k,\cdot)$ is continuous on $\mathbb{R}_0^+$. Moreover, if a family of measures $\{\mu_j\} \subset \mathcal{M}_{\mathbb{C}}(M)$ is tight and uniformly bounded, then $\{(\mathcal{F}\mu_j)(k,\cdot)\}$ is equicontinuous on $\mathbb{R}_0^+$.*
- ***(ii)*** *Each measure $\mu \in \mathcal{M}_{\mathbb{C}}(M)$ is uniquely determined by $\mathcal{F}\mu$.*
- ***(iii)*** *If $\{\mu_n\}$ is a sequence of measures belonging to $\mathcal{M}_+(M)$, $\mu \in \mathcal{M}_+(M)$, and $\mu_n \overset{w}{\longrightarrow} \mu$, then for each $k \in \mathbb{Z}$ we have*

$$(\mathcal{F}\mu_n)(k,\cdot) \xrightarrow[n\to\infty]{} (\mathcal{F}\mu)(k,\cdot) \qquad \textit{uniformly on compact sets.}$$

- ***(iv)*** *Suppose that $\lim_{x\to\infty} A(x) = \infty$. If $\{\mu_n\}$ is a sequence of measures belonging to $\mathcal{M}_+(M)$ whose Δ-Fourier transforms are such that*

$$(\mathcal{F}\mu_n)(k,\lambda) \xrightarrow[n\to\infty]{} f(k,\lambda) \qquad \textit{pointwise in } (k,\lambda) \in \mathbb{Z} \times \mathbb{R}_0^+$$

for some real-valued function f such that $f(0,\cdot)$ is continuous at a neighbourhood of zero, then $\mu_n \overset{w}{\longrightarrow} \mu$ for some measure $\mu \in \mathcal{M}_+(M)$ such that $\mathcal{F}\mu \equiv f$.

Proof

- ***(i)*** We have $(\mathcal{F}\mu)(k,\lambda) = (\mathcal{F}_{\ell_k}\widehat{\mu}_k)(\lambda)$ where $\mathcal{F}_{\ell_k}$ is the Sturm–Liouville transform of measures determined by ℓ_k (Definition 4.3), thus the result follows from Proposition 4.7(i).
- ***(ii)*** Let $\mu \in \mathcal{M}_{\mathbb{C}}(M)$ be such that $(\mathcal{F}\mu)(k,\lambda) = 0$ for all $k \in \mathbb{Z}$ and $\lambda \geq 0$. Let $f \in \mathrm{C_c}(\mathbb{R}_0^+)$ and $g \in \mathrm{C}^1(\mathbb{T})$. Recalling that the Fourier series $g(\theta) = \sum_{k\in\mathbb{Z}} \langle g, e^{-i2k\pi\cdot}\rangle e^{i2k\pi\theta}$ converges absolutely and uniformly [50, Theorem 1.4.2], we get

$$\begin{aligned}\int_M f(x)\, g(\theta)\, \mu(d(x,\theta)) &= \int_M f(x) \sum_{k\in\mathbb{Z}} \langle g, e^{-i2k\pi\cdot}\rangle\, e^{i2k\pi\theta}\, \mu(d(x,\theta)) \\ &= \sum_{k\in\mathbb{Z}} \langle g, e^{-i2k\pi\cdot}\rangle \int_{\mathbb{R}_0^+} f(x)\, \widehat{\mu}_{-k}(dx) \\ &= 0,\end{aligned}$$

where the last equality holds because, by Proposition 4.7(ii), $(\mathcal{F}\mu)(k, \cdot) \equiv 0$ implies that $\widehat{\mu}_k = 0$. By the Stone-Weierstrass theorem (see [169, Section 38] and also [102, Corollary 15.3]), this implies that μ is the zero measure.

(iii) This follows directly from Proposition 4.7(iii).

(iv) By the same argument in the proof of Proposition 3.4(iv), it is enough to show that $\{\mu_n\}$ is tight. Fix $\varepsilon > 0$. Given that $f(0, \cdot)$ is continuous near zero, we can choose $\delta > 0$ such that

$$\left| \frac{1}{\delta} \int_0^{2\delta} \bigl(f(0,0) - f(0,\lambda)\bigr) d\lambda \right| < \varepsilon.$$

Furthermore, we have $\lim_{x\to\infty} \widetilde{w}_{0,\lambda}(x) = 0$ for all $\lambda > 0$ (Corollary 4.1), and thus we can pick $0 < \beta < \infty$ such that

$$\int_0^{2\delta} \bigl(1 - V_{0,\lambda}(\boldsymbol{\xi})\bigr) d\lambda \geq \delta \qquad \text{for all } (x,\theta) \in (\beta,\infty) \times \mathbb{T}.$$

We now compute

$$\begin{aligned} \mu_n\bigl([\beta,\infty) \times \mathbb{T}\bigr) &\leq \frac{1}{\delta} \int_{[\beta,\infty)\times\mathbb{T}} \int_0^{2\delta} \bigl(1 - V_{0,\lambda}(\boldsymbol{\xi})\bigr) d\lambda \, \mu_n(d\boldsymbol{\xi}) \\ &\leq \frac{1}{\delta} \int_0^{2\delta} \bigl((\mathcal{F}\mu_n)(0,0) - (\mathcal{F}\mu_n)(0,\lambda)\bigr) d\lambda \end{aligned}$$

and from this inequality we obtain

$$\begin{aligned} \limsup_{n\to\infty} \mu_n([\beta,\infty) \times \mathbb{T}) &\leq \frac{1}{\delta} \limsup_{n\to\infty} \int_0^{2\delta} \bigl((\mathcal{F}\mu_n)(0,0) - (\mathcal{F}\mu_n)(0,\lambda)\bigr) d\lambda \\ &= \frac{1}{\delta} \int_0^{2\delta} \bigl(f(0,0) - f(0,\lambda)\bigr) d\lambda \\ &< \varepsilon, \end{aligned}$$

where ε is arbitrary, showing that $\{\mu_n\}$ is tight.

In the remainder of this section we will always assume that $\lim_{x\to\infty} A(x) = \infty$.

Corollary 5.5 *For each $k \in \mathbb{Z}$, the mapping $(\mu, \nu) \mapsto \mu \underset{k}{*} \nu$ is continuous in the weak topology.*

Proof We have $\bigl(\mathcal{F}(\delta_{\xi_1} \underset{k}{*} \delta_{\xi_2})\bigr)(j,\lambda) = e^{-i2j\pi(\theta_1+\theta_2)} (\mathcal{F}_{\ell_j}(\delta_{x_1} \underset{k}{\diamond} \delta_{x_2}))(\lambda)$, and it follows from Proposition 4.9 that the right-hand side is a continuous function of $(\boldsymbol{\xi}_1, \boldsymbol{\xi}_2)$. Using Proposition 5.15(iv) we conclude that $(\boldsymbol{\xi}_1, \boldsymbol{\xi}_2) \mapsto \delta_{\xi_1} \underset{k}{*} \delta_{\xi_2}$ is continuous in the weak topology, which readily implies our claim.

The operator $\mathcal{T}_k^\mu$ defined by the integral

$$(\mathcal{T}_k^\mu h)(\boldsymbol{\xi}) := \int_M h \, d(\delta_{\boldsymbol{\xi}} \underset{k}{*} \mu)$$

is said to be the Δ_k-*translation* by the measure $\mu \in \mathcal{M}_{\mathbb{C}}(M)$. The next result summarizes its mapping properties. For brevity, we write $L_k^p := L^p(M, B_k(x)dxd\theta)$.

Proposition 5.16

(a) *If $h \in \mathrm{C_b}(M)$, then $\mathcal{T}_k^\mu h \in \mathrm{C_b}(M)$ for all $\mu \in \mathcal{M}_{\mathbb{C}}(M)$.*
(b) *If $h \in \mathrm{C_0}(M)$, then $\mathcal{T}_k^\mu h \in \mathrm{C_0}(M)$ for all $\mu \in \mathcal{M}_{\mathbb{C}}(M)$.*
(c) *Let $1 \leq p \leq \infty$, $\mu \in \mathcal{M}_+(M)$ and $h \in L_k^p$. The Δ_k-translation $(\mathcal{T}_k^\mu h)(x)$ is a Borel measurable function of $x \in M$, and we have*

$$\|\mathcal{T}_k^\mu h\|_{L_k^p} \leq \|\mu\| \cdot \|h\|_{L_k^p} \tag{5.45}$$

(d) *Let $p_1, p_2 \in [1, \infty]$ such that $\frac{1}{p_1} + \frac{1}{p_2} \geq 1$, and write $\mathcal{T}_k^{\boldsymbol{\xi}} := \mathcal{T}_k^{\delta_{\boldsymbol{\xi}}}$ ($\boldsymbol{\xi} \in M$). For $h \in L_k^{p_1}$ and $g \in L_k^{p_2}$, the Δ_k-convolution*

$$(h \underset{k}{*} g)(\boldsymbol{\xi}) = \int_M (\mathcal{T}_k^{\boldsymbol{\xi}_1} h)(\boldsymbol{\xi}) \, g(\boldsymbol{\xi}_1) \, B_k(x_1) dx_1 d\theta_1$$

is well-defined and, for $s \in [1, \infty]$ defined by $\frac{1}{s} = \frac{1}{p_1} + \frac{1}{p_2} - 1$, it satisfies

$$\|h \underset{k}{*} g\|_{L_k^s} \leq \|h\|_{L_k^{p_1}} \|g\|_{L_k^{p_2}}$$

(in particular, $h \underset{k}{*} g \in L_k^s$).

Proof

(a) Immediate consequence of Corollary 5.5.
(b) By Proposition 4.10(iv) and a dominated convergence argument we have $\big(\mathcal{F}(\delta_{\boldsymbol{\xi}} \underset{k}{*} \mu)\big)(j, \lambda) = \int_M e^{-i2j\pi(\theta+\theta_1)} (\mathcal{F}_{\ell_j}(\delta_x \underset{k}{\diamond} \delta_{x_1}))(\lambda) \mu(d\boldsymbol{\xi}_1) \longrightarrow 0$ as $x \to \infty$. An argument similar to that of the proof of 4.10(iv) then yields that $\mathcal{T}_k^\mu h \in \mathrm{C_0}(M)$ for all $\mu \in \mathcal{M}_{\mathbb{C}}(M)$.
(c) In the case $p = \infty$, the proof is straightforward. Let $1 \leq p < \infty$. Suppose first that $h(x, \theta) = f(x)g(\theta)$ and observe that

$$(\mathcal{T}_k^{(x_2,\theta_2)} h)(x_1, \theta_1) = (\mathcal{T}_{\ell_k}^{x_2} f)(x_1) \cdot (\mathcal{T}_{\mathbb{T}}^{\theta_2} g)(\theta_1),$$

where $\mathcal{T}_{\ell_k}^x$ is the generalized translation associated with the Sturm–Liouville operator ℓ_k and $(\mathcal{T}_{\mathbb{T}}^{\theta_2} g)(\theta_1) := g(\theta_1 + \theta_2)$ is the ordinary translation on the torus.

By Proposition 4.10(i) we have $\|\mathcal{T}^x_{\ell_k} f\|_{L^p(\mathbb{R}^+,\, B_k(x)dx)} \leq \|f\|_{L^p(\mathbb{R}^+,\, B_k(x)dx)}$, and therefore

$$\begin{aligned}\|\mathcal{T}_k^{(x_2,\theta_2)} h\|_{L_k^p} &= \|\mathcal{T}^x_{\ell_k} f\|_{L^p(\mathbb{R}^+,\, B_k(x)dx)} \|\mathcal{T}^{\theta_2}_{\mathbb{T}} g\|_{L^p(\mathbb{T})}\\ &\leq \|f\|_{L^p(\mathbb{R}^+,\, B_k(x)dx)} \|g\|_{L^p(\mathbb{T})}\\ &= \|h\|_{L_k^p}.\end{aligned}$$

A continuity argument then yields that $\|\mathcal{T}_k^{(x,\theta)} h\|_{L_k^p} \leq \|h\|_{L_k^p}$ for all $h \in L_k^p$ and $(x,\theta) \in M$, showing that (5.45) holds for Dirac measures $\mu = \delta_{(x,\theta)}$. We can then extend the result to all $\mu \in \mathcal{M}_+(M)$ by using Minkowski's integral inequality.

(d) Identical to that of Proposition 3.18.

In the next statement we show that if a heat kernel exists for the heat semigroup $\{e^{t\Delta_N}\}$, then the functions $e^{t\Delta_N} V_{k,\lambda} = e^{-t\lambda} V_{k,\lambda}$ also admit a product formula whose measures do not depend on the spectral parameter λ and, moreover, are absolutely continuous with respect to Ω_g.

Proposition 5.17 *Assume that the action of $e^{t\Delta_N}$ on $L^2(M)$ is given by a symmetric heat kernel satisfying conditions I and II of Proposition 5.11. Let $\boldsymbol{\gamma}_{t,k,\xi_1,\xi_2}$ be the positive measure defined by*

$$\boldsymbol{\gamma}_{t,k,\xi_1,\xi_2}(d\xi_3) = \int_M \boldsymbol{\gamma}_{k,\xi_4,\xi_2}(d\xi_3)\, p(t,\xi_1,\xi_4)\Omega_g(d\xi_4).$$

Then, the product $e^{-t\lambda} V_{k,\lambda}(\xi_1)\, V_{k,\lambda}(\xi_2)$ admits, for $t > 0$ the integral representation

$$e^{-t\lambda} V_{k,\lambda}(\xi_1)\, V_{k,\lambda}(\xi_2) = \int_M V_{k,\lambda}(\xi_3)\, \boldsymbol{\gamma}_{t,k,\xi_1,\xi_2}(d\xi_3) \qquad \big(\xi_1, \xi_2 \in M,\ \lambda \in \operatorname{supp}(\boldsymbol{\rho}_k)\big).$$

Proof By direct calculation we get

$$\begin{aligned}\int_M V_{k,\lambda}(\xi_3)\, \boldsymbol{\gamma}_{t,k,\xi_1,\xi_2}(d\xi_3) &= \int_M \int_M V_{k,\lambda}(\xi_3) \boldsymbol{\gamma}_{k,\xi_4,\xi_2}(d\xi_3)\, p(t,\xi_1,\xi_4)\Omega_g(d\xi_4)\\ &= e^{-t\lambda} V_{k,\lambda}(\xi_1)\, V_{k,\lambda}(\xi_2),\end{aligned}$$

where, by Proposition 5.13 and Corollary 5.4, the second equality holds for $t \geq 0$, $\lambda \in \operatorname{supp}(\boldsymbol{\rho}_k)$, $\xi_2 \in M$ and Ω_g-almost every $\xi_1 \in M$. Using the symmetry relation $\int_M V_{k,\lambda}(\xi_3)\, \boldsymbol{\gamma}_{t,k,\xi_1,\xi_2}(d\xi_3) = \int_M V_{k,\lambda}(\xi_3)\, \boldsymbol{\gamma}_{t,k,\xi_2,\xi_1}(d\xi_3)$, the identity extends by continuity to all $\xi_1, \xi_2 \in M$. (The given symmetry can be deduced by noting that, by (4.43) and Propositions 5.10–5.11, we have for $g \in \mathrm{C}_\mathrm{c}^2(\mathbb{R}^+)$

$$\begin{aligned}\int_M & e^{i2k\pi\theta_3} g(x_3)\boldsymbol{\gamma}_{t,k,\boldsymbol{\xi}_1,\boldsymbol{\xi}_2}(d\boldsymbol{\xi}_3)\\ &= \int_M \int_{\mathbb{R}_0^+} V_{k,\lambda}(\boldsymbol{\xi}_4)\, V_{k,\lambda}(\boldsymbol{\xi}_2)\, (\mathcal{F}_{\Delta_k} g)(\lambda)\, \boldsymbol{\rho}_k(d\lambda)\, p(t,\boldsymbol{\xi}_1,\boldsymbol{\xi}_4)\Omega_g(d\boldsymbol{\xi}_4)\\ &= \int_{\mathbb{R}_0^+} e^{-t\lambda} V_{k,\lambda}(\boldsymbol{\xi}_1)\, V_{k,\lambda}(\boldsymbol{\xi}_2)\, (\mathcal{F}_{\Delta_k} g)(\lambda)\, \boldsymbol{\rho}_k(d\lambda)\end{aligned}$$

and, therefore, $(\widehat{\boldsymbol{\gamma}_{t,k,\boldsymbol{\xi}_1,\boldsymbol{\xi}_2}})_{-k} = (\widehat{\boldsymbol{\gamma}_{t,k,\boldsymbol{\xi}_2,\boldsymbol{\xi}_1}})_{-k}\cdot)$

5.4.3 Infinitely Divisible Measures and Convolution Semigroups

The basic notions of divisibility and probabilistic harmonic analysis for the Δ_k-convolution can be defined in the usual way (cf. Sects. 3.5 and 4.5):

Definition 5.4

- The set $\mathcal{P}_{k,\mathrm{id}}$ of Δ_k-*infinitely divisible measures* is defined by

$$\mathcal{P}_{k,\mathrm{id}} = \bigl\{\mu \in \mathcal{P}(M) \bigm| \text{for all } n \in \mathbb{N} \text{ there exists } \nu_n \in \mathcal{P}(M) \text{ such that } \mu = (\nu_n)^{*_k n}\bigr\}, \tag{5.46}$$

  where $(\nu_n)^{*_k n}$ denotes the n-fold Δ_k-convolution of the measure ν_n with itself.
- The Δ_k-*Poisson measure* associated with $\nu \in \mathcal{M}_+(M)$ is

$$\mathbf{e}_k(\nu) := e^{-\|\nu\|} \sum_{n=0}^{\infty} \frac{\nu^{*_k n}}{n!}$$

  (the infinite sum converging in the weak topology).
- A measure $\mu \in \mathcal{P}(M)$ is called a Δ_k-*Gaussian measure* if $\mu \in \mathcal{P}_{k,\mathrm{id}}$ and

$$\mu = \mathbf{e}_k(\nu) \underset{k}{*} \vartheta \quad \bigl(\nu \in \mathcal{M}_+(M),\ \vartheta \in \mathcal{P}_{k,\mathrm{id}}\bigr) \qquad \Longrightarrow \qquad \mathbf{e}_k(\nu) = \delta_{(0,0)}.$$

It is easy to check that, for $\nu \in \mathcal{M}_+(M)$, $(j,\lambda) \in \mathbb{Z} \times \mathbb{R}_0^+$,

$$\int_M e^{-i2j\pi\theta}\, \widetilde{w}_{k,\lambda}(x)\, \mathbf{e}_k(\nu)(d\boldsymbol{\xi}) = \exp\biggl(\int_M \bigl[e^{-i2j\pi\theta}\, \widetilde{w}_{k,\lambda}(x) - 1\bigr]\nu(d\boldsymbol{\xi})\biggr). \tag{5.47}$$

(This is an equivalent characterization of Δ_k-Poisson measures, because by [19, Theorem 2.2.4] each measure $\mu \in \mathcal{M}_{\mathbb{C}}(M)$ is characterized by the integrals $\int_M e^{-i2j\pi\theta}\, \widetilde{w}_{k,\lambda}(x)\mu(d\boldsymbol{\xi})$.) More generally, if the positive measure ν is (possibly) unbounded and the equality (5.47) holds for some measure $\mathbf{e}_k(\nu) \in \mathcal{P}(M)$, then we will also say that $\mathbf{e}_k(\nu)$ is a Δ_k-*Poisson measure* associated with ν.

Definition 5.5 A family $\{\mu_t\}_{t\geq 0} \subset \mathcal{P}(M)$ is a Δ_k-*convolution semigroup* if it satisfies the conditions

$$\mu_s \underset{k}{*} \mu_t = \mu_{s+t} \text{ for all } s,t \geq 0, \qquad \mu_0 = \delta_{(0,0)} \qquad \text{and} \quad \mu_t \xrightarrow{w} \delta_{(0,0)} \text{ as } t \downarrow 0.$$

The Δ_k-convolution semigroup $\{\mu_t\}_{t\geq 0}$ is said to be *Gaussian* if μ_1 is a Δ_k-Gaussian measure.

A measure $\mu \in \mathcal{M}_{\mathbb{C}}(M)$ is said to be *symmetric* if $\mu(B) = \mu(\check{B})$ for all Borel subsets $B \subset M$, where $\check{B}$ is the image of B under the mapping $(x,\theta) \mapsto (x, 1-\theta)$. One can show that for each symmetric measure $\mu \in \mathcal{P}_{k,\mathrm{id}}$ there exists a unique Δ_k-convolution semigroup $\{\mu_t\}_{t\geq 0}$ such that $\mu_1 = \mu$; consequently, there is a one-to-one correspondence between symmetric Δ_k-infinitely divisible measures and symmetric Δ_k-convolution semigroups. (The proof is similar to that of the corresponding result for the ordinary convolution on the torus, see also [19, Theorem 5.3.4].)

Proposition 5.18 (Lévy–Khintchine Type Representation) *Any symmetric measure $\mu \in \mathcal{P}_{k,\mathrm{id}}$ can be represented as*

$$\mu = \gamma \underset{k}{*} \mathbf{e}_k(\nu),$$

where $\mathbf{e}_k(\nu)$ is the Δ_k-Poisson measure associated with the σ-finite positive measure $\nu = \lim_{t\downarrow 0}(\frac{1}{t}\mu_t)|_{M\setminus(0,0)}$ and γ is a Δ_k-Gaussian measure.

The representation is unique, i.e. if $\mu = \widetilde{\gamma} \underset{k}{*} \mathbf{e}_k(\widetilde{\nu})$ for a σ-finite positive measure $\widetilde{\nu}$ and a Gaussian measure $\widetilde{\gamma}$, then $\nu = \widetilde{\nu}$ and $\gamma = \widetilde{\gamma}$.

Proof This is a particular case of Proposition 2.17. (Note also that by [196, Theorem 3.1] the notion of Gaussian measures introduced in Proposition 2.17 is equivalent to the definition of Δ_k-Gaussian measures presented in Definition 5.4.)

Exactly as in the previous chapters (cf. Propositions 3.8 and 4.12), every convolution semigroup determines a transition semigroup with the expected Feller-type properties:

Proposition 5.19 *If $\{\mu_t\}_{t\geq 0}$ is a Δ_k-convolution semigroup, then the family $\{T_t\}_{t\geq 0}$ defined by*

$$(T_t h)(\boldsymbol{\xi}) := (\mathcal{T}_k^{\mu_t} h)(\boldsymbol{\xi}) = \int_M h \, d(\delta_{\boldsymbol{\xi}} \underset{k}{*} \mu_t)$$

is a conservative Feller semigroup such that the identity $T_t \mathcal{T}_k^\nu f = \mathcal{T}_k^\nu T_t f$ *holds for all* $t \geq 0$ *and* $\nu \in \mathcal{M}_\mathbb{C}(M)$*. The restriction* $\{T_t|_{\mathrm{C_c}(M)}\}$ *can be extended to a strongly continuous contraction semigroup* $\{T_t^{(p)}\}$ *on the space* $L^p(M)$ $(1 \leq p < \infty)$*. Moreover, the operators* $T_t^{(p)}$ *are given by* $T_t^{(p)} f = \mathcal{T}_k^{\mu_t} f$ $(f \in L^p(M))$*.*

The next result shows the heat semigroup generated by the Neumann Laplacian has the convolution semigroup property, in the sense that its action can be represented in terms of integrals with respect to Δ_k-convolution semigroups:

Proposition 5.20 *For* $k \in \mathbb{Z}$*, let* $\mathrm{m}_0 \in \mathcal{M}_\mathbb{C}(M)$ *be an absolutely continuous measure with respect to* Ω_g *whose density function* q_{m_0} *belongs to* $L^2(M) \cap L^1(M, \zeta_k \cdot \Omega_g)$*, and such that* $(\widehat{\mathrm{m}_0})_j = 0$ *for each* $j \neq k$*. Then there exists a Gaussian* Δ_k*-convolution semigroup* $\{\mu_t^k\}_{t\geq 0}$ *such that*

$$\int_M (e^{t\Delta_N} h)(\boldsymbol{\xi})\, \mathrm{m}_0(d\boldsymbol{\xi}) = \int_M \frac{h(\boldsymbol{\xi})}{\zeta_k(x)} \bigl(\mu_t^k \underset{k}{*} (\zeta_k \cdot \mathrm{m}_0)\bigr)(d\boldsymbol{\xi}) \qquad \bigl(h \in L^2(M),\ t \geq 0\bigr). \tag{5.48}$$

Proof For $t > 0$, let $\mu_t^k = \alpha_t^k \otimes \delta_0$, where $\alpha_t^k(dx) = p_{\ell_k}(t, 0, x) B_k(x)dx$ are the transition probabilities of the diffusion process generated by the Sturm–Liouville operator ℓ_k (cf. Proposition 4.14). We recall from the proof of Corollary 5.4 that we have $e^{-t\lambda} \in L^2(\boldsymbol{\rho_k})$ and $\alpha_t^k(dx) = (\mathcal{F}_{\ell_k}^{-1} e^{-t\cdot})(x) B_k(x)dx$, where $\mathcal{F}_{\ell_k}^{-1} e^{-t\cdot} \in L^1(\mathbb{R}^+, B_k(x)dx)$.

Our first claim is that the measure $\frac{1}{\zeta_k}(\mu_t^k \underset{k}{*} (\zeta_k \cdot \mathrm{m}_0))$ is absolutely continuous with respect to Ω_g and that its density function $q_{\mu_t^k, \mathrm{m}_0}$ belongs to $L^2(M)$. Note first that, by assumption, $(\widehat{\mathrm{m}_0})_j = 0$ for $j \neq k$, and therefore (e.g. by Proposition 5.15(ii)) $\mathrm{m}_0 = (\widehat{\mathrm{m}_0})_k \otimes \varkappa_k$, where $\varkappa_k$ is the measure on $\mathbb{T}$ defined by $\varkappa_k(d\theta) = e^{i2k\pi\theta} d\theta$. We thus have

$$\mu_t^k \underset{k}{*} (\zeta_k \cdot \mathrm{m}_0) = (\alpha_t^k \underset{k}{\diamond} (\zeta_k \cdot (\widehat{\mathrm{m}_0})_k)) \otimes \varkappa_k.$$

The absolute continuity assumption on m_0 implies that $(\widehat{\mathrm{m}_0})_k(dx) = (\widehat{q_{\mathrm{m}_0}})_k(x) A(x)dx$ with $(\widehat{q_{\mathrm{m}_0}})_k \in L^2(A)$, so we can now use the properties of the convolution $\underset{k}{\diamond}$ (see Proposition 4.19) to conclude that $\frac{1}{\zeta_k}(\alpha_t^k \underset{k}{\diamond} (\zeta_k \cdot (\widehat{\mathrm{m}_0})_k))$ is also absolutely continuous with respect to $A(x)dx$ with density belonging to $L^2(A)$, and this proves the claim.

Let $h \in L^2(M)$. Combining the above with Proposition 5.10, we may now compute

$$\begin{aligned} \int_M (e^{t\Delta_N} h)(\boldsymbol{\xi})\, \mathrm{m}_0(d\boldsymbol{\xi}) &= \bigl\langle e^{t\Delta_N} h, \overline{q_{\mathrm{m}_0}} \bigr\rangle_{L^2(M)} \\ &= \sum_{j\in\mathbb{Z}} \bigl\langle \mathcal{F}(e^{t\Delta_N} h)_j, (\mathcal{F}\,\overline{q_{\mathrm{m}_0}})_j \bigr\rangle_{L^2(\boldsymbol{\rho}_j)} \end{aligned}$$

$$= \langle e^{-t\cdot}(\mathcal{F}h)_{-k}, (\mathcal{F}\overline{q_{\mathrm{m}_0}})_{-k}\rangle_{L^2(\rho_k)}$$

$$= \Big\langle (\mathcal{F}h)_{-k}, (\mathcal{F}\mu_t^k)(-k,\cdot)\,(\mathcal{F}\overline{q_{\mathrm{m}_0}})_{-k}\Big\rangle_{L^2(\rho_k)}$$

$$= \sum_{j\in\mathbb{Z}} \langle (\mathcal{F}h)_j, (\mathcal{F}\overline{q_{\mu_t^k,\mathrm{m}_0}})_j\rangle_{L^2(\rho_j)} = \langle h, \overline{q_{\mu_t^k,\mathrm{m}_0}}\rangle_{L^2(M)}$$

$$= \int_M \frac{h(\boldsymbol{\xi})}{\zeta_k(x)} \big(\mu_t^k \underset{k}{*} (\zeta_k\cdot \mathrm{m}_0)\big)(d\boldsymbol{\xi}),$$

so that (5.48) holds.

One can check that the Dirichlet form $(\mathcal{E}, H^1(M))$ is regular (see [24, 66]). As observed in Sect. 5.1, it follows that there exists a Hunt process $\{W_t\}_{t\geq 0}$ with state space M whose transition semigroup $\{P_t\}_{t\geq 0}$ is a quasi-continuous version of $\{e^{t\Delta_N}\}$. The process $\{W_t\}$ is called the *reflected Brownian motion* on the manifold (M, g).

The convolution semigroup property (5.48) can be rewritten as

$$\mathbb{E}_{\mathrm{m}_0}[h(W_t)] = \int_M h\, d(\widetilde{\mu}_t^k \underset{k}{\star} \mathrm{m}_0), \qquad \big(h \in L^2(M),\ t\geq 0\big), \tag{5.49}$$

where:

- $\mathbb{E}_{\mathrm{m}_0}$ is the expectation operator of the reflected Brownian motion with initial distribution $\mathrm{m}_0 \in \mathcal{M}_{\mathbb{C}}(M)$ (defined as $\mathbb{E}_{\mathrm{m}_0}[h(W_t)] := \int_M \mathbb{E}_{\xi}[h(W_t)]\mathrm{m}_0(d\boldsymbol{\xi})$, where $\mathbb{E}_{\xi}$ is the usual expectation operator for the process started at the point $\boldsymbol{\xi}$);
- The convolution $\underset{k}{\star}$ is defined by $\nu_1 \underset{k}{\star} \nu_2 = \frac{1}{\zeta_k}\big((\zeta_k\cdot\nu_1)\underset{k}{*}(\zeta_k\cdot\nu_2)\big)$ or, equivalently, by $(\nu_1 \underset{k}{\star} \nu_2)(\cdot) = \int_M\int_M \boldsymbol{\gamma}_{k,\xi_1,\xi_2}(\cdot)\,\nu_1(d\boldsymbol{\xi_1})\,\nu_2(d\boldsymbol{\xi_2})$, with $\boldsymbol{\gamma}_{k,\xi_1,\xi_2}$ given as in (5.43);
- $\widetilde{\mu}_t^k := \frac{\mu_t^k}{\zeta_k}$ (so that $\widetilde{\mu}_t^k$ satisfies the convolution semigroup property with respect to $\underset{k}{\star}$).

Corollary 5.6 *Let* $\mathrm{m}_0 \in \mathcal{M}_{\mathbb{C}}(M)$ *be an absolutely continuous measure with respect to* Ω_g *whose density function* q_{m_0} *belongs to* $L^2(M)\cap\big(\bigcap_{k=0}^{\infty} L^1(M, \zeta_k\cdot\Omega_g)\big)$. *Then there exist Gaussian* Δ_k*-convolution semigroups* $\{\mu_t^k\}_{t\geq 0}$ *such that*

$$\begin{aligned}\int_M (e^{t\Delta_N}h)(\boldsymbol{\xi})\,\mathrm{m}_0(d\boldsymbol{\xi}) &= \\ = \sum_{k\in\mathbb{Z}}\int_M e^{i2k\pi\theta}\frac{\widehat{h}_k(x)}{\zeta_k(x)}&\big(\mu_t^k \underset{k}{*} (\zeta_k\cdot \mathbf{m}_{0,-k})\big)(d\boldsymbol{\xi})\end{aligned} \qquad \big(h\in L^2(M),\ t\geq 0\big),$$

where $\mathbf{m}_{0,k} = (\widehat{\mathrm{m}_0})_k \otimes \varkappa_k$ *and* $\widehat{h}_k$ *is given as in* (5.37).

Proof We have

$$\begin{aligned}&\int_M e^{t\Delta_N}\biggl(\sum_{k\in\mathbb{Z}} e^{i2k\pi\theta}\,\widehat{h}_k(x)\biggr)\,\mathrm{m}_0(d\boldsymbol{\xi})\\ &\qquad = \sum_{k\in\mathbb{Z}}\int_M e^{t\Delta_N}\bigl(e^{i2k\pi\theta}\,\widehat{h}_k(x)\bigr)\,\bigl((\widehat{\mathrm{m}_0})_{-k}\otimes\varkappa_{-k}\bigr)(d\boldsymbol{\xi}).\end{aligned}$$

Since each measure $(\widehat{\mathrm{m}_0})_{-k}\otimes\varkappa_{-k}$ satisfies $\bigl((\widehat{\mathrm{m}_0})_{-k}\otimes\varkappa_{-k}\widehat{\bigr)}_j = 0$ for $j\neq -k$, the corollary follows by applying Proposition 5.20 to each term in the right-hand side.

The result of Proposition 5.20 can be extended to other Markovian semigroups whose generators are functions of the Neumann Laplacian:

Proposition 5.21 *For $k\in\mathbb{Z}$, let $\mathrm{m}_0\in\mathcal{M}_{\mathbb{C}}(M)$ be an absolutely continuous measure with respect to Ω_g whose density function q_{m_0} belongs to $L^2(M)\cap L^1(M,\zeta_k\cdot\Omega_g)$, and such that $(\widehat{\mathrm{m}_0})_j = 0$ for each $j\neq k$. Let ψ_k be a function of the form*

$$\psi_k(\lambda) = c\lambda + \int_{\mathbb{R}^+}(1-\widetilde{w}_{k,\lambda}(x))\,\tau(dx)\qquad(\lambda\geq 0),\tag{5.50}$$

where $c\geq 0$ and τ is a σ-finite measure on $\mathbb{R}^+$ which is finite on the complement of any neighbourhood of 0 and such that $\int_{\mathbb{R}^+}(1-\widetilde{w}_{k,\lambda}(x))\,\tau(dx)<\infty$ for $\lambda\geq 0$. Assume also that $e^{-t\psi_k(\cdot)}\in L^2(\boldsymbol{\rho_k})$ for all $t>0$. Then there exists a Δ_k-convolution semigroup $\{\mu_t^{\psi_k}\}_{t\geq 0}$ such that

$$\begin{aligned}&\int_M (e^{-t\psi_k(-\Delta_N)}h)(\boldsymbol{\xi})\,\mathrm{m}_0(d\boldsymbol{\xi}) =\\ &\quad = \int_M \frac{h(\boldsymbol{\xi})}{\zeta_k(x)}\,\bigl(\mu_t^{\psi_k}\underset{k}{*}(\zeta_k\cdot\mathrm{m}_0)\bigr)(d\boldsymbol{\xi})\end{aligned}\qquad\bigl(h\in L^2(M),\ t\geq 0\bigr),\tag{5.51}$$

where $e^{-t\psi_k(-\Delta_N)}$ is defined via the usual spectral calculus.

We observe that, since $e^{-t\lambda}\in L^2(\boldsymbol{\rho_k})$ for all $t>0$, the assumption $e^{-t\psi_k(\cdot)}\in L^2(\boldsymbol{\rho_k})$ is automatically satisfied whenever $c>0$ in the right hand side of (5.50).

Proof By Theorem 4.5, there exists a $\underset{k}{\diamond}$-convolution semigroup $\{\alpha_t^{\psi_k}\}_{t\geq 0}$ such that $(\mathcal{F}_{\ell_k}\alpha_t^{\psi_k})(\lambda) = e^{-t\psi_k(\lambda)}$. Using Theorem 2.5 and the assumption $e^{-t\psi_k(\cdot)}\in L^2(\boldsymbol{\rho_k})$, we deduce that $\alpha_t^{\psi_k}(dx) = (\mathcal{F}_{\ell_k}^{-1}e^{-t\psi_k(\cdot)})(x)\,B_k(x)dx$, where $\mathcal{F}_{\ell_k}^{-1}e^{-t\psi_k(\cdot)}\in L^1(\mathbb{R}^+,\ B_k(x)dx)$. The result can now be proved using the same argument as in Proposition 5.20 above.

The sesquilinear form $\mathcal{E}^{\psi_k}$ associated with the Markovian self-adjoint operator $-\psi_k(-\Delta_N)$, defined as

$$\mathcal{D}(\mathcal{E}^{\psi_k}) = \mathcal{D}\big(\sqrt{\psi_k(-\Delta_N)}\big), \qquad \mathcal{E}^{\psi_k}(u, v) = \big\langle \sqrt{\psi_k(-\Delta_N)}\, u,\, \sqrt{\psi_k(-\Delta_N)}\, v \big\rangle_{L^2(M)},$$

is a regular Dirichlet form on $L^2(M)$. (We can prove this claim using Proposition 4.11 and the proof of Proposition 3.1 of [136].) Accordingly, there exists a Hunt process $\{X_t\}_{t\geq 0}$ with state space M such that $(e^{-t\psi_k(-\Delta_N)}h)(\xi) = \mathbb{E}_\xi[h(X_t)]$, and therefore the convolution semigroup property (5.51) translates into the Lévy-like representation

$$\mathbb{E}_{\mathrm{m}_0}[h(X_t)] = \int_M h\, d(\widetilde{\mu}_t^{\psi_k} \underset{k}{\star} \mathrm{m}_0) \qquad (h \in L^2(M),\ t \geq 0)$$

for the law of the process $\{X_t\}$. (Here m_0 is any complex measure satisfying the assumptions in Proposition 5.21.) The representation (5.49) for the law of reflected Brownian motion on (M, g) is a particular case of this result.

Remark 5.3 In general we cannot state a counterpart of Corollary 5.6 for semigroups generated by functions of the Neumann Laplacian. Indeed, in order to derive such a result for the semigroup $\{e^{-t\psi(-\Delta_N)}\}$ one would need that the function $\psi(\lambda)$ could be written, for each $k = 0, 1, \ldots$, as $c_k\lambda + \int (1 - \widetilde{v}_{k,\lambda})\, d\tau_k$ with $c_k \geq 0$ and τ_k measures satisfying the conditions stated in Proposition 5.21, but there are no reasons to expect that this is possible other than in the trivial case $\psi(\lambda) = c\lambda$. (See [211] for a related investigation on the set of stable infinitely divisible measures for the hypergroup $(\mathbb{R}_0^+, \diamond)$, which was shown to depend nontrivially on the eigenfunctions of the operator ℓ_k^k.)

5.4.4 *Special Cases*

We now present some special cases in which the theory of special functions provides further information on the eigenfunction expansion of Δ_N and the associated convolution structure. We start with an example where the solutions $V_{k,\lambda}$ can be expressed in terms of the Whittaker function of the second kind, and the Fourier decomposition gives rise to a family of Sturm–Liouville operators which are generators of drifted Bessel processes.

Example 5.10 Consider the case $A(x) = \sqrt{x}$, so that the Riemannian metric on $M = \mathbb{R}^+ \times \mathbb{T}$ is $g = dx^2 + x\, d\theta^2$, the volume form is $d\Omega_g = \sqrt{x}\, dx d\theta$ and the Laplace–Beltrami operator on (M, g) is $\Delta = \partial_x^2 + \frac{1}{2x}\partial_x + \frac{1}{x}\partial_\theta^2$.

(i) *The unique solution of the boundary value problem* (5.29) *is given by*

$$V_{k,\lambda}(x,\theta) = e^{i2k\pi\theta}\,(2ix\sqrt{\lambda})^{-\frac{1}{4}} M_{\frac{2(k\pi)^2 i}{\sqrt{\lambda}},-\frac{1}{4}}(2ix\sqrt{\lambda}), \tag{5.52}$$

where $M_{\alpha,\nu}(z)$ *is the Whittaker function of the first kind.*

One can verify this by noting that the Sturm–Liouville equation $-\Delta_k\, w = \lambda w$ is equivalent, up to a change of variables, to $\mathfrak{L}_k v = \widetilde{\lambda} v$, where $\mathfrak{L}_k = -\frac{d^2}{dx^2} - \bigl(\frac{1}{2x} + (4k\pi)^2\bigr)\frac{d}{dx}$ and $\widetilde{\lambda} = \lambda + 4(2k\pi)^4$. The solutions of $\mathfrak{L}_k v = \widetilde{\lambda} v$ were described in Example 2.3.

(ii) *Let* $\sigma_k(\lambda) := 2^{-3/2}\pi^{-2}\lambda^{-1/4}\exp\bigl(-\frac{2k^2\pi^3}{\sqrt{\lambda}}\bigr)\bigl|\Gamma\bigl(\frac{1}{4} - \frac{2(k\pi)^2 i}{\sqrt{\lambda}}\bigr)\bigr|^2$. *The integral operator* $\mathcal{F} : L^2(M) \to \bigoplus_{k\in\mathbb{Z}} L^2(\mathbb{R}^+, \sigma_k(\lambda)d\lambda)$ *defined by*

$$(\mathcal{F}h)_k(\lambda) := \int_0^\infty\!\!\int_0^1 h(x,\theta)\, e^{-i2k\pi\theta} d\theta\,(2ix\sqrt{\lambda})^{-\frac{1}{4}} M_{\frac{2(k\pi)^2 i}{\sqrt{\lambda}},-\frac{1}{4}}(2ix\sqrt{\lambda})\, x^{1/2} dx$$

is a spectral representation of the Laplace–Beltrami operator (cf. Proposition 5.10), and its inverse is given by

$$(\mathcal{F}^{-1}\{\varphi_k\})(\boldsymbol{\xi}) = \sum_{k\in\mathbb{Z}} \int_0^\infty \varphi_k(\lambda)\, e^{i2k\pi\theta}\,(2ix\sqrt{\lambda})^{-\frac{1}{4}} M_{\frac{2(k\pi)^2 i}{\sqrt{\lambda}},-\frac{1}{4}}(2ix\sqrt{\lambda})\,\sigma_k(\lambda)d\lambda.$$

To obtain this result we just need to recall from Proposition 5.10 that $(\mathcal{F}h)_k(\lambda) \equiv (\mathcal{F}_{\Delta_k}\widehat{h}_k)(\lambda)$ and $(\mathcal{F}^{-1}\{\varphi_k\})(\boldsymbol{\xi}) \equiv \sum_{k\in\mathbb{Z}} e^{i2k\pi\theta}(\mathcal{F}_{\Delta_k}^{-1}\varphi_k)(x)$, and then observe that the eigenfunction expansion of Δ_k can be deduced from that of $\mathfrak{L}_k$ (Example 2.3) by elementary changes of variable.

(iii) *For each* $k \in \mathbb{N}_0$ *and* $\boldsymbol{\xi}_j = (x_j, \theta_j) \in M$ $(j = 1, 2)$ *there exists a positive measure* $\boldsymbol{\gamma}_{k,\boldsymbol{\xi}_1,\boldsymbol{\xi}_2}$ *on* M *such that for all* $\tau \in \mathbb{C}$ *the solutions* (5.52) *satisfy*

$$\begin{aligned} &e^{i2k\pi(\theta_1+\theta_2)}(2i\tau x_1 x_2)^{-\frac{1}{4}}\, M_{\frac{2(k\pi)^2 i}{\tau},-\frac{1}{4}}(2ix_1\tau)\, M_{\frac{2(k\pi)^2 i}{\tau},-\frac{1}{4}}(2ix_2\tau) \\ &\qquad = \int_M e^{i2k\pi\theta_3}\, x_3^{-\frac{1}{4}}\, M_{\frac{2(k\pi)^2 i}{\tau},-\frac{1}{4}}(2ix_3\tau)\,\boldsymbol{\gamma}_{k,\boldsymbol{\xi}_1,\boldsymbol{\xi}_2}(d\boldsymbol{\xi}_3). \end{aligned}$$

The support of measure $\boldsymbol{\gamma}_{k,\boldsymbol{\xi}_1,\boldsymbol{\xi}_2}$ *is the set* $[|x_1 - x_2|, x_1 + x_2] \times \{\theta_1 + \theta_2\}$.

This product formula is a particular case of Proposition 5.13. One should notice that $\boldsymbol{\gamma}_{k,\boldsymbol{\xi}_1,\boldsymbol{\xi}_2}(d\boldsymbol{\xi}) = e^{\mu(x_1+x_2-x_3)}\,\boldsymbol{\nu}_{x_1,x_2}(dx_3)\,\delta_{\theta_1+\theta_2}(d\theta_3)$, where $\boldsymbol{\nu}_{x_1,x_2}$ is the measure of the product formula (4.53) (with parameters $\alpha = \frac{1}{2}$ and $\mu = 2(2k\pi)^4$).

It was noted by Boscain and Prandi [24] that one can formally interpret the manifold (M, g) as a cone-like surface of revolution $\mathcal{S} = \{(t, r(t)\cos\theta, r(t)\sin\theta) \mid t > 0,\, \theta \in \mathbb{T}\}$ with profile $r(t) \sim \sqrt{t}$ as $t \downarrow 0$. The properties of self-adjoint extensions of the Laplace–Beltrami operator (and the corresponding Markovian semigroups) on such cone-like manifolds have been widely studied, see [24] and references therein. As a particular case of Corollary 5.6, we obtain the following convolution semigroup property for the heat semigroup generated by the Neumann realization of the Laplace–Beltrami operator on (M, g):

(iv) *If* $\mathrm{m}_0 \in \mathcal{M}_{\mathbb{C}}(M)$ *satisfies the absolute continuity assumption of Corollary 5.6, then the transition probabilities of the reflected Brownian motion* $\{W_t\}$ *on the manifold* (M, g) *with initial distribution* m_0 *can be written as*

$$\mathbb{E}_{\mathrm{m}_0}[h(W_t)] \equiv \int_M (e^{t\Delta_N} h)(\boldsymbol{\xi})\, \mathrm{m}_0(d\boldsymbol{\xi}) = \sum_{k\in\mathbb{Z}} \int_M e^{i2k\pi\theta}\, \widehat{h}_k(x)\, \bigl(\widetilde{\mu}_t^k \underset{k}{\star} \mathbf{m}_{0,-k}\bigr)(d\boldsymbol{\xi}), \tag{5.53}$$

where $\{\widetilde{\mu}_t^k\}_{t\geq 0}$ *is a convolution semigroup with respect to the convolution* $\underset{k}{\star}$ *defined by* $(\mu \underset{k}{\star} \nu)(\cdot) = \int_M \int_M \boldsymbol{\gamma}_{k,\boldsymbol{\xi}_1,\boldsymbol{\xi}_2}(\cdot)\, \mu(d\boldsymbol{\xi}_1)\, \nu(d\boldsymbol{\xi}_2)$ *and the measures* $\mathbf{m}_{0,-k}$ *are defined as in Corollary 5.6.*

The convolution semigroups $\{\widetilde{\mu}_t^k\}$ are, by definition, of the form $(\frac{\alpha_t^k}{\zeta_k}) \otimes \delta_0$, where $\{\alpha_t^k\}$ is the law of the one-dimensional diffusion process (started at $x = 0$) generated by the Sturm–Liouville operator ℓ_k defined in (5.32). The equations $\ell_k u = \lambda u$ and $\mathfrak{L}_k v = \widetilde{\lambda} v$ are related via a change of variables; therefore, the identity (5.53) can be intepreted as a decomposition of the law of $\{W_t\}$ in terms of transition probabilities of (one-dimensional) drifted Bessel processes.

Finally, we call attention to the following convolution semigroup representation for Markovian semigroups generated by fractional powers of the Laplace–Beltrami operator:

(v) *Let* $\mathrm{m}_0 \in \mathcal{M}_{\mathbb{C}}(M)$ *satisfy the assumptions of Proposition 5.21 and* $(\widehat{\mathrm{m}_0})_j = 0$ *for each* $j \neq 0$. *Let* $0 < q < 1$. *Then the Markovian semigroup generated by the operator* $-(-\Delta_N)^q$ *is such that*

$$\int_M (e^{-t(-\Delta_N)^q} h)(\boldsymbol{\xi})\, \mathrm{m}_0(d\boldsymbol{\xi}) = \int_M h(\xi)\, \bigl(\nu_{q,t} \underset{0}{\star} \mathrm{m}_0\bigr)(d\boldsymbol{\xi}),$$

where $\{\nu_{q,t}\}_{t\geq 0}$ *is a* $\underset{0}{\star}$*-convolution semigroup.*

This representation can be obtained from Proposition 5.21 after observing that the convolution $\underset{0}{\star}$ is (modulo the product with the trivial convolution on the torus) the Kingman convolution with parameter $\eta = -\frac{1}{4}$, so that by Theorem 2.7 the function $\psi_0(\lambda) = \lambda^q$ belongs to the set of admissible functions of the form (5.50).

Example 5.11 Consider now the more general case $A(x) = x^\beta$ with $0 < \beta < 1$. The corresponding Riemannian metric, $g = dx^2 + x^{2\beta} d\theta^2$, endows the space $M = \mathbb{R}^+ \times \mathbb{T}$ with a metric structure which, like in the previous example, can be formally interpreted as that of a surface of revolution with profile $r(t) \sim t^{\beta}$ as $t \downarrow 0$.

If $\beta \neq \frac{1}{2}$, the solution of the boundary value problem (5.29) and the spectral measures $\boldsymbol{\rho_k}$ $(k = 1, 2, \ldots)$ can no longer be written in closed form. Notwithstanding, it is clear that the convolution semigroup property of the Laplace–Beltrami operator $\Delta = \partial_x^2 + \frac{\beta}{x}\partial_x + \frac{1}{x^{2\beta}}\partial_\theta^2$ on the cone-like manifold (M, g), stated in property *(iv)* of the previous example, continues to hold here. Moreover, the convolution $\underset{0}{\star}$ is now the Kingman convolution with parameter $\eta = \frac{\beta-1}{2}$, and therefore the convolution semigroup representation for $\{e^{-t(-\Delta_N)^q}\}_{t\geq 0}$, formulated in property *(v)* of the previous example, extends to the present setting without any essential change.

In the latter example, the limiting case $\beta = 0$ corresponds to the trivial (product) metric on the cylinder $\mathbb{R}^+ \times \mathbb{T}$, for which the convolutions introduced in the previous subsections have a particularly simple structure:

Example 5.12 If $A \equiv 1$, the Fourier decomposition (5.28) yields the Sturm–Liouville operators $\Delta_k = \partial_x^2 - (2k\pi)^2$. The eigenfunction expansion (5.33)–(5.34) is simply a composition of a Fourier series in the variable θ and a cosine Fourier transform in the variable x,

$$(\mathcal{F}h)_k(\lambda) = \int_0^\infty \int_0^1 h(\boldsymbol{\xi}) e^{-i2k\pi\theta} d\theta \, \cos(x\widetilde{\lambda}_k)\, dx \qquad (\widetilde{\lambda}_k = \sqrt{\lambda - (2k\pi)^2}),$$

$$(\mathcal{F}^{-1}\{\varphi_k\})(\boldsymbol{\xi}) = \frac{1}{\pi} \sum_{k\in\mathbb{Z}} \int_{(2k\pi)^2}^\infty \varphi_k(\lambda) e^{i2k\pi\theta} \cos(x\widetilde{\lambda}_k)\, \widetilde{\lambda}_k^{-1} d\lambda,$$

and the product formula $V_{k,\lambda}(\boldsymbol{\xi_1})\, V_{k,\lambda}(\boldsymbol{\xi_2}) = \int_M V_{k,\lambda}\, d\boldsymbol{\gamma}_{k,\boldsymbol{\xi}_1,\boldsymbol{\xi}_2}$ (where $V_{k,\lambda}(\boldsymbol{\xi_1}) = e^{-i2k\pi\theta}\cos(x\widetilde{\lambda}_k)$) holds for the measures $\boldsymbol{\gamma}_{k,\boldsymbol{\xi}_1,\boldsymbol{\xi}_2} = \frac{1}{2}(\delta_{|x_1-x_2|} + \delta_{x_1+x_2}) \otimes \delta_{\theta_1+\theta_2}$, which do not depend on k. Accordingly, the convolution $\star \equiv \underset{k}{\star}$ has the product structure

$$\delta_{\boldsymbol{\xi}_1} \star \delta_{\boldsymbol{\xi}_2} = (\delta_{x_1} \underset{\text{sym}}{\diamond} \delta_{x_2}) \otimes (\delta_{\theta_1} \underset{\mathbb{T}}{\diamond} \delta_{\theta_2}),$$

where $\underset{\text{sym}}{\diamond}$ is the symmetric convolution (Example 4.1(a)) and $\underset{\mathbb{T}}{\diamond}$ is the ordinary convolution on the torus. In turn, the convolutions $\underset{k}{*}$ of Definition 5.2 are, modulo the product with the convolution on $\mathbb{T}$, identical to the convolutions of the so-called *cosh hypergroups* (as defined in [19, Example 3.5.71]).

We proceed with another example where the family of Δ_k-convolutions on (M, g) yields a generalization of a well-known one-dimensional generalized convolution:

Example 5.13 Consider $A(x) = (\sinh x)^{2\alpha+1}(\cosh x)^{2\beta+1}$, which satisfies condition (5.24) provided that $-\frac{1}{2} \leq \beta \leq \alpha < 0$ with $\alpha \neq -\frac{1}{2}$. The first component in the Fourier decomposition of the Laplace–Beltrami operator is the Sturm–Liouville operator $\Delta_0 = \partial_x^2 + [(2\alpha+1)\coth(x) + (2\beta+1)\tanh(x)]\partial_x$, hence the convolution $\underset{0}{\star}$ is the product of the convolution of the Jacobi hypergroup (Example 4.1(e)) with the ordinary convolution on the torus.

The results of the previous subsections show that the Sturm–Liouville solutions determined by the operators $\Delta_k = \Delta_0 - (2k\pi)^2(\sinh x)^{-4\alpha-2}(\cosh x)^{-4\beta-2}$ also admit a product formula, whose measures are also supported on $[|x_1 - x_2|, x_1 + x_2]$. The convolution structure associated with the Laplace–Beltrami operator (5.25) on (M, g) is therefore a natural two-dimensional extension of the Jacobi hypergroup.

In all the examples above, the support of the convolution $\delta_{\xi_1} \underset{k}{*} \delta_{\xi_2} = \nu_{k,\xi_1,\xi_2}$ does not depend on the parameter k. Our final example shows that this is not always the case:

Example 5.14 Let $\phi \in \mathrm{C}_\mathrm{c}^\infty(\mathbb{R}_0^+)$ be a nonnegative decreasing function with $\mathrm{supp}(\phi) = [0, S]$ and let $A(x) = \exp(\int_0^x \phi(y)dy)$. By definition we have $\delta_{\xi_1} \underset{k}{*} \delta_{\xi_2} = \pi_{x_1,x_2}^{[k]} \otimes \delta_{\theta_1+\theta_2}$, where $\pi_{x_1,x_2}^{[k]}$ is the measure of the product formula for the Sturm–Liouville solutions determined by $\ell_k = -\frac{1}{B_k}\frac{d}{dx}(B_k \frac{d}{dx})$. It follows from Proposition 4.17 that the supports of the measures $\pi_{x_1,x_2}^{[k]}$ are given by

$$\mathrm{supp}(\pi_{x_1,x_2}^{[0]})$$
$$= \begin{cases} [|x_1 - x_2|, x_1 + x_2], & \min\{x_1, x_2\} \le 2S, \\ [|x_1 - x_2|, 2S + |x_1 - x_2|] \cup [x_1 + x_2 - 2S, x_1 + x_2], & \min\{x_1, x_2\} > 2S. \end{cases}$$

and

$$\mathrm{supp}(\pi_{x_1,x_2}^{[k]}) = [|x_1 - x_2|, x_1 + x_2], \qquad k \ge 1.$$

5.4.5 *Product Formulas and Convolutions Associated with Elliptic Operators on Subsets of* $\mathbb{R}^2$

The goal of this subsection is to show that the techniques used above also allow us to construct families of convolution-like operators for elliptic differential operators on $\mathbb{R}^+ \times I \subset \mathbb{R}^2$ of the general form

$$\mathcal{G}^{\wp} = \partial_x^2 + \frac{A'(x)}{A(x)}\partial_x + \frac{1}{A(x)^2}\wp_z \qquad (x \in \mathbb{R}^+,\ z \in I),$$

where $\wp_z = \frac{1}{\mathfrak{r}(z)}\big(\mathfrak{p}(z)\partial_z^2 + \mathfrak{p}'(z)\partial_z\big)$ is a Sturm–Liouville operator which admits an associated generalized convolution. As in Sect. 2.2 we assume that the coefficients of $\wp$ are such that $\mathfrak{p}, \mathfrak{r} > 0$ on (a,b), $\mathfrak{p}, \mathfrak{r}$ are locally absolutely continuous on (a,b) and $\int_a^c \int_y^c \frac{dz}{\mathfrak{p}(z)}\, \mathfrak{r}(y) dy < \infty$. The coefficient $A(x)$ is assumed to satisfy the conditions (5.24) and $\lim_{x\to\infty} A(x) = \infty$.

Let us fix some notation: set $I = [a,b)$ if b is an exit or natural endpoint of $\wp$ and $I = [a,b]$ if the endpoint b is regular or entrance. We shall write $M = \mathbb{R}_0^+ \times I$ and $\Omega^\wp\bigl(d(x,z)\bigr) = A(x)dx\, \mathfrak{r}(z)dz$. We denote by ψ_η the unique solution of $-\wp_z(u) = \eta u$, $u(a) = 1$, $(\mathfrak{p}u')(a) = 0$ (cf. Lemma 2.1), and the eigenfunction expansion for the Neumann realization of $\wp$ (cf. Theorem 2.5) will be denoted as

$$(\mathcal{J}_\wp g)(\eta) := \int_a^b g(z)\, \psi_\eta(z)\, \mathfrak{r}(z) dz,$$

$$(\mathcal{J}_\wp^{-1}\varphi)(z) := \int_{\mathbb{R}_0^+} \varphi(\eta)\, \psi_\eta(z)\, \boldsymbol{\sigma}(d\eta).$$

If the endpoint b is regular, entrance or exit, then the inverse transform is written as $(\mathcal{J}_\wp^{-1}\varphi)(z) = \sum_{k=1}^\infty \frac{1}{\|\psi_{\eta_k}\|^2} \varphi(\eta_k)\, \psi_{\eta_k}(z)$, where the η_k are eigenvalues of $\wp$ (cf. Proposition 2.11). In these conditions, an application of the eigenfunction expansion to functions $h(x,z) \in L^2(M, \Omega^\wp)$ yields the decomposition

$$L^2(M, \Omega^\wp) = \bigoplus_{k=1}^\infty H_{\eta_k}^\wp, \qquad H_\eta^\wp := \{\psi_\eta(z) v(x) \mid v \in L^2(A)\}.$$

A similar expansion also holds if b is natural, with the direct sums replaced by direct integrals [63, §7.4]. Note also that if $u \in H_\eta^\wp \cap \mathrm{C}_\mathrm{c}^\infty(M)$ then $\mathcal{G}^\wp u = \mathcal{G}_{\eta_k} u$, where $\mathcal{G}_\eta := \partial_x^2 + \frac{A'(x)}{A(x)} \partial_x - \frac{\eta}{A(x)^2}$.

The following result is a counterpart of Proposition 5.10:

Proposition 5.22 *For each $(\lambda, \eta) \in \mathbb{C} \times \mathbb{R}_0^+$, there exists a unique solution $V_{\lambda,\eta}^\wp \in \{\psi_\eta(z) w(x) \mid w \in \mathrm{C}(\mathbb{R}_0^+)\}$ of the boundary value problem*

$$-\mathcal{G}^\wp w = \lambda w, \qquad w(0,z) = \psi_\eta(z), \qquad w^{[1]}(0,z) = 0.$$

There exists a locally finite positive Borel measure $\boldsymbol{\rho}^\wp$ on $(\mathbb{R}_0^+)^2$ such that the map $h \mapsto \mathcal{F}_\wp h$, where

$$(\mathcal{F}_\wp h)(\lambda, \eta) := \int_M h(x,z)\, V_{\lambda,\eta}^\wp(x,z)\, \Omega^\wp(d(x,z)) \qquad \bigl(\lambda, \eta \geq 0\bigr) \tag{5.54}$$

is an isometric isomorphism $\mathcal{F}_\wp : L^2(M, \Omega^\wp) \longrightarrow L^2((\mathbb{R}_0^+)^2, \boldsymbol{\rho}^\wp)$ whose inverse is given by

$$(\mathcal{F}_\wp^{-1}\Phi)(x,z) = \int_{(\mathbb{R}_0^+)^2} \Phi(\lambda, \eta)\, V_{\lambda,\eta}^\wp(x,z)\, \boldsymbol{\rho}^\wp(d(\lambda, \eta)). \tag{5.55}$$

The integrals in (5.54) *and* (5.55) *are understood as limits in* $L^2((\mathbb{R}_0^+)^2, \boldsymbol{\rho}^{\wp})$ *and* $L^2(M, \Omega^{\wp})$ *respectively.*

If b *is regular, entrance or exit, then* $\boldsymbol{\rho}^{\wp}(\Lambda_1 \times \Lambda_2) = \sum_{\eta_k \in \Lambda_2} \frac{1}{\|\psi_{\eta_k}\|^2} \boldsymbol{\rho}^{\wp}_{\eta_k}(\Lambda_1)$, *where* $\boldsymbol{\rho}^{\wp}_{\eta}$ *is the spectral measure of (the Neumann realization of) the Sturm–Liouville operator* $\mathcal{G}_\eta$, *and the expansion* (5.54)–(5.55) *reduces to*

$$\mathcal{F}_{\wp} : L^2(M, \Omega^{\wp}) \longrightarrow \bigoplus_{k=1}^{\infty} L^2(\mathbb{R}_0^+, \boldsymbol{\rho}^{\wp}_{\eta_k}),$$

$$\mathcal{F}_{\wp} h \equiv \bigl((\mathcal{F}_{\wp} h)(\cdot, \eta_1), (\mathcal{F}_{\wp} h)(\cdot, \eta_2), \ldots\bigr),$$

$$(\mathcal{F}_{\wp}^{-1}\{\varphi_k\})(x, z) = \sum_{k=1}^{\infty} \frac{1}{\|\psi_{\eta_k}\|^2} \int_{\mathbb{R}_0^+} \varphi_k(\lambda)\, V^{\wp}_{\lambda,\eta_k}(x, z)\, \boldsymbol{\rho}^{\wp}_{\eta_k}(d\lambda).$$

Proof The result for b regular, entrance or exit can be proved in a direct way using the same method as in Proposition 5.10.

If b is natural, start by considering the operator $\mathcal{G}^{\wp}$ on the restricted domain $M_N = [0, N] \times [a, N]$, where $\max\{0, a\} < N < \infty$. Applying first the eigenfunction expansion of the Sturm–Liouville operator $\wp$ on the interval $[a, N]$ (with boundary condition $u'(N) = 0$) and then the eigenfunction expansion of the Sturm–Liouville operators $\mathcal{G}_\eta$ on $[0, N]$ (also with $u'(N) = 0$), we obtain a discrete eigenfunction expansion of the form

$$(\mathcal{F}_{\wp,N}\, h)(\lambda_{k,N}, \eta_{k,N}) = \int_{M_N} h(x, z)\, V^{\wp}_{\lambda_{k,N},\eta_{k,N}}(x, z)\, \Omega^{\wp}(d(x, z)),$$

$$(\mathcal{F}_{\wp,N}^{-1}\{c_k\})(x, z) = \sum_{k=1}^{\infty} \frac{c_k}{\|V_{\lambda_{k,N},\eta_{k,N}}\|^2}\, V_{\lambda_{k,N},\eta_{k,N}}(x, z).$$

Using the techniques of [26], one can show that in the limit $N \to \infty$ this discrete expansion gives rise to an eigenfunction expansion of the general form (5.54)–(5.55), where $\boldsymbol{\rho}^{\wp}$ is the limiting spectral measure.

Proposition 5.23 (Product Formula for $V^{\wp}_{\lambda,\eta}$**)** *Suppose that there exists a family* $\{\pi^{\wp}_{z_1,z_2}\}_{z_1,z_2 \in I} \subset \mathcal{P}(I)$ *such that*

$$\psi_\eta(z_1)\, \psi_\eta(z_2) = \int_I \psi_\eta\, d\pi^{\wp}_{z_1,z_2} \qquad \bigl(z_1, z_2 \in I,\ \eta \in \operatorname{supp}(\boldsymbol{\sigma})\bigr). \tag{5.56}$$

Then for each $\eta \in \operatorname{supp}(\sigma)$, $\boldsymbol{\xi}_1 = (x_1, z_1) \in M$ *and* $\boldsymbol{\xi}_2 = (x_2, z_2) \in M$ *there exists a positive measure* $\boldsymbol{\gamma}^{\wp}_{\eta,\boldsymbol{\xi}_1,\boldsymbol{\xi}_2}$ *on M such that the product* $V^{\wp}_{\lambda,\eta}(\boldsymbol{\xi}_1)\, V^{\wp}_{\lambda,\eta}(\boldsymbol{\xi}_2)$ *admits the integral representation*

$$V^{\wp}_{\lambda,\eta}(\boldsymbol{\xi}_1)\, V^{\wp}_{\lambda,\eta}(\boldsymbol{\xi}_2) = \int_M V^{\wp}_{\lambda,\eta}(\boldsymbol{\xi}_3)\, \boldsymbol{\gamma}^{\wp}_{\eta,\boldsymbol{\xi}_1,\boldsymbol{\xi}_2}(d\boldsymbol{\xi}_3) \qquad \big(\lambda \in \mathbb{C},\ \eta \in \operatorname{supp}(\sigma)\big). \tag{5.57}$$

Notice that the assumption on the existence of the product formula (5.56) obviously holds if $\wp$ belongs to the family of Sturm–Liouville operators satisfying Assumption MP of Chap. 4. It also holds for many Sturm–Liouville operators with discrete spectrum defined on compact intervals of $\mathbb{R}$ (sufficient conditions are given in [15, pp. 312–314], [19, pp. 242–245]).

Proof It is straightforward that $V^{\wp}_{\lambda,\eta}(x,z) = \psi_\eta(z)\, \zeta_\eta(x)\, \widetilde{w}_{\lambda,\eta}(x)$, where $\zeta_\eta(x) := \cosh\big(\sqrt{\eta} \int_0^x \frac{dy}{A(y)}\big)$ and $\widetilde{w}_{\lambda,\eta}$ is the solution of

$$-\frac{1}{B_\eta}(B_\eta \widetilde{w}')' = \lambda \widetilde{w}, \qquad \widetilde{w}(0) = 1, \qquad (B_\eta \widetilde{w}')(0) = 0,$$

where $B_\eta(x) = A(x)\zeta_\eta(x)^2$. Arguing as in the proof of Proposition 5.13, we deduce that the product formula (5.57) holds for the positive measures

$$\boldsymbol{\gamma}^{\wp}_{\eta,\boldsymbol{\xi}_1,\boldsymbol{\xi}_2}(d\boldsymbol{\xi}_3) = \frac{\zeta_\eta(x_1)\zeta_\eta(x_2)}{\zeta_\eta(x_3)}\, \boldsymbol{\nu}^{\wp}_{\eta,\boldsymbol{\xi}_1,\boldsymbol{\xi}_2}(d\boldsymbol{\xi}_3),$$

where $\boldsymbol{\nu}^{\wp}_{\eta,\boldsymbol{\xi}_1,\boldsymbol{\xi}_2} := \pi^{[\eta]}_{x_1,x_2} \otimes \pi^{\wp}_{z_1,z_2}$ and $\pi^{[\eta]}_{x_1,x_2}$ is the measure of the product formula for $\widetilde{v}_{\lambda,\eta}$.

Definition 5.6 Suppose that there exists $\{\pi^{\wp}_{z_1,z_2}\}_{z_1,z_2 \in I} \subset \mathcal{P}(I)$ such that (5.56) holds, and let $\eta \in \operatorname{supp}(\sigma)$, $\lambda \geq 0$ and $\mu, \nu \in \mathcal{M}_{\mathbb{C}}(M)$. The measure

$$(\mu \underset{\eta,\wp}{*} \nu)(\cdot) = \int_M \int_M \boldsymbol{\nu}^{\wp}_{\eta,\boldsymbol{\xi}_1,\boldsymbol{\xi}_2}(\cdot)\, \mu(d\boldsymbol{\xi}_1)\, \nu(d\boldsymbol{\xi}_2)$$

is called the $\mathcal{G}_\eta$*-convolution* of the measures μ and ν. The functions

$$(\mathcal{F}_\wp \mu)(\lambda,\eta) = \int_M \frac{V^{\wp}_{\lambda,\eta}(\boldsymbol{\xi})}{\zeta_\eta(x)}\, \mu(d\boldsymbol{\xi}) \qquad \text{and} \qquad (\mathcal{T}^{\mu}_{\eta,\wp} h)(\boldsymbol{\xi}) := \int_M h\, d(\delta_{\boldsymbol{\xi}} \underset{\eta,\wp}{*} \mu)$$

are, respectively, the $\boldsymbol{\mathcal{G}}^{\wp}$*-Fourier transform* of μ and the $\mathcal{G}_\eta$*-translation* of a function h by μ.

Unsurprisingly, the $\mathcal{G}_\eta$-*convolution* shares many properties with the Δ_k-convolution studied in the previous sections, among which the following:

Proposition 5.24 *Assume that there exists* $\{\pi^{\wp}_{z_1,z_2}\}_{z_1,z_2\in I} \subset \mathcal{P}(I)$ *such that* (5.56) *holds. Assume also that* $e^{-t\cdot} \in L^2(\mathbb{R}_0^+, \boldsymbol{\sigma})$ *for all* $t > 0$.

(a) *For each* $\eta \in \mathrm{supp}(\boldsymbol{\sigma})$, *the space* $(\mathcal{M}_{\mathbb{C}}(M), \underset{\eta,\wp}{*})$, *equipped with the total variation norm, is a commutative Banach algebra over* $\mathbb{C}$ *whose identity element is the Dirac measure* $\delta_{(0,a)}$. *Moreover, the subset* $\mathcal{P}(M)$ *is closed under the* $\mathcal{G}_\eta$-*convolution.*

(b) $\big(\mathcal{F}_\wp(\mu \underset{\eta,\wp}{*} \nu)\big)(\lambda,\eta) = (\mathcal{F}_\wp \mu)(\lambda,\eta)\cdot(\mathcal{F}_\wp \nu)(\lambda,\eta)$ *for all* $\lambda \geq 0$ *and* $\eta \in \mathrm{supp}(\boldsymbol{\sigma})$.

(c) *Each measure* $\mu \in \mathcal{M}_{\mathbb{C}}(M)$ *is uniquely determined by* $\mathcal{F}_\wp \mu$.

Set $\Sigma := \mathrm{supp}(\boldsymbol{\sigma})$ *if* $I = [a,b]$ *and set* $\Sigma := \mathbb{R}_0^+$ *if* $I = [a,b)$. *In the latter case, assume also that* $\lim_{z\uparrow b} \psi_\eta(z) = 0$ *for all* $\eta > 0$. *Then:*

(d) *Let* $\{\mu_n\}$ *be a sequence of measures belonging to* $\mathcal{M}_+(M)$ *whose* $\mathcal{G}^\wp$-*Fourier transforms are such that*

$$(\mathcal{F}_\wp \mu_n)(\lambda,\eta) \xrightarrow[n\to\infty]{} f(\lambda,\eta) \qquad \text{pointwise in } (\lambda,\eta) \in \mathbb{R}_0^+ \times \Sigma,$$

where the function f *is such that*

$$\begin{cases} f(\cdot,0) \text{ is continuous at a neighbourhood of zero} & \text{if } b \text{ is regular or entrance,} \\ f \text{ is continuous at a neighbourhood of } (0,0) & \text{if } b \text{ is exit or natural.} \end{cases}$$

Then $\mu_n \xrightarrow{w} \mu$ *for some measure* $\mu \in \mathcal{M}_+(M)$ *such that* $\mathcal{F}_\wp \mu \equiv f$.

(e) *For each* $\eta \in \Sigma$ *the mapping* $(\mu,\nu) \mapsto \mu \underset{\eta,\wp}{*} \nu$ *is continuous in the weak topology.*

(f) *If* $h \in \mathrm{C_b}(M)$ *(respectively* $\mathrm{C_0}(M)$*) then* $\mathcal{T}^\mu_{\eta,\wp} h \in \mathrm{C_b}(M)$ *(resp.* $\mathrm{C_0}(M)$*) for all* $\mu \in \mathcal{M}_{\mathbb{C}}(M)$.

(g) *Let* $1 \leq p \leq \infty$, $\mu \in \mathcal{M}_+(M)$ *and* $h \in L^p_{\eta,\wp} := L^p\big(M, B_\eta(x)dx\, \mathfrak{r}(z)dz\big)$. *The* $\mathcal{G}_\eta$-*translation* $(\mathcal{T}^\mu_{\eta,\wp} h)(x)$ *is a Borel measurable function of* $x \in M$, *and we have*

$$\|\mathcal{T}^\mu_{\eta,\wp} h\|_{L^p_{\eta,\wp}} \leq \|\mu\| \cdot \|h\|_{L^p_{\eta,\wp}}.$$

We omit the proofs as they contain no new ideas.

Notions such as infinite divisibility and convolution semigroups with respect to the $\mathcal{G}_\eta$-convolution can also be defined like in the previous subsection, giving rise to a Lévy–Khintchine type representation and to Feller semigroups on $\mathrm{C_0}(M)$. The details are left to the reader.

Remark 5.4 The above result on the existence of a family of convolutions associated with the functions $V_{\lambda,\eta}^{\wp}$ can be interpreted in the context of the theory of multiparameter Sturm–Liouville spectral problems.

First we recall some known results. Consider the system of Sturm–Liouville equations

$$-(p_m u_m')'(x_m) + (q_m u_m)(x_m) = \sum_{n=1}^{N} \lambda_n (r_{mn} u_m)(x_m) \qquad (m = 1, \dots, N), \tag{5.58}$$

where $N \in \mathbb{N}$ and $a_m \leq x_m \leq b_m$, together with boundary conditions at the endpoints a_m and b_m of the form

$$u_m(a_m)\cos(\vartheta_m) = u_m'(a_m)\sin(\vartheta_m), \qquad u_m(b_m)\cos(\vartheta_m') = u_m'(b_m)\sin(\vartheta_m') \tag{5.59}$$

($m = 1, \dots, N$). Let us assume that the intervals $I_m = [a_m, b_m]$ are bounded, the functions p_m, q_m, r_{mn} are sufficiently well-behaved and $r(x) = \det\{r_{mn}(x_m)\} > 0$ for $x = (x_1, \dots, x_N) \in \prod_{m=1}^N I_m$. If $\lambda = (\lambda_1, \dots, \lambda_N)$ is chosen such that for each m there exists a nontrivial solution $u_m(x_m; \lambda)$ of (5.58)–(5.59), then the function $u(x; \lambda) = \prod_{i=1}^N u_m(x_m; \lambda)$ is said to be an eigenfunction of the system (5.58)–(5.59) corresponding to the eigenvalue λ.

By the completeness theorem for multiparameter eigenvalue problems [59], the following Fourier-like expansion holds:

$$h(x) = \sum_k (\boldsymbol{F}h)(\lambda^{(k)})\, u(x; \lambda^{(k)}), \quad \text{where } (\boldsymbol{F}h)(\lambda^{(k)}) := \int_a^b h(x)\, u(x; \lambda^{(k)})\, r(x) dx, \tag{5.60}$$

$\lambda^{(k)}$ are the eigenvalues of (5.58)–(5.59), and $\int_a^b = \int_{a_1}^{b_1} \cdots \int_{a_N}^{b_N}$.

Similar results have been established for (singular) systems where some of the intervals $[a_m, b_m]$ are unbounded; in this case, the sum in (5.60) is, in general, replaced by a Stieltjes integral with respect to a spectral function [26, 27]. However, compared to one-dimensional Sturm–Liouville operators, much less is known regarding the spectral properties of such singular systems [2, 171].

A comparison of this general formulation of the multiparameter Sturm–Liouville eigenvalue problem with the eigenfunction expansion for the operator $\boldsymbol{\mathcal{G}}^{\wp}$ shows that the transformation (5.54)–(5.55) can be reinterpreted as a Fourier-like expansion for the system of differential equations (5.58) with $N = 2$, $x_1 \in \mathbb{R}_0^+$, $x_2 \in [a, b]$, $\lambda_1 = \lambda$, $\lambda_2 = \eta$, $p_1 = r_{11} = A$, $p_2 = p$, $r_{22} = r$, $r_{12} = \frac{1}{A}$ and $q_1 = q_2 = r_{21} = 0$.

As we saw in Chap. 4, a crucial requirement in the theory of product formulas and convolutions associated with one-dimensional Sturm–Liouville equations $-(pu')' + qu = \lambda r u$ is that the measures of the product formula should not depend on the spectral parameter λ. Similarly, the measures of product formula (5.57) for the generalized eigenfunctions $w^{\wp}_{\lambda,\eta}$ do not depend on one of the spectral parameters (the measures $\gamma_{\eta,\xi_1,\xi_2}$ are independent of λ); this is a fundamental property which (as we saw above) enables us to develop the theory of $\mathcal{G}_\eta$-convolutions. This suggests that the natural way to introduce the notion of a product formula for a general Sturm–Liouville system (5.58) (regular or singular, with suitable boundary conditions) is as follows:

Let $1 \leq s \leq N$. *The system* (5.58) *is said to admit a* $(\lambda_1, \ldots, \lambda_s)$-product formula *if for each* $x^{(1)}, x^{(2)} \in I := \prod_{m=1}^N I_m$ *there exists a positive measure* $\boldsymbol{\gamma}^{\lambda_{s+1},\ldots,\lambda_N}_{x^{(1)},x^{(2)}}$ *on* I *such that the product* $u(x^{(1)}; \lambda)\, u(x^{(2)}; \lambda)$ *admits the representation*

$$u(x^{(1)}; \lambda)\, u(x^{(2)}; \lambda) = \int_I u(x; \lambda)\, \boldsymbol{\gamma}^{\lambda_{s+1},\ldots,\lambda_N}_{x^{(1)},x^{(2)}}(dx), \qquad \lambda_1, \ldots, \lambda_N \geq 0. \tag{5.61}$$

As far as we are aware, no general results are available on the existence of such product formulas for nontrivial Sturm–Liouville systems of the form (5.58). (Here the word 'nontrivial' means that $r_{mn} \neq 0$ for some $m \neq n$.) Proposition 5.23 can be interpreted as an initial step in this direction. Developing a general theory of product formulas for nontrivial systems of Sturm–Liouville equations is an interesting problem which remains open for further investigation.

Appendix A
Some Open Problems

The following list is a summary of the open problems that arise from the present work, part of which have already been mentioned in the main text.

(a) Prove or disprove the existence of Whittaker convolution semigroups whose log-Whittaker transforms are of the form $\psi(\lambda) = \lambda^\beta$ where $0 < \beta < 1$ (cf. Remark 3.9).

(b) What is the support of the measures π_x of the Laplace-type representation for Sturm–Liouville solutions established in Theorem 4.1? (On the one hand, we know from [19, Theorem 3.5.58] that $\text{supp}(\pi_x)$ is compact when the Sturm–Liouville operators satisfies the assumptions of Theorem 4.1; on the other hand, Theorem 3.2 shows that $\text{supp}(\pi_x)$ is noncompact in the case of the Whittaker operator. A general result is missing.)

(c) Establish necessary and sufficient conditions for:

(c1) The measures of the Sturm–Liouville product formula (4.21) to be absolutely continuous with respect to the Lebesgue measure. (See [210] for known results on Sturm–Liouville hypergroups satisfying the assumptions of Theorem 4.1.)

(c2) The measures of the product formula to be of the form $\boldsymbol{\nu}_{x,y}(d\xi) = k(x, y, \xi)r(\xi)d\xi$, where $k(x, y, \xi) = \mathcal{F}^{-1}[w_{(\cdot)}(x)w_{(\cdot)}(y)](\xi)$. (This is known to hold on some Sturm–Liouville hypergroups [30], and we saw in (3.55) that it also holds for the Whittaker convolution.)

(d) Let $([a, b), *)$ be a Sturm–Liouville convolution constructed as in Chap. 4. Prove or disprove that the log $\mathcal{L}$-transform of any $\mathcal{L}$-Gaussian measure μ is of the form $\psi_\mu(\lambda) = c\lambda$ for some $c > 0$. (As noted in Remark 3.7, this result is known to hold on the hypergroups studied by Zeuner.)

(e) Is it possible to extend the characterization of weak convergence stated in Remark 4.27.I and the theory of infinite divisibility of Sects. 4.5.1–4.5.3 to Sturm–Liouville convolutions not satisfying Assumption MP_∞?

R. Sousa et al., *Convolution-like Structures, Differential Operators and Diffusion Processes*, Lecture Notes in Mathematics 2315,
https://doi.org/10.1007/978-3-031-05296-5

(f) Generalize the Lévy-type characterization for the associated one-dimensional diffusion (stated in Remark 4.4) to a larger class of Sturm–Liouville convolutions. In particular, can the Lévy-type characterization be extended to the drifted Bessel process?

(g) Determine a closed-form expression for the measures of the product formula (4.53) for the Whittaker function of the first kind. Can this be achieved using techniques similar to those used in Sect. 3.2?

(h) Provide examples of Sturm–Liouville convolutions (other than the Whittaker convolution, cf. Corollary 3.6) for which Theorem 4.9 yields an explicit expression for the solution of integral equations with special functions in the kernel. In particular, can we find explicit solutions for convolution-type integral equations with respect to the convolutions of the Bessel-Kingman and Jacobi hypergroups?

(i) Prove the existence of probabilistic product formulas for Sturm–Liouville operators with nondifferentiable coefficients and extend the theory developed in Chap. 4 to the induced convolution operators. (The reader should note that most of the Sturm–Liouville theory presented in Sect. 2.2 extends to differential operators with measure-valued coefficients, cf. [53, 107].)

(j) Do the techniques used in Chap. 4 also allow us to generalize the known results on existence of Sturm–Liouville hypergroups of compact type to operators for which the associated hyperbolic Cauchy problem is degenerate? Does this give rise to a notion of degenerate Sturm–Liouville hypergroups of compact type?

(k) Can one construct positivity-preserving convolution-like operators associated with diffusion processes with absorbing (Dirichlet) boundary conditions? In the one-dimensional case, can this be accomplished by adapting the PDE approach of Chap. 4?

(l) Establish a nonexistence theorem (similar to Theorem 5.4) for convolutions associated with diffusions on unbounded domains of $\mathbb{R}^d$ ($d \geq 2$), Brownian motions on noncompact Riemannian manifolds, or other multidimensional diffusions whose generator does not have a discrete spectrum.

(m) Can we take advantage of other known results on the asymptotic behavior of solutions of second-order differential equations (cf. [52]) to extend the nonexistence result of Theorem 5.9 to other families of Sturm–Liouville operators? In particular, does the conclusion of Theorem 5.9 also hold if $\lim_{\xi\to\infty} \frac{A'(\xi)}{2A(\xi)} = -\infty$?

(n) Find nontrivial examples of Sturm–Liouville systems (other than those studied in Sect. 5.4) whose generalized eigenfunctions admit a product formula of the form (5.61). Study how such examples could be unified into a general theory of product formulas for multiparameter Sturm–Liouville spectral problems.

References

1. D. Applebaum, *Probability on compact Lie groups* (Springer, Cham, 2014)
2. F.V. Atkinson, A.B. Mingarelli, *Multiparameter Eigenvalue Problems—Sturm–Liouville Theory* (CRC Press, Boca Raton, 2011)
3. F. Avram, N.N. Leonenko, N. Šuvak, On spectral analysis of heavy-tailed Kolmogorov-Pearson diffusions. Markov Process. Related Fields **19**(2), 249–298 (2013)
4. R. Bañuelos, K. Burdzy, On the "hot spots" conjecture of J. Rauch. J. Funct. Anal. **164**(1), 1–33 (1999)
5. M.T. Barlow, R.F. Bass, Z.-Q. Chen, M. Kassmann, Non-local Dirichlet forms and symmetric jump processes. Trans. Am. Math. Soc. **361**, 1963–1999 (2009)
6. O.E. Barndorff-Nielsen, T.Mikosch, S.I. Resnick (eds.), *Lévy Processes: Theory and Applications* (Birkhäuser, Boston, 2001)
7. C. Barrabès, V. Frolov, R. Parentani, Metric fluctuation corrections to Hawking radiation. Phys. Rev. D **59**, 124010 (1999)
8. R.F. Bass, P. Hsu, Some potential thoery for reflecting Brownian motion in Hölder and Lipschitz domains. Ann. Probab. **19**(2), 486–508 (1991)
9. H. Bauer, *Probability Theory* (Walter De Gruyter, Berlin, 1996)
10. H. Bauer, *Measure and Integration Theory* (Walter De Gruyter, Berlin, 2001)
11. C. Bennewitz, W. N. Everitt, The Titchmarsh-Weyl eigenfunction expansion theorem for Sturm–Liouville differential equations, in *Sturm–Liouville Theory: Past and Present*, ed. by W.O. Amrein, A.M. Hinz, D.B. Pearson (Birkhäuser, Basel, 2005), pp. 137–171
12. N. Ben Salem, Convolution semigroups and central limit theorem associated with a dual convolution structure. J. Theor. Probab. **7**(2), 417–436 (1994)
13. Y. M. Berezansky, S. G. Krein, Continuous algebras. Dokl. Akad. Nauk SSSR **72**, 5–8 (1950)
14. Y.M. Berezansky, S.G. Krein, Hypercomplex systems with continual basis. Uspekhi Mat. Nauk **12**(1), 147–152 (1957)
15. Y.M. Berezansky, A.A. Kalyuzhnyi, *Harmonic Analysis in Hypercomplex Systems* (Kluwer Academic Publishers, Dordrecht, 1998)
16. C. Berg, G. Forst, *Potential Theory on Locally Compact Abelian Groups* (Springer, Berlin, 1975)
17. G. Birkhoff, G.-C. Rota, *Ordinary Differential Equations*, 4th edn. (Wiley, New York, 1989)
18. A.V. Bitsadze, *Equations of the Mixed Type* (Pergamon Press, Oxford, 1964)
19. W.R. Bloom, H. Heyer, *Harmonic Analysis of Probability Measures on Hypergroups* (Walter de Gruyter, Berlin, 1994)
20. V. Bobkov, On exact Pleijel's constant for some domains. Doc. Math. **23**, 799–813 (2018)

R. Sousa et al., *Convolution-like Structures, Differential Operators and Diffusion Processes*, Lecture Notes in Mathematics 2315,
https://doi.org/10.1007/978-3-031-05296-5

21. A.N. Borodin, *Stochastic Processes* (Birkhäuser, Cham, 2017)
22. A. N. Borodin, P. Salminen, *Handbook of Brownian Motion: Facts and Formulae* (Springer, Basel, 2002)
23. M. Borowiecka-Olszewska, B.H. Jasiulis-Gołdyn, J.K. Misiewicz, J. Rosiński, Lévy processes and stochastic integrals in the sense of generalized convolutions. Bernoulli **21**(4), 2513–2551 (2015)
24. U. Boscain, D. Prandi, Self-adjoint extensions and stochastic completeness of the Laplace–Beltrami operator on conic and anticonic surfaces. J. Differ. Equ. **260**, 3234–3269 (2016)
25. B. Böttcher, R. Schilling, J. Wang, Lévy-type processes: construction, approximation and sample path properties, in *Lévy Matters III*. Lecture Notes in Mathematics, vol. 2099 (Springer, Berlin, 2014)
26. P.J. Browne, A singular multi-parameter eigenvalue problem in second order ordinary differential equations. J. Differ. Equ. **12**, 81–94 (1972)
27. P.J. Browne, Abstract multiparameter theory I. J. Math. Anal. Appl. **60**(1), 259-273 (1977)
28. K. Burdzy, The hot spots problem in planar domains with one hole. Duke Math. J. **129**(3), 481–502 (2005)
29. R. W. Carroll, *Transmutation Theory and Applications* (North-Holland, Amsterdam, 1985)
30. H. Chebli, Opérateus de translation généralisée et semi-groupes de convolution, in *Théorie du Potentiel et Analyse Harmonique*, ed. by J. Faraut (Springer, Berlin, 1974), pp. 35–59
31. J. Cheeger, On the Hodge theory of Riemannian pseudomanifolds, in *Geometry of the Laplace Operator*, ed. by R. Osserman, A. Weinstein. Proc. Sympos. Pure Math. vol. XXXVI (American Mathematical Society, Providence, 1980), pp. 91–146
32. Z.-Q. Chen, W.-T. (L.) Fan, Functional central limit theorem for Brownian particles in domains with Robin boundary condition. J. Funct. Anal. **269**(12), 3765–3811 (2015)
33. Z.-Q. Chen, M. Fukushima, *Symmetric Markov Processes, Time Change and Boundary Theory* (Princeton University Press, Princeton, 2011)
34. Z.-Q. Chen, T. Zhang, A probabilistic approach to mixed boundary value problems for elliptic operators with singular coefficients. Proc. Am. Math. Soc. **142**(6), 2135–2149 (2014)
35. F.M. Cholewinski, A Hankel convolution complex inversion theory. Mem. Am. Math. Soc. **58**, 67 (1965)
36. M. Choulli, L. Kayser, E. M. Ouhabaz, Observations on Gaussian upper bounds for Neumann heat kernels. Bull. Aust. Math. Soc. **92**, 429–439 (2015)
37. E.A. Coddington, N. Levinson, *Theory of Ordinary Differential Equations* (Mc-Graw Hill, New York, 1955)
38. W.C. Connett, C. Markett, A.L. Schwartz, Convolution and hypergroup structures associated with a class of Sturm–Liouville systems. Trans. Am. Math. Soc. **332**(1), 365–390 (1992)
39. W.C. Connett, A.L. Schwartz, Subsets of $\mathbb{R}$ which support hypergroups with polynomial characters. J. Comput. Appl. Math. **65**, 73–84 (1995)
40. R. Courant, *Methods of Mathematical Physics—Vol. II: Partial Differential Equations* (Wiley, New York, 1962)
41. W. Craig, S. Weinstein, On determinism and well-posedness in multiple time dimensions. Proc. R. Soc. A **465**, 3023–3046 (2009)
42. E.B. Davies, *One-parameter Semigroups* (Academic Press, London, 1980)
43. E.B. Davies, Heat Kernels and Spectral Theory (Cambridge University Press, Cambridge, 1989)
44. L. Debnath, D. Bhatta, *Integral Transforms and their Applications*, 2nd edn. (CRC Press, Boca Raton, 2007)
45. J. Delsarte, Sur une extension dc la formule de Taylor. J. Math. Pures Appl. **17**, 213–231 (1938)
46. I.H. Dimovski, *Convolutional Calculus* (Publ. House Bulg. Acad. Sci., Sofia, 1982)
47. C. Donati-Martin, R. Ghomrasni, M. Yor, On certain Markov processes attached to exponential functionals of Brownian motion; application to Asian options. Rev. Mat. Iberoam. **17**(1), 179–193 (2001)

48. N. Dunford, J.T. Schwartz, *Linear Operators—Part II: Spectral Theory* (Wiley, New York, 1963)
49. C.F. Dunkl, The measure algebra of a locally compact hypergroup. Trans. Am. Math. Soc. **179**, 331–348 (1973)
50. H. Dym, H.P. McKean, *Fourier Series and Integrals* (Academic Press, New York, 1972)
51. H. Dym, H.P. McKean, *Gaussian Processes, Function Theory, and the Inverse Spectral Problem* (Academic Press, New York, 1976)
52. M.S.P. Eastham, *The Asymptotic Solution of Linear Differential Systems—Applications of the Levinson Theorem* (Oxford University Press, Oxford, 1989)
53. J. Eckhardt, G. Teschl, Sturm–Liouville operators with measure-valued coefficients. J. Anal. Math. **120**, 151–224 (2013)
54. J. Elliott, Eigenfunction expansions associated with singular differential operators. Trans. Am. Math. Soc. **78**, 406–425 (1955)
55. A. Erdélyi, W. Magnus, F. Oberhettinger, F. G. Tricomi, *Higher Transcendental Functions*, vol. I (McGraw-Hill, New York, 1953)
56. A. Erdélyi, W. Magnus, F. Oberhettinger, F.G. Tricomi, *Higher Transcendental Functions*, vol. II (McGraw–Hill, New York, 1953)
57. S.N. Ethier, T.G. Kurtz, *Markov Processes—Characterization and Convergence* (Wiley, New York, 1986)
58. W.N. Everitt, On the transformation theory of ordinary second-order linear symmetric differential expressions. Czech. Math. J. **32**(2), 275-306 (1982)
59. M. Faierman, The completeness and expansion theorems associated with the multi-parameter eigenvalue problem in ordinary differential equations. J. Differ. Equ. **5**, 197–213 (1969)
60. W. Feller, *An Introduction to Probability Theory and its Applications*, 2nd edn. (Wiley, New York, 1971)
61. M. Flensted-Jensen, T. Koornwinder, The convolution structure for Jacobi function expansions. Ark. Mat. **11**(1–2), 245–262 (1973)
62. G.B. Folland, *Real Analysis: Modern Techniques and Their Applications* (Wiley, New York, 1999)
63. G.B. Folland, *A Course in Abstract Harmonic Analysis*, 2nd edn. (CRC Press, Boca Raton, 2016)
64. F. Früchtl, Sturm–Liouville hypergroups and asymptotics. Monatsh. Math. **186**(1), 11–36 (2018)
65. M. Fukushima, On general boundary conditions for one-dimensional diffusions with symmetry. J. Math. Soc. Jpn. **66**(1), 289–316 (2014)
66. M. Fukushima, Y. Oshima, M. Takeda, *Dirichlet Forms and Symmetric Markov Processes* (Walter De Gruyter, Berlin, 2011)
67. C. Fulton, H. Langer, Sturm–Liouville operators with singularities and generalized Nevanlinna functions. Complex Anal. Oper. Theory **4**, 179–243 (2010)
68. L. Gallardo, K. Trimèche, Lie theorems for one-dimensional hypergroups. Integr. Transf. Spec. Funct. **13**(1), 71–92 (2002)
69. L. Gårding, *Application of the Theory of Direct Integrals of Hilbert Spaces to some Integral and Differential Operators*. Lect. Ser. Inst. Fluid Dynam. Appl. Math., vol. 11 (1954)
70. G. Gasper, Banach algebras for Jacobi series and positivity of a kernel. Ann. Math. **95**(2), 261–280 (1972)
71. I. Gelfand, D. Raikov, G. Shilov, *Commutative Normed Rings* (Chelsea, New York, 1964)
72. R.K. Getoor, Markov operators and their associated semi-groups. Pac. J. Math. **9**, 449–472 (1959)
73. H.-J. Glaeske, A.P. Prudnikov, K.A. Skòrnik, *Operational Calculus and Related Topics* (CRC Press, Boca Raton, 2006)
74. D.S. Grebenkov, B.-T. Nguyen, Geometrical structure of Laplacian eigenfunctions. SIAM Rev. **55**(4), 601–667 (2013)
75. A. Grigor'yan, *Heat Kernel and Analysis on Manifolds* (American Mathematical Society, Providence; International Press, Boston, 2009)

76. A. Grigor'yan, A. Telcs, Two-sided estimates of heat kernels on metric measure spaces. Ann. Probab. **40**(3), 1212–1284 (2012)
77. E. Grosswald, *Bessel Polynomials* (Springer, Berlin, 1978)
78. D.L. Guy, Hankel multiplier transformations and weighted p-norms. Trans. Am. Math. Soc. **95**(1), 137–189 (1960)
79. B. Helffer, M. Hoffmann-Ostenhof, T. Hoffmann-Ostenhof, N. Nadirashvili, Spectral theory for the dihedral group. Geom. Funct. Anal. **12**(5), 989–1017 (2002)
80. B. Helffer, M.P. Sundqvist, On nodal domains in Euclidean balls. Proc. Am. Math. Soc. **144**(11), 4777–4791 (2016)
81. D. Henry, *Perturbation of the Boundary in Boundary-Value Problems of Partial Differential Equations* (Cambridge University Press, Cambridge, 2005)
82. H. Heyer, Probability theory on hypergroups: a survey, in *Probability Measures on Groups VII*, ed. by H. Heyer (Springer, Berlin, 1984), pp. 481–550
83. H. Heyer, G. Pap, Martingale characterizations of increment processes in a commutative hypergroup. Adv. Pure Appl. Math. **1**(1), 117–140 (2010)
84. I.I. Hirschman, Variation diminishing Hankel transforms. J. Anal. Math. **8**, 307–336 (1960)
85. P.V. Hoang, T. Tuan, N.X. Thao, V.K. Tuan, Boundedness in weighted L_p spaces for the Kontorovich–Lebedev–Fourier generalized convolutions and applications. Integr. Transf. Spec. Funct. **28**(8), 590–604 (2017)
86. N.T. Hong, P.V. Hoang, V.K. Tuan, The convolution for the Kontorovich-Lebedev transform revisited. J. Math. Anal. Appl. **440**(1), 369–378 (2016)
87. E.P. Hsu, Estimates of derivatives of the heat kernel on a compact Riemannian manifold. Proc. Am. Math. Soc. **127**(12), 3739–3744 (1999)
88. K. Itô, *Essentials of Stochastic Processes* (American Mathematical Society, Providence, 2006)
89. K. Itô, H.P. McKean, *Diffusion Processes and Their Sample Paths* (Springer, Berlin, 1996)
90. S.R. Jain, R. Samajdar, Nodal portraits of quantum billiards: domains, lines and statistics. Rev. Mod. Phys. **89**, 045005 (2017)
91. B.H. Jasiulis-Gołdyn, Kendall random walks. Probab. Math. Statist. **36**(1), 165–185 (2016)
92. B.H. Jasiulis-Gołdyn, K. Łukaszewicz, J.K. Misiewicz, E. Omey, Renewal theory for extremal Markov sequences of the Kendall type. Probab. Math. Statist. **36**(1), 165–185 (2020)
93. B.H. Jasiulis-Gołdyn, J.K. Misiewicz, Kendall random walk, Williamson transform, and the corresponding Wiener–Hopf factorization. Lith. Math. J. **57**(4), 479–489 (2017)
94. R.I. Jewett, Spaces with an Abstract Convolution of Measures. Adv. Math. **18**(1), 1–101 (1975)
95. C. Judge, S. Mondal, Euclidean triangles have no hot spots. Ann. Math. **191**(2), 167–211 (2020)
96. I.S. Kac, The existence of spectral functions of generalized second order differential systems with boundary conditions at the singular end. Am. Math. Soc. Transl. (2) **62**, 204–262 (1967)
97. V.A. Kakichev, On the convolution for integral transforms. Izv. Vyssh. Uchebn. Zaved. Mat. **2**, 53–62 (1967) (in Russian)
98. V.A. Kakichev, N.X. Thao, On a method for the construction of generalized integral convolutions. Izv. Vyssh. Uchebn. Zaved. Mat. **1**, 31–40 (1998) (in Russian)
99. I. Karatzas, S.E. Shreve, *Brownian Motion and Stochastic Calculus*, 2nd edn. (Springer, New York, 1991)
100. S. Karlin, H.M. Taylor, *A Second Course in Stochastic Processes* (Academic Press, New York, 1981)
101. J.F.C. Kingman, Random walks with spherical symmetry. Acta Math. **109**, 11–53 (1963)
102. A. Klenke, *Probability Theory—A Comprehensive Course*, 3rd edn. (Springer, London, 2020)
103. V.N. Kolokoltsov, *Markov Processes, Semigroups and Generators* (Walter De Gruyter, Berlin, 2011)
104. V. Komornik, P. Loreti, *Fourier Series in Control Theory* (Springer, New York, 2005)

105. T.H. Koornwinder, Jacobi functions and analysis on noncompact semisimple Lie groups, in *Special Functions: Group Theoretical Aspects and Applications*, ed. by R.A. Askey, T.H. Koornwinder, W. Schempp (Reidel, Dordrecht, 1984), pp. 1–85
106. T.H. Koornwinder, A.L. Schwartz, Product formulas and associated hypergroups for orthogonal polynomials on the simplex and on a parabolic biangle. Constr. Approx. **13**(4), 537–567 (1997)
107. S. Kotani, On a generalized Sturm–Liouville operator with a singular boundary. J. Math. Kyoto Univ. **15**(2), 423–454 (1975)
108. S. Kotani, Krein's strings with singular left boundary. Rep. Math. Phys. **59**(3), 305–316 (2007)
109. H.L. Krall, O. Frink, A new class of orthogonal polynomials: the Bessel polynomials. Trans. Am. Math. Soc. **65**, 100–115 (1949)
110. M.G. Krein, On some cases of the effective determination of the density of a nonhomogeneous string from its spectral function. Dokl. Akad. Nauk SSSR **93**, 617–620 (1953) (in Russian)
111. M.G. Krein, Integral equations on a half-line with kernel depending on the difference of the arguments. Am. Math. Soc. Transl. (2) **22**, 163–288 (1962)
112. H. Kunita, Stochastic differential equations and stochastic flows of diffeomorphisms, in *École d'été de probabilités de Saint-Flour XII–1982*, ed. by R.M. Dudley, H. Kunita, F. Ledrappier (Springer, Berlin, 1984), pp. 143–303
113. B.J. Laurenzi, Derivatives of Whittaker functions $W_{\kappa,1/2}$ and $M_{\kappa,1/2}$ with respect to order κ. Math. Comp. **27**(121), 129–132 (1973)
114. N.N. Lebedev, Some singular integral equations connected with the integral representations of mathematical physics. Dokl. Akad. Nauk SSSR **65**, 621–624 (1949) (in Russian)
115. N.N. Leonenko, N. Šuvak, Statistical inference for reciprocal gamma diffusion process. J. Stat. Plann. Infer. **140**, 30–51 (2010)
116. N.N. Leonenko, N. Šuvak, Statistical inference for Student diffusion process. Stoch. Anal. Appl. **28**(6), 972–1002 (2010)
117. B.M. Levitan, Die Verallgemeinerung der Operation der Verschiebung im Zusammenhang mit fastperiodischen Funktionen. Mat. Sb. **7**(49), 449–478 (1940)
118. B.M. Levitan, Normed rings generated by the generalized operation of translation. C. R. (Doklady) Acad. Sci. URSS (N. S.) **47**, 3 6 (1945) (in Russian)
119. B.M. Levitan, *The application of generalized displacement operators to linear differential equations of the second order*. Uspekhi Mat. Nauk **4**(1), 3–112 (1949) (in Russian)
120. B.M. Levitan, On a class of solutions of the Kolmogorov-Smoluchowski equation. Vestnik Leningrad. Univ. **15**(7), 81–115 (1960) (in Russian)
121. B.M. Levitan, *Generalized Translation Operators and Some of Their Applications* (Israel Program for Scientific Translations, Jerusalem, 1964)
122. V. Linetsky, The spectral decomposition of the option value. Int. J. Theor. Appl. Financ **7**(3), 337–384 (2004)
123. V. Linetsky, The spectral representation of Bessel processes with constant drift: applications in queuing and finance. J. Appl. Probab. **41**(2), 327–344 (2004)
124. V. Linetsky, Spectral expansions for Asian (average price) options. Oper. Res. **52**(6), 856–867 (2004)
125. J.V. Linnik, I.V. Ostrovskiĭ, *Decomposition of Random Variables and Vectors* (American Mathematical Society, Providence, 1977)
126. L.L. Littlejohn, A.M. Krall, Orthogonal polynomials and singular Sturm–Liouville systems I. Rocky Mount. J. Math. **16**(3), 435–479 (1986)
127. G.L. Litvinov, Hypergroups and hypergroup algebras. J. Soviet Math. **38**(2), 1734–1761 (1987)
128. Y.L. Luke, *The Special Functions and Their Approximations*, vol. I (Academic Press, New York and London, 1969)
129. G.W. Mackey, Harmonic analysis as the exploitation of symmetry—a historical survey. Bull. Am. Math. Soc. **3**, 543–698 (1980)

130. N.K. Mamadaliev, On representation of a solution to a modified Cauchy problem. Sib. Math. J. **41**(5), 889–899 (2000)
131. C. Markett, Norm estimates for generalized translation operators associated with a singular differential operator. Indag. Math. **46**, 299–313 (1984)
132. P. Maroni, Une théorie algébrique des polynômes orthogonaux. Application aux polynômes orthogonaux semi-classiques, in *Orthogonal Polynomials and Their Applications*. IMACS Ann. Comput. Appl. Math., vol. 9 (Baltzer, Basel, 1991), pp. 95–130
133. P. Maroni, Variations around classical orthogonal polynomials. Connected problems. J. Comput. Appl. Math. **48**(1–2), 133–155 (1993)
134. M.A.M. Marrocos, A.L. Pereira, Eigenvalues of the Neumann Laplacian in symmetric regions. J. Math. Phys. **56**, 111502 (2015)
135. L. Mattner, Complex differentiation under the integral. Nieuw Arch. Wiskd. (5) **2**(1), 32–35 (2001)
136. I. McGillivray, A recurrence condition for some subordinated strongly local Dirichlet forms. Forum Math. **9**(2), 229–246 (1997)
137. H. McKean, Elementary solutions for certain parabolic partial differential equations. Trans. Am. Math. Soc. **82**, 519–548 (1956)
138. P. Medvegyev, *Stochastic Integration Theory* (Oxford University Press, Oxford, 2007)
139. J.K. Misiewicz, Lévy process with substable increments via generalized convolution. Lith. Math. J. **55**(2), 255–262 (2015)
140. J.K. Misiewicz, Generalized convolutions and the Levi-Civita functional equation. Aequat. Mat. **92**, 911–933 (2018)
141. J.K. Misiewicz Every symmetric weakly-stable random vector is pseudo-isotropic. J. Math. Anal. Appl. **483**, 939–965 (2020)
142. Y. Miyamoto, The hot spots conjecture for a certain class of planar convex domains. J. Math. Phys. **50**(10), 103530 (2009)
143. M. Musiela, M. Rutkowski, *Martingale Methods in Financial Modelling* (Springer, Berlin, 2005)
144. M.A. Naimark, *Linear Differential Operators. Part II: Linear Differential Operators in Hilbert Space* (Frederick Ungar Publishing, New York, 1968)
145. F.W.J. Olver, D.W. Lozier, R.F. Boisvert, C.W. Clark (eds.), *NIST Handbook of Mathematical Functions* (Cambridge University Press, Cambridge, 2010)
146. K.R. Parthasarathy, *Probability Measures on Metric Spaces* (American Mathematical Society, Providence, 1967)
147. K.R. Parthasarathy, R. Ranga Rao, S.R.S. Varadhan, Probability distributions on locally compact abelian groups. Illinois J. Math. **7**(2), 337–369 (1963)
148. M.N. Pascu, Scaling coupling of reflecting Brownian motions and the hot spots problem. Trans. Am. Math. Soc. **354**(11), 4681–4702 (2002)
149. G. Peskir, On the fundamental solution of the Kolmogorov–Shiryaev equation, in *From Stochastic Calculus to Mathematical Finance—The Shiryaev Festschrift*, ed. by Yu. Kabanov, R.Liptser, J. Stoyanov (Springer, Berlin, 2006), pp. 536–546
150. I.G. Petrovsky, *Lectures on Partial Differential Equations* (Interscience Publishers, 1954)
151. M. Pivato, *Linear Partial Differential Equations and Fourier Theory* (Cambridge University Press, Cambridge, 2010)
152. A.S. Polunchenko, G. Sokolov, An analytic expression for the distribution of the generalized Shiryaev–Roberts diffusion. Methodol. Comput. Appl. Probab. **18**(4), 1153–1195 (2016)
153. A.D. Polyanin, V.F. Zaitsev, *Handbook of Exact Solutions for Ordinary Differential Equations* (CRC Press, Boca Raton, 2002)
154. A. Prasad, U.K. Mandal, Composition of pseudo-differential operators associated with Kontorovich–Lebedev transform. Integr. Transf. Spec. Funct. **27**(11), 878–892 (2016)
155. A.P. Prudnikov, Y.A. Brychkov, O.I. Marichev, *Integrals and Series, Vol. I: Elementary Functions* (Gordon and Breach, New York and London, 1986)
156. A.P. Prudnikov, Y.A. Brychkov, O.I. Marichev, *Integrals and Series, Vol. II: Special Functions* (Gordon and Breach, New York and London, 1986)

157. A.P. Prudnikov, Y.A. Brychkov, O.I. Marichev, *Integrals and Series, Vol. III: More Special Functions* (Gordon and Breach, New York and London, 1989)
158. A.P. Prudnikov, Y.A. Brychkov, O.I. Marichev, *Integrals and Series, Vol. V: Inverse Laplace Transforms* (Gordon and Breach, New York and London, 1992)
159. E.V. Radkevich, Equations with nonnegative characteristic form II. J. Math. Sci. **158**, 453–604 (2009)
160. J. Rauch, *Partial Differential Equations* (Springer, New York, 1991)
161. M. Renardy, R.C. Rogers, *An Introduction to Partial Differential Equations*, 2nd edn. (Springer, Berlin, 2004)
162. C. Rentzsch, A Lévy Khintchine type representation of convolution semigroups on commutative hypergroups. Probab. Math. Statist. **18**(1), 185–198 (1998)
163. C. Rentzsch, M. Voit, Lévy Processes on Commutative Hypergroups, in *Probability on Algebraic Structures: AMS Special Session on Probability on Algebraic Structures, March 12–13, 1999, Gainesville, Florida*, ed. by G. Budzban, P. Feinsilver, A. Mukherjea (American Mathematical Society, Providence, 2000), pp. 83–105
164. D. Revuz, M. Yor, *Continuous Martingales and Brownian Motion* (Springer, Berlin, 1999)
165. M.Rösler, Convolution algebras which are not necessarily positivity-preserving, in *Applications of Hypergroups and Related Measure Algebras—A Joint Summer Research Conference, July 31-August 6, 1993, Seattle WA*, ed. by W.C. Connett, M.-O. Gebuhrer, A.L. Schwartz (American Mathematical Society, Providence, 1995), pp. 299–318
166. K. Sato, *Lévy Processes and Infinitely Divisible Distributions* (Cambridge University Press, Cambridge, 1999)
167. K. Schmüdgen, *Unbounded Self-Adjoint Operators on Hilbert Space* (Springer, Dordrecht, 2012)
168. A.N. Shiryaev, The problem of the most rapid detection of a disturbance in a stationary process. Soviet Math. Dokl. **2**, 795–799 (1961)
169. G.F. Simmons, *Introduction to Topology and Modern Analysis* (McGraw-Hill, New York, 1963)
170. B. Simon, *Operator Theory—A Comprehensive Course in Analysis, Part 4* (American Mathematical Society, Providence, 2015)
171. B.D. Sleeman, Multiparameter spectral theory and separation of variables. J. Phys. A **41**(1), 015209 (2008)
172. R. Sousa, M. Guerra, S. Yakubovich, Lévy processes with respect to the index Whittaker convolution. Trans. Am. Math. Soc. **374**(4), 2383–2419 (2021)
173. R. Sousa, M. Guerra, S. Yakubovich, A unified construction of product formulas and convolutions for Sturm–Liouville operators. Anal. Math. Phys. **11**(2), article 87 (2021)
174. R. Sousa, M. Guerra, S. Yakubovich, On the product formula and convolution associated with the index Whittaker transform. J. Math. Anal. Appl. **475**(1), 939–965 (2019)
175. R. Sousa, M. Guerra, S. Yakubovich, The hyperbolic maximum principle approach to the construction of generalized convolutions, in *Special Functions and Analysis of Differential Equations*, ed. by P. Agarwal, R.P. Agarwal, M. Ruzhansky (CRC Press, Boca Raton, 2020)
176. R. Sousa, M. Guerra, S. Yakubovich, Product formulas and convolutions for Laplace–Beltrami operators on product spaces: beyond the trivial case (2020). Preprint, arXiv:2006.14522. [Submitted for publication on a peer-reviewed journal]
177. R. Sousa, S. Yakubovich, The spectral expansion approach to index transforms and connections with the theory of diffusion processes. Commun. Pure Appl. Anal. **17**(6), 2351–2378 (2018)
178. R. Spector, Aperçu de la théorie des hypergroupes, in *Analyse harmonique sur les groupes de Lie – Séminaire Nancy-Strasbourg 1973–75*, ed. by P. Eymard, J. Faraut, G. Schiffmann, R. Takahashi (Springer, Berlin, 1975), pp. 643–673
179. H.M. Srivastava, V.K. Tuan, The Cherry transform and its relationship with a singular Sturm–Liouville problem. Q. J. Math. **51**(3), 371–383 (2000)
180. H.M. Srivastava, Y.V. Vasil'ev, S. Yakubovich, A class of index transforms with Whittaker's function as the kernel. Q. J. Math. **49**(2), 375–394 (1998)

181. D.W. Stroock, S.R.S. Varadhan, *Multidimensional Diffusion Processes* (Springer, Berlin, 1979)
182. K.-T. Sturm, How to construct diffusion processes on metric spaces. Potential Anal. **8**(2), 149–161 (1998)
183. K.-T. Sturm, Diffusion processes and heat kernels on metric spaces. Ann. Probab. **26**(1), 1–55 (1998)
184. L. Székelyhidi, *Functional Equations on Hypergroups* (World Scientific, Singapore, 2013)
185. G. Teschl, Mathematical methods in quantum mechanics. With applications to Schrödinger operators, 2nd edn. (American Mathematical Society, Providence, 2014)
186. N. Thanh Hai, S. Yakubovich, *The Double Mellin-Barnes Type Integrals and their Applications to Convolution Theory* (World Scientific, Singapore, 1996)
187. E.C. Titchmarsh, *Eigenfunction Expansions Associated with Second-Order Differential Equations* (Clarendon, Oxford, 1962)
188. K. Trimèche, *Generalized Wavelets and Hypergroups* (Gordon and Breach, Amsterdam, 1997)
189. K. Uhlenbeck, Generic properties of eigenfunctions. Am. J. Math. **98**(4), 1059–1078 (1976)
190. K. Urbanik, Generalized convolutions. Stud. Math. **23**, 217–245 (1964)
191. K. Urbanik, Generalized convolutions II. Stud. Math. **45**, 57–70 (1973)
192. K. Urbanik, Generalized convolutions III. Stud. Math. **80**, 167–189 (1984)
193. K. Urbanik, Generalized convolutions IV. Stud. Math. **83**, 57–95 (1986)
194. K. Urbanik, Analytical methods in probability theory, in *Transactions of the Tenth Prague Conference on Information Theory, Statistical Decision Functions, Random Processes*, vol. A (Reidel, Dordrecht, 1988), pp. 151–163
195. N. Van Thu, Generalized independent increments processes. Nagoya Math. J. **133**, 155–175 (1994)
196. V.E. Volkovich, Infinitely divisible distributions in algebras with stochastic convolution. J. Sov. Math. **40**(4), 459–467 (1988)
197. V.E. Volkovich, Quasiregular stochastic convolutions. J. Sov. Math. **47**(5), 2685–2699 (1989)
198. V.E. Volkovich, On symmetric stochastic convolutions. J. Theoret. Probab. **5**(3), 417–430 (1992)
199. F.-Y. Wang, L. Yan, Gradient estimate on convex domains and applications. Proc. Am. Math. Soc. **141**(3), 1067–1081 (2013)
200. G.N. Watson, *A Treatise on the Theory of Bessel Functions*, 2nd edn. (Cambridge University Press, Cambridge, 1944)
201. J. Weidmann, *Spectral Theory of Ordinary Differential Operators* (Springer, Berlin, 1987)
202. J. Wimp, A class of integral transforms. Proc. Edinb. Math. Soc. **14**(2), 33–40 (1964)
203. J. Wloka, *Partial Differential Equations* (Cambridge University Press, Cambridge, 1987)
204. S. Yakubovich, *Index Transforms* (World Scientific, Singapore, 1996)
205. S. Yakubovich, On the least values of L_p-norms for the Kontorovich–Lebedev transform and its convolution. J. Approx. Theory **131**(2), 231–242 (2004)
206. S. Yakubovich, On the Plancherel theorem for the Olevskii transform. Acta Math. Vietnam. **31**(3), 249–260 (2006)
207. S. Yakubovich, Y.F. Luchko, *The Hypergeometric Approach to Integral Transforms and Convolutions* (Kluwer Academic Publishers, Dordrecht, 1994)
208. H. Zeuner, Laws of Large Numbers of Hypergroups on $\mathbb{R}^+$. Math. Ann. **283**(4), 657–678 (1989)
209. H. Zeuner, One-dimensional hypergroups. Adv. Math. **76**(1), 1–18 (1989)
210. H. Zeuner, Moment functions and laws of large numbers on hypergroups. Math. Z. **211**(1), 369–407 (1992)
211. H. Zeuner, Domains of attraction with inner norming on Sturm–Liouville hypergroups. J. Appl. Anal. **1**(2), 213–221 (1995)

Index

R. Sousa et al., *Convolution-like Structures, Differential Operators and Diffusion Processes*, Lecture Notes in Mathematics 2315,
https://doi.org/10.1007/978-3-031-05296-5

LECTURE NOTES IN MATHEMATICS

Springer

Editorial Policy

1. Lecture Notes aim to report new developments in all areas of mathematics and their applications – quickly, informally and at a high level. Mathematical texts analysing new developments in modelling and numerical simulation are welcome.

   Manuscripts should be reasonably self-contained and rounded off. Thus they may, and often will, present not only results of the author but also related work by other people. They may be based on specialised lecture courses. Furthermore, the manuscripts should provide sufficient motivation, examples and applications. This clearly distinguishes Lecture Notes from journal articles or technical reports which normally are very concise. Articles intended for a journal but too long to be accepted by most journals, usually do not have this "lecture notes" character. For similar reasons it is unusual for doctoral theses to be accepted for the Lecture Notes series, though habilitation theses may be appropriate.

2. Besides monographs, multi-author manuscripts resulting from SUMMER SCHOOLS or similar INTENSIVE COURSES are welcome, provided their objective was held to present an active mathematical topic to an audience at the beginning or intermediate graduate level (a list of participants should be provided).

   The resulting manuscript should not be just a collection of course notes, but should require advance planning and coordination among the main lecturers. The subject matter should dictate the structure of the book. This structure should be motivated and explained in a scientific introduction, and the notation, references, index and formulation of results should be, if possible, unified by the editors. Each contribution should have an abstract and an introduction referring to the other contributions. In other words, more preparatory work must go into a multi-authored volume than simply assembling a disparate collection of papers, communicated at the event.

3. Manuscripts should be submitted either online at www.editorialmanager.com/lnm to Springer's mathematics editorial in Heidelberg, or electronically to one of the series editors. Authors should be aware that incomplete or insufficiently close-to-final manuscripts almost always result in longer refereeing times and nevertheless unclear referees' recommendations, making further refereeing of a final draft necessary. The strict minimum amount of material that will be considered should include a detailed outline describing the planned contents of each chapter, a bibliography and several sample chapters. Parallel submission of a manuscript to another publisher while under consideration for LNM is not acceptable and can lead to rejection.

4. In general, **monographs** will be sent out to at least 2 external referees for evaluation.

   A final decision to publish can be made only on the basis of the complete manuscript, however a refereeing process leading to a preliminary decision can be based on a pre-final or incomplete manuscript.

   Volume Editors of **multi-author works** are expected to arrange for the refereeing, to the usual scientific standards, of the individual contributions. If the resulting reports can be forwarded to the LNM Editorial Board, this is very helpful. If no reports are forwarded or if other questions remain unclear in respect of homogeneity etc, the series editors may wish to consult external referees for an overall evaluation of the volume.

5. Manuscripts should in general be submitted in English. Final manuscripts should contain at least 100 pages of mathematical text and should always include

   – a table of contents;
   – an informative introduction, with adequate motivation and perhaps some historical remarks: it should be accessible to a reader not intimately familiar with the topic treated;
   – a subject index: as a rule this is genuinely helpful for the reader.
   – For evaluation purposes, manuscripts should be submitted as pdf files.

6. Careful preparation of the manuscripts will help keep production time short besides ensuring satisfactory appearance of the finished book in print and online. After acceptance of the manuscript authors will be asked to prepare the final LaTeX source files (see LaTeX templates online: https://www.springer.com/gb/authors-editors/book-authors-editors/manuscriptpreparation/5636) plus the corresponding pdf- or zipped ps-file. The LaTeX source files are essential for producing the full-text online version of the book, see http://link.springer.com/bookseries/304 for the existing online volumes of LNM). The technical production of a Lecture Notes volume takes approximately 12 weeks. Additional instructions, if necessary, are available on request from lnm@springer.com.

7. Authors receive a total of 30 free copies of their volume and free access to their book on SpringerLink, but no royalties. They are entitled to a discount of 33.3 % on the price of Springer books purchased for their personal use, if ordering directly from Springer.

8. Commitment to publish is made by a *Publishing Agreement*; contributing authors of multiauthor books are requested to sign a *Consent to Publish form*. Springer-Verlag registers the copyright for each volume. Authors are free to reuse material contained in their LNM volumes in later publications: a brief written (or e-mail) request for formal permission is sufficient.

Addresses:

Professor Jean-Michel Morel, CMLA, École Normale Supérieure de Cachan, France
E-mail: moreljeanmichel@gmail.com

Professor Bernard Teissier, Equipe Géométrie et Dynamique,
Institut de Mathématiques de Jussieu – Paris Rive Gauche, Paris, France
E-mail: bernard.teissier@imj-prg.fr

Springer: Ute McCrory, Mathematics, Heidelberg, Germany,
E-mail: lnm@springer.com

Printed in the United States
by Baker & Taylor Publisher Services